Transcriptional Regulation in Bacteria

Guest Editor

Tomohiro Shimada

MDPI

Basel • Beijing • Wuhan • Barcelona • Belgrade • Novi Sad • Cluj • Manchester

Guest Editor
Tomohiro Shimada
Department of Agricultural Chemistry
Meiji University
Kawasaki-shi
Japan

Editorial Office
MDPI AG
Grosspeteranlage 5
4052 Basel, Switzerland

This is a reprint of the Special Issue, published open access by the journal *Microorganisms* (ISSN 2076-2607), freely accessible at: www.mdpi.com/journal/microorganisms/special_issues/C5YZSI1YJR.

For citation purposes, cite each article independently as indicated on the article page online and using the guide below:

Lastname, A.A.; Lastname, B.B. Article Title. *Journal Name* **Year**, *Volume Number*, Page Range.

ISBN 978-3-7258-2916-3 (Hbk)
ISBN 978-3-7258-2915-6 (PDF)
https://doi.org/10.3390/books978-3-7258-2915-6

Contents

microorganisms

MDPI

Editorial

Transcriptional Regulation in Bacteria

Tomohiro Shimada

School of Agriculture, Meiji University, 1-1-1 Kawasaki-Shi, Kanagawa 214-8571, Japan; tomoshimada@meiji.ac.jp

1. Introduction

The goal of research in the post-genomic era, now that the full extent of the genes encoded on microbial genomes is known [https://www.ncbi.nlm.nih.gov/genome (accessed on 5 November 2024)], is to elucidate the whole transcriptional mechanism by which microorganism express genes through transcriptional regulation with transcriptional regulatory factors. The genome sequence of microorganisms reveals their gene set and the set of transcriptional regulators that regulate the expression of those genes, providing a complete overview of the transcriptional regulatory mechanisms in microorganisms [1,2]. Before genome sequencing was available, the analysis of the transcriptional regulation of transcriptional regulators occurred at the gene level, whereas it is now possible to analyze this at a genome-wide scale [3–5]. To understand the intrinsic transcriptional regulatory mechanisms, it is necessary to note that transcriptional regulation forms a hierarchical network structure and to understand the distinction between direct regulation and indirect effects [6–11].

The new Special Issue, entitled "Transcriptional Regulation in Bacteria", in the *Microorganisms* journal, includes a total of ten original articles, providing new information about the transcriptional regulation of various functions in a wide variety of microorganisms. Recent studies were collected, ranging from detailed and precise studies of the transcriptional regulatory mechanisms of a single gene or via a single transcription factor [contributions 1–3] to studies of the genome regulatory networks of whole genomes and newly revealed transcriptional regulations [contributions 4–7]. It is also revealed that transcription factors comprehensively regulate multiple genes involved in multiple seemingly unrelated biological functions [contributions 8–10]. The analysis of transcriptional regulation revealed not only the molecular mechanisms of transcriptional regulation, but also the functional network of genes, leading to the elucidation of new biological functions of these microorganisms.

check for updates

Citation: Shimada, T. Transcriptional Regulation in Bacteria. *Microorganisms* **2024**, *12*, 2514. https://doi.org/10.3390/microorganisms12122514

Received: 29 November 2024
Accepted: 5 December 2024
Published: 6 December 2024

2. An Overview of Published Articles

Transcriptional regulation in microorganisms is carried out via the interaction of the transcription machinery, RNA polymerase, with the sigma subunit and transcription factors. Therefore, an elucidation of the function of functionally unknown transcription factors will provide new insights into novel transcriptional regulatory mechanisms and the environmental adaptation of microorganisms. Saito et al. identified a genomic binding site of a functionally unknown transcription factor, YegW, in *Escherichia coli*. Moreover, they showed that the effector is ADP–glucose, a precursor to glycogen synthesis, and that it functions as a repressor of the *yegTUV* operon, which is involved in glycogen accumulation. Its demonstrated effects on glycogen accumulation revealed its mechanism for accumulating a carbon source suitable for cell proliferation, and YegW was renamed GgaR (repressor of glycogen accumulation) [contribution 1]. Zhang et al. analyzed the *Bacillus thuringiensis* function unknown transcription factor HD73_5014 and found that it increased the transcript level of *pepV*, which encodes a dipeptidase. They proposed that HD73_5014 is renamed PepR (dipeptidase regulator) [contribution 2]. Belin et al. analyzed the interaction of the arabinose-responsive transcription factor AraC of *Escherichia coli* with the α subunit of RNA polymerase in detail. It became clear that the N-terminal domain of AraC exhibits at least three distinct activities: dimerization, arabinose binding, and transcriptional activation [contribution 3]. These research results focus on the functions of the transcription factors, revealing new physiological roles of transcription factors and new molecular mechanisms of transcription factors.

The function of a transcription factor has often been understood from the perspective of the function of a specific target gene(s). Thus, even for transcription factors thought to have a known function, their precise role in the genome as a whole remains unknown Therefore, once again, a genome-wide comprehensive analysis could contribute to the elucidation of new target gene(s) and their precise role. Xiao et al. analyzed the culture conditions under which PurR, a known transcription factor for purine biosynthesis in *Yersinia pestis*, is affected, as well as providing a quantitative RNA analysis. They found novel targets, including the type VI secretion system, and discovered a new physiological role for PurR [contribution 4]. In *Acinetobacter baumannii*, Huang et al. analyzed EmaSR, a two-component regulatory system of ethanol metabolism, and clarified its role in ethanol and the acetate metabolism, as well as its regulation of the acetate:succinyl-CoA transferase gene [contribution 5]. Huang et al. analyzed the mechanism by which the two-component regulatory system LiaSR induces Chlorhexidine resistance in *Streptococcus mutans*, suggesting that this resistance ability is achieved via the regulation of the *lmrB* efflux pump [contribution 6]. Xu et al. analyzed HigBA2, one of the TA systems in the main persister formation factors in *Mycobacterium tuberculosis*, and identified a novel set of target genes based on the HigA2-binding motif. This provided insights into the mechanism of parsister formation [contribution 7]. This research reveals a new transcriptional regulatory network based on a genome-wide analysis of transcription factors whose functions were previously known. Approaches such as these are useful not only for understanding the intrinsic role of transcription factors in the genome, but also for understanding how microorganisms adapt to changes in the environment.

To understand the molecular mechanisms of phenomena in which the involvement of specific transcription factors is unknown, approaches that analyze gene expression across the genome, such as transcriptome analysis, are useful. Song et al. studied the mechanism of feather degradation by *Pseudomonas aeruginosa* Gxun-7 for bioremediation. The RNA-seq analysis inferred a set of genes involved in feather degradation and proposed a model for the molecular mechanism [contribution 8]. Sanders et al. analyzed the mechanism of silver ion resistance in *Escherichia coli*. They identified a group of genes that responded to ionic silver and revealed that two-component regulatory systems such as CusSR and HprSR were involved in the major response [contribution 9]. Schlüter et al. improved acarbose yield by regulating the expression level of osmotic stress sigma factor σH^{As} in *Actinoplanes* sp. SE50/110. Transcriptome analysis identified a set of genes affected by σH^{As} and contributed to our understanding of its genome-wide role [contribution 10]. These approaches identified the transcriptional regulators involved in these important phenomena.

3. Conclusions

This Special Issue addresses the understanding of diverse transcriptional regulatory mechanisms in various microbial species. Although the genome sequences of microorganisms are rapidly being revealed, experimental analyses of individual transcriptional regulatory mechanisms are necessary because these sequences do not tell us how, when, and to what extent genes are expressed [12–14]. In addition, although the recent research tends to emphasize the analysis of some genes that directly benefit human society, it is necessary to understand the regulatory mechanisms of the entire genome because the genes of an organism form a network structure and multiple groups of genes function in a coordinated manner [15,16]. One of our major goals is to understand the entire transcriptional regulatory mechanism based on the functional elucidation of all transcriptional regulators in a single organism.

Funding: This research received no external funding.

Acknowledgments: As Guest Editor of the Special Issue "Transcriptional Regulation in Bacteria", I would like to express my deep appreciation to all authors whose valuable work was published under this issue and thus contributed to the success of the edition.

Conflicts of Interest: The author declares no conflict of interest.

List of Contributions

1. Saito, S.; Kobayashi, I.; Hoshina, M.; Uenaka, E.; Sakurai, A.; Imamura, S.; Shimada, T. Regulatory Role of GgaR (YegW) for Glycogen Accumulation in *Escherichia coli* K-12. *Microorganisms* **2024**, *12*, 115.
2. Zhang, X.; Wang, H.; Yan, T.; Chen, Y.; Peng, Q.; Song, F. A Novel Regulator PepR Regulates the Expression of Dipeptidase Gene *pepV* in *Bacillus thuringiensis*. *Microorganisms* **2024**, *12*, 579.
3. Belin, D.; Costafrolaz, J.; Silva, F. AraC Functional Suppressors of Mutations in the C-Terminal Domain of the RpoA Subunit of the *Escherichia coli* RNA Polymerase. *Microorganisms* **2024**, *12*, 1928.
4. Xiao, L.; Jin, J.; Song, K.; Qian, X.; Wu, Y.; Sun, Z.; Xiong, Z.; Li, Y.; Zhao, Y.; Shen, L.; et al. Regulatory Functions of PurR in *Yersinia pestis*: Orchestrating Diverse Biological Activities. *Microorganisms* **2023**, *11*, 2801.
5. Huang, Y.W.; Shu, H.Y.; Lin, G.H. Gene Expression of Ethanol and Acetate Metabolic Pathways in the *Acinetobacter baumannii* EmaSR Regulon. *Microorganisms* **2024**, *12*, 331.
6. Huang, S.; Huang, J.; Du, J.; Li, Y.; Wu, M.; Chen, S.; Zhan, L.; Huang, X. The LiaSR Two-Component System Regulates Resistance to Chlorhexidine in *Streptococcus mutans*. *Microorganisms* **2024**, *12*, 468.
7. Xu, M.; Liu, M.; Liu, T.; Pan, X.; Ren, Q.; Han, T.; Gou, L. HigA2 (Rv2021c) Is a Transcriptional Regulator with Multiple Regulatory Targets in *Mycobacterium tuberculosis*. *Microorganisms* **2024**, *12*, 1244.
8. Song, C.; Liu, R.; Yin, D.; Xie, C.; Liang, Y.; Yang, D.; Jiang, M.; Zhang, H.; Shen, N. A Comparative Transcriptome Analysis Unveils the Mechanisms of Response in Feather Degradation by *Pseudomonas aeruginosa* Gxun-7. *Microorganisms* **2024**, *12*, 841.
9. Sanders, B.R.; Thomas, L.S.; Lewis, N.M.; Ferguson, Z.A.; Graves, J.L., Jr.; Thomas, M.D. It Takes Two to Make a Thing Go Right: Epistasis, Two-Component Response Systems, and Bacterial Adaptation. *Microorganisms* **2024**, *12*, 2000.
10. Schlüter, L.; Busche, T.; Bondzio, L.; Hütten, A.; Niehaus, K.; Schneiker-Bekel, S.; Pühler, A.; Kalinowski, J. Sigma Factor Engineering in *Actinoplanes* sp. SE50/110: Expression of the Alternative Sigma Factor Gene *ACSP50_0507* (σH^{As}) Enhances Acarbose Yield and Alters Cell Morphology. *Microorganisms* **2024**, *12*, 1241.

References

1. Ishihama, A. Building a complete image of genome regulation in the model organism *Escherichia coli*. *J. Gen. Appl. Microbiol.* **2018**, *63*, 311–324. [CrossRef] [PubMed]
2. Lara, P.; Gama-Castro, S.; Salgado, H.; Rioualen, C.; Tierrafría, V.H.; Muñiz-Rascado, L.J.; Bonavides-Martínez, C.; Collado-Vides, J. Flexible gold standards for transcription factor regulatory interactions in *Escherichia coli* K-12: Architecture of evidence types. *Front. Genet.* **2024**, *15*, 1353553. [CrossRef] [PubMed]
3. Shimada, T.; Ogasawara, H.; Kobayashi, I.; Ishihama, A. Genomic SELEX screening of regulatory targets of transcription factors. *Methods Mol. Biol.* **2024**, *2819*, 77–102. [PubMed]
4. Grainger, D.C.; Hurd, D.; Harrison, M.; Holdstock, J.; Busby, S.J. Studies of the distribution of *Escherichia coli* cAMP-receptor protein and RNA polymerase along the *E. coli* chromosome. *Proc. Natl. Acad. Sci. USA* **2005**, *102*, 17693–17698. [CrossRef] [PubMed]
5. Chumsakul, O.; Nakamura, K.; Fukamachi, K.; Ishikawa, S.; Oshima, T. GeF-seq: A Simple Procedure for Base-Pair Resolution ChIP-seq. *Methods Mol. Biol.* **2024**, *2819*, 39–53. [PubMed]
6. Ishihama, A.; Shimada, T. Hierarchy of transcription factor network in *Escherichia coli* K-12: H-NS-mediated silencing and Anti-silencing by global regulators. *FEMS Microbiol. Rev.* **2021**, *45*, fuab032. [CrossRef] [PubMed]
7. Shimada, T.; Ogasawara, H.; Ishihama, A. Single-target regulators form a minor group of transcription factors in *Escherichia coli*. *Nucleic Acids Res.* **2018**, *46*, 3921–3936. [CrossRef] [PubMed]

8. Browning, D.F.; Busby, S.J. Local and global regulation of transcription initiation in bacteria. *Nat. Rev. Microbiol.* **2016**, *14*, 638–650. [CrossRef] [PubMed]
9. Busby, S.J.W.; Browning, D.F. Transcription activation in *Escherichia coli* and *Salmonella*. *EcoSal Plus.* **2024**, eesp-0039-2020. [CrossRef] [PubMed]
10. Gagarinova, A.; Hosseinnia, A.; Rahmatbakhsh, M.; Istace, Z.; Phanse, S.; Moutaoufik, M.T.; Zilocchi, M.; Zhang, Q.; Aoki, H.; Jessulat, M.; et al. Auxotrophic and prototrophic conditional genetic networks reveal the rewiring of transcription factors in *Escherichia coli*. *Nat. Commun.* **2022**, *13*, 4085. [CrossRef] [PubMed]
11. Femerling, G.; Gama-Castro, S.; Lara, P.; Ledezma-Tejeida, D.; Tierrafría, V.H.; Muñiz-Rascado, L.; Bonavides-Martínez, C.; Collado-Vides, J. Sensory Systems and Transcriptional Regulation in *Escherichia coli*. *Front. Bioeng. Biotechnol.* **2022**, *10*, 823240. [CrossRef] [PubMed]
12. Gao, Y.; Yurkovich, J.T.; Seo, S.W.; Kabimoldayev, I.; Dräger, A.; Chen, K.; Sastry, A.V.; Fang, X.; Mih, N.; Yang, L.; et al. Systematic discovery of uncharacterized transcription factors in *Escherichia coli* K-12 MG1655. *Nucleic Acids Res.* **2018**, *46*, 10682–10696. [CrossRef] [PubMed]
13. Gao, Y.; Lim, H.G.; Verkler, H.; Szubin, R.; Quach, D.; Rodionova, I.; Chen, K.; Yurkovich, J.T.; Cho, B.K.; Palsson, B.O. Unraveling the functions of uncharacterized transcription factors in *Escherichia coli* using ChIP-exo. *Nucleic Acids Res.* **2021**, *49*, 9696–9710. [CrossRef] [PubMed]
14. Rodionova, I.A.; Gao, Y.; Monk, J.; Hefner, Y.; Wong, N.; Szubin, R.; Lim, H.G.; Rodionov, D.A.; Zhang, Z.; Saier, M.H., Jr.; et al. A systems approach discovers the role and characteristics of seven LysR type transcription factors in *Escherichia coli*. *Sci. Rep.* **2022**, *12*, 7274. [CrossRef] [PubMed]
15. Duarte-Velázquez, I.; de la Mora, J.; Ramírez-Prado, J.H.; Aguillón-Bárcenas, A.; Tornero-Gutiérrez, F.; Cordero-Loreto, E.; Anaya-Velázquez, F.; Páramo-Pérez, I.; Rangel-Serrano, Á.; Muñoz-Carranza, S.R.; et al. *Escherichia coli* transcription factors of unknown function: Sequence features and possible evolutionary relationships. *Peer J.* **2022**, *10*, e13772. [CrossRef] [PubMed]
16. Moore, L.R.; Caspi, R.; Boyd, D.; Berkmen, M.; Mackie, A.; Paley, S.; Karp, P.D. Revisiting the y-ome of *Escherichia coli*. *Nucleic Acids Res.* **2024**, *52*, 12201–12207. [CrossRef] [PubMed]

microorganisms

Article

Regulatory Role of GgaR (YegW) for Glycogen Accumulation in *Escherichia coli* K-12

Shunsuke Saito [1], Ikki Kobayashi [1], Motoki Hoshina [2,†], Emi Uenaka [1,3], Atsushi Sakurai [3], Sousuke Imamura [3] and Tomohiro Shimada [1,*,†]

1 School of Agriculture, Meiji University, 1-1-1 Kawasaki-Shi, Kanagawa 214-8571, Japan
2 Research and Development Section, Diagnostics Division, YAMASA Corporation, 2-10-1 Araoicho, Choshi, Chiba 288-0056, Japan
3 Space Environment and Energy Laboratories, Nippon Telegraph and Telephone Corporation, Musashino-shi, Tokyo 180-8585, Japan
* Correspondence: tomoshimada@meiji.ac.jp
† Previous Affiliation: Micro-Nano Technology Research Center, Hosei-University, Koganei 184-0003, Tokyo, Japan.

Abstract: Glycogen, the stored form of glucose, accumulates upon growth arrest in the presence of an excess carbon source in *Escherichia coli* and other bacteria. Chromatin immunoprecipitation screening for the binding site of a functionally unknown GntR family transcription factor, YegW, revealed that the *yegTUV* operon was a single target of the *E. coli* genome. Although none of the genes in the *yegTUV* operon have a clear function, a previous study suggested their involvement in the production of ADP-glucose (ADPG), a glycogen precursor. Various validation through in vivo and in vitro experiments showed that YegW is a single-target transcription factor that acts as a repressor of *yegTUV*, with an intracellular concentration of consistently approximately 10 molecules, and senses ADPG as an effector. Further analysis revealed that YegW repressed glycogen accumulation in response to increased glucose concentration, which was not accompanied by changes in the growth phase. In minimal glucose medium, *yegW*-deficient *E. coli* promoted glycogen accumulation, at the expense of poor cell proliferation. We concluded that YegW is a single-target transcription factor that senses ADPG and represses glycogen accumulation in response to the amount of glucose available to the cell. We propose renaming YegW to GgaR (repressor of glycogen accumulation).

Keywords: transcription factor; glycogen; ADP-glucose; *Escherichia coli*; gene regulation

Citation: Saito, S.; Kobayashi, I.; Hoshina, M.; Uenaka, E.; Sakurai, A.; Imamura, S.; Shimada, T. Regulatory Role of GgaR (YegW) for Glycogen Accumulation in *Escherichia coli* K-12. *Microorganisms* **2024**, *12*, 115. https://doi.org/10.3390/microorganisms12010115

Academic Editor: Vincenzo Scarlato

Received: 4 December 2023
Revised: 31 December 2023
Accepted: 3 January 2024
Published: 5 January 2024

1. Introduction

Glycogen accumulation occurs in *Escherichia coli* and many other bacteria. Glycogen is formed when there is an excess of carbon under conditions in which growth is limited because of the lack of a growth nutrient, such as a nitrogen, sulfur, or phosphate source [1]. Glycogens play a role in prolonging the viability of microorganisms in the stationary phase by providing a source of carbon and energy [2,3]. The accumulated glycogen also plays an important role as the primary source of glucose for energy production during regrowth in the early inducible phase [4]. Recent reports have also indicated that glycogen metabolism is involved in various stress conditions, such as starvation, biofilm formation, cold stress, desiccation, and oxidative stress in *E. coli* [5].

In *E. coli*, the enzymes for glycogen synthesis and degradation are encoded by genes that constitute the *glgBXCAP* single operon: *glgC* (glucose-1-phosphate adenylyltransferase (AGPase)), *glgA* (glycogen synthase), *glgB* (glycogen branching enzyme), *glgP* (glycogen phosphorylase), and *glgX* (glycogen debranching enzyme) [6,7]. ADP-glucose (ADPG) is a glucosyl donor for bacterial glycogen synthesis, and *E. coli* glycogen synthases are specific for the sugar nucleotide ADPG [8]. Therefore, the modulation of AGPase activity is an important allosteric regulator of bacterial glycogen synthesis. The primary activator of *E. coli* AGPase is fructose 1,6-bisphosphate, and AMP is its major inhibitor [1,9,10] Therefore, under conditions of limited growth with excess carbon in the medium, the accumulation of glycolytic intermediates, such as fructose-1,6-bisphosphate, is thought to occur and could be a signal for the activation of ADPG synthesis. Although there are many reports on the allosteric regulation of enzymes involved in glycogen metabolism, gene expression is reportedly induced during the stationary phase [7]. The rate of glycogen synthesis is inversely related to the growth rate when growth is limited by the availability of certain nutrients, such as nitrogen. Consistent with this relationship, the levels of glycogen biosynthetic enzymes in *E. coli* increase as cultures enter the stationary phase [2,3].

In transcriptional regulation, full-length *glgBXCAP* is transcribed from the *glgB* promoter, and the transcription unit of *glgCAP* is transcribed from the promoter of *glgC*, both of which are recognized by the major sigma subunit, sigma-70, encoded by *rpoD*, which plays a major role in the transcription of growth-related genes. In addition, there is another promoter upstream of *glgB* [11], which is recognized by sigma-38, encoded by *rpoS*, and plays a role in the transcription of stationary-phase-specific genes. Glycogen accumulates during the stationary phase in an RpoS-dependent manner [12]. Other transcription factors, such as the cAMP receptor protein (CRP), which senses glucose uptake activity, activate *glgCAP* but do not affect *glgBX* [11], and the PhoP-PhoQ two-component system activates *glgBXCAP* (sensing environmental Mg^{2+}) [13]. At the translational level, CsrA (a carbon storage regulator) binds to *glgCAP* mRNA and inhibits its translation [14]. In addition to a set of *glg* genes, it has recently been reported that the inner membrane nucleoside transporters NupC and NupG incorporate ADPG, depending on CRP regulation [15]. Previously, the KO collection, a collection of 3985 non-essential gene-disrupted mutants of *E. coli* K-12, and the ASKA library, a gene expression library of 4123 genes, were used to screen for genes involved in glycogen metabolism [16–18]. This has led to the proposal of various functions related to glycogen metabolism; however, so far, the genes that directly act on glycogen metabolism are limited to the *glg* operon and its regulators, and glycogen is thought to accumulate during the stationary phase. Because glycogen production is based on glucose availability, it should be regulated not only by the growth phase, but also by the amount of glucose and similar sugars; however, such mechanisms and regulations have not yet been revealed.

E. coli contains approximately 4500 protein-coding sequences in its genome. *E. coli* has seven sigma factors and approximately 300 transcription factors (TFs) to precisely regulate the genes required for responses to various environmental changes [19,20]. However, even within the model bacterium *E. coli*, the function of one-fifth of this species' genes remains unknown. Owing to the advancement of genome-wide research technologies, such as transcriptomics and proteomics, the regulatory role of each TF has been identified mainly based on transcription patterns in the absence of a test regulator or after its overexpression. A large amount of knowledge of TFs is assembled in databases such as EcoCyc [21] and RegulonDB [22], but with the use of these gene expression analyses it is difficult to distinguish between the direct regulation of test TF and their indirect influence. The regulatory targets of TFs often include genes encoding other TFs that form the hierarchy of the TF network [20]. To identify the direct regulatory targets of TFs, Chromatin Immunoprecipitataion (ChIP) [23], and Genomic SELEX analysis [24] have been widely used in vivo and in vitro, respectively. TFs bind to DNA and regulate nearby genes. Thus, predicting the regulation of target promoters, genes, and operons is possible based on the location of TF-binding sites. Using these genome-wide methods to identify the direct regulatory network of a TF and its regulated gene(s), it is possible to infer the function or role of unknown regulated gene(s) or unknown TFs based on the function of either the TF feature or the regulated gene.

To understand the overall picture of genomic transcriptional regulatory mechanisms, it is necessary to understand the functions of a set of *E. coli* TFs. As a part of our research strategy, in this study we sought to characterize YegW, which has been recognized as one of the uncharacterized TFs in *E. coli* and belongs to the GntR family. YegW has a DNA-binding domain at the N-terminus and a signal receiving domain, which is involved in interactions with the ligand at the C-terminus [19,25]. A systematic search was performed to identify the target genes controlled by YegW using the ChIP-chip method [26]. As a result, the binding of YegW was only observed at two closely located sites within the *yegTUV* upstream region on the *E. coli* K-12 genome. Although the function of *yegTUV* is unknown, its possible association with ADPG has been predicted [15], so we tested the effect of YegW on glycogen accumulation in this study. YegW is a repressor of *yegTUV*, sensing ADPG and suppressing glycogen accumulation. Therefore, we propose renaming YegW to GgaR (a repressor of glycogen accumulation).

2. Materials and Methods

2.1. *E. coli Strains and Plasmids*

E. coli BW25113 [27], its *ggaR* single-gene knockout mutant JW2088 [28], and the expression plasmid from the ASKA clone library were obtained from the *E. coli* Stock Center (National Bio-Resource Center, Chiba, Japan). Cells were grown in Luria broth (LB) medium, Kornberg medium (KM; 1.1% K_2HPO_4, 0.85% KH_2PO_4, and 0.6% yeast extract) [29] or M9 minimal medium supplemented with various concentrations of glucose at 37 °C with constant shaking at 150 rpm. When necessary, kanamycin (20 $\mu g \cdot mL^{-1}$) or chloramphenicol (30 $\mu g \cdot mL^{-1}$) was added to the medium. Cell growth was monitored by measuring turbidity at 600 nm.

2.2. *ChIP-Chip Analysis*

A ChIP assay was used to measure the chromosome-wide DNA-binding profile of GgaR using the experimental protocols described by Shimada et al. [26]. Briefly, cultures of *E. coli* BW25113 and, as a control, JW2088 Δ*ggaR* were grown to mid-log phase at 37 °C. The cells were then treated with 1% formaldehyde and broken open by sonication, which also fragmented the cross-linked nucleoproteins. Cross-linked GgaR-DNA complexes were immunoprecipitated from cleared lysates of BW25113 cells using anti-GgaR rabbit polyclonal anti-serum, and parallel samples were isolated from control JW2088 Δ*ggaR* cells. The cross-links were then reversed, and the immunoprecipitated DNA was purified. DNA samples isolated from BW25113 and control Δ*ggaR* cells were labelled with Cy5 and Cy3, respectively. To identify the DNA segments specifically associated with GgaR, the two labelled samples were combined and hybridized to an *E. coli* DNA tiling microarray [26]. For each probe, the Cy5/Cy3 ratio was measured and plotted against the corresponding position on the *E. coli* BW25113 chromosome to create a GgaR-binding profile.

2.3. *Purification of the GgaR Protein*

The plasmid (pGgaR) for the purification of GgaR was constructed as described by Shimada et al. [30]. Briefly, GgaR coding sequences were polymerase chain reaction (PCR)-amplified using *E. coli* K-12 W3110 genomic DNA as a template and inserted into the pET21a (+) vector (Novagen, Darmstadt, Germany) between the NdeI and NotI sites. The pGgaR expression plasmid was transformed into *E. coli* BL21 (DE3) cells. Transformants were grown in LB medium, and GgaR expression was induced using isopropyl β-D-thiogalactopyranoside (IPTG) in the middle of the exponential phase. The GgaR protein was purified by affinity purification using a Ni-nitrilotriacetic acid (NTA) agarose column. The affinity-purified GgaR protein was stored and frozen in the storage buffer at −80 °C until further use. Protein purity was greater than 95% as determined by sodium dodecyl sulfate-polyacrylamide gel electrophoresis (SDS-PAGE).

2.4. *Gel Shift Assay*

A gel shift assay was performed according to standard procedures [31]. Probes for the GgaR-binding target sequences were generated using PCR amplification using a pair of primers (Table S1a) and Ex Taq DNA polymerase (TaKaRa, Kyoto, Japan). A mixture of each probe and GgaR was incubated at 37 °C for 30 min in the binding buffer. When the effectors were added, the mixture was incubated for an additional 30 min. After addition of the DNA loading solution, the mixture was subjected to 5% polyacrylamide gel electrophoresis (PAGE). The DNA in the gels was stained with GelRed (Biotium, Fremont, CA, USA) and detected using LuminoGraph I (Atto, Tokyo, Japan).

2.5. *Northern Blot Analysis*

Total RNA was extracted from *E. coli* cells using ISOGEN solution (Nippon Gene, Tokyo, Japan). RNA purity was verified using electrophoresis on a 1.2% agarose gel with formaldehyde, and followed by GelRed-staining. Northern blot analysis was performed as described previously [32]. Digoxigenin (DIG)-labelled probes were prepared using PCR amplification using BW25113 genomic DNA as a template with a pair of primers (Table S1b), DIG-11-dUTP (Roche, Basel, Switzerland), dNTP, gene-specific pair primers, and Ex Taq DNA polymerase. An amount of 3 μg of total RNAs were incubated in formaldehyde-MOPS (morpholinepropanesulfonic acid) loading buffer for 5 min at 65 °C for denaturation. The samples were subjected to electrophoresis on formaldehyde-containing 1.5% agarose gel, and then transferred onto a nylon membrane (Roche). Hybridization was performed using DIG Easy Hyb system (Roche) at 50 °C overnight with a DIG-labelled probe. The membranes were treated with anti-DIG-AP Fab fragments and CDP-Star (Roche) to detect the DIG-labelled probe and the images were scanned using LuminoGraph I (Atto).

2.6. *Observation and Measurement of Glycogen*

The intracellular accumulation of glycogen was qualitatively detected using staining cell pellets with an iodine solution (0.01 M I_2, 0.03 M KI) [29]. The same number of *E. coli* cells (OD_{600} × culture volume (mL) = 2.0) was stained with an iodine solution. Intracellular glycogen was observed using transmission electron microscopy, following the method described by Eydallin et al. [16] with some modifications. Cells were fixed with 2.5% glutaraldehyde in 0.01 M phosphate buffer (pH 7.3) for 30 min at 4 °C, rinsed in 0.1 M phosphate buffer, and post-fixed 2% OsO_4 in phosphate buffer (pH 7.3) for 1 h at 4 °C. After three washes with pure water, the cells were resuspended several times in an aqueous solution with gradually increasing ethanol concentrations for dehydration. The cells were further resuspended several times in an ethanol solution with gradually increasing acetone concentration. The cells were then resuspended in acetone solution with gradually increasing concentrations of Quetol651 and, finally, the cells were embedded in 100% Quetol651 solution at 60 °C for two days. The ultrathin sections were cut using a diatome diamond knife. The sections were stained with 2% uranyl acetate and lead acetate, and inspected using a transmission electron microscope at 120 kV (JEM-2010).

Glycogen measurements were performed according to the method described by Iijima et al. (2021) [33] with some modifications. Equal amounts of cells (OD_{600} × cell culture (mL) = 10) were collected using centrifugation (8000× *g* at 25 °C for 5 min) and the pellets were resuspended in 100 μL of 3.5% (*w*/*v*) sulfuric acid. The cell suspensions were incubated at 100 °C for 160 min and centrifuged at 20,500× *g* at 4 °C for 1 min. Glucose levels in the supernatants were quantified using LabAssay Glucose (Wako, Osaka, Japan) and a Multiskan FC microplate reader (Thermo Scientific, Waltham, MA, USA). The total protein content of the cell suspensions was quantified using a Bradford protein assay kit (Bio-Rad) and a Multiskan FC microplate reader (Thermo Scientific).

2.7. *Biofilm Formation Assay*

Crystal violet staining was performed as described previously [34]. *E. coli* cells were grown in a KM medium with 1% (*w*/*v*) glucose at 37 °C in a 2.0 mL tube. Following incubation for 24 h, planktonic cells were discarded and the tubes were washed twice with phosphate-buffered saline (PBS) (-) and subsequently stained with 0.1% crystal violet for 20 min at room temperature. After extensive washing with H_2O, the biofilm-bound crystal violet was extracted with 500 μL of 70% ethanol and measured at OD_{595}.

2.8. Starvation Viability Assay

A single *E. coli* colony was picked up from KM agar plate for inoculating KM medium for overnight, which was then cultured fresh 5 mL KM medium with 1% (*w*/*v*) glucose at 37 °C with 150 rpm shaking rate until OD_{600} reached to 1.0. The inoculated culture was centrifuged at 6000× *g* at 25 °C for 3 min, and the cell pellets were washed three times with PBS buffer. The cells were then resuspended in 1 mL PBS buffer and placed in incubator at 25 °C for 0, 2, 4, 6, and 12 days, and were then spread on KM agar plates at each time point. The number of colonies formed on the plate after overnight incubation was counted and defined as viable cells compared to those on day 0. The experiments were independently repeated three times for the wild-type and Δ*ggaR* strains.

2.9. Real-Time (RT)-qPCR Analysis

RT-qPCR was performed according to standard procedures [35]. *E. coli* cells were inoculated into KM supplemented with glucose (0.25% or 1%) at 37 °C under aeration with constant shaking at 150 rpm. Total RNA was extracted from exponential phase *E. coli* cells (OD_{600} = 0.5) using ISOGEN solution (Nippon Gene, Tokyo, Japan). Total RNA was transcribed to cDNA using the THUNDERBIRD SYBR qPCR RT Set (TOYOBO, Osaka, Japan). Quantitative PCR (qPCR) was conducted using the THUNDERBIRD SYBR qPCR Mix (TOYOBO) and a LightCycler 96 system (Roche). The primer pairs are listed in Table S1c. The cDNA templates were serially diluted four-fold and used for qPCR analysis. The qPCR mixtures, containing 10 μL of THUNDERBIRD SYBR qPCR Mix (TOYOBO), 1 μL of each primer (5 μM stock), 7 μL water, and 1 μL cDNA, were amplified under the following thermal cycling conditions: 95 °C treatment for 2 min, 45 cycles of 10 s at 95 °C and 20 s at 55 °C, and then incubated for 20 s at 72 °C. The 16S rRNA expression level was used to normalize the varying levels of the test samples, and the relative expression levels were quantified using Relative Quantification software provided by Roche. The results are presented as the average of three independent experiments.

2.10. Western Blot Analysis

Western blot analysis was performed using a standard method, as described previously [36]. Briefly, *E. coli* cells grown in KM supplemented with 1% (*w*/*v*) glucose or M9 minimal medium supplemented with 0.4% (*w*/*v*) glucose were harvested in time course using centrifugation and resuspended in lysis buffer (50 mM Tris-HCl, pH 7.5, 50 mM NaCl, 5% glycerol, and 1 mM dithiothreitol), and lysozyme was added to a final concentration of 20 μg·mL^{-1}. After sonication, 3 μg of cell extracts was subjected to 10% SDS-PAGE and blotted on to polyvinylidene difluoride membranes using semi-dry transfer apparatus. Membranes were first immunoreacted with rabbit anti-GgaR or anti-RpoA anti-serum, and then immunoreacted with anti-rabbit IgG and HRP conjugate. The membranes were developed using an enhanced chemiluminescence kit (Amersham Pharmacia Biotech, Buckinghamshire, UK). The intensity of the bands was analyzed using an LAS-4000 IR multicolor (Fuji Film, Tokyo, Japan).

3. Results

3.1. Search for GgaR-Binding Locations Using ChIP-Chip Screening

To reveal the regulatory role of the transcription factor GgaR (renamed YegW), we used ChIP to identify the binding of GgaR across chromosomes of growing *E. coli* cells. Thus, wild-type BW25113 and the Δ*ggaR* derivative JW2088 were grown aerobically in LB medium to an OD_{600} of 0.4. The cells were then treated with formaldehyde, and cellular DNA was extracted and sonicated to yield DNA fragments of approximately 500 bp. After immunoprecipitation with anti-GgaR antibodies, DNA fragments from BW25113 and control JW2088 Δ*ggaR* cells were purified, labelled with Cy5 and Cy3, respectively, mixed, and hybridized to the microarray. After washing and scanning, the fluorescence intensities of the Cy5 and Cy3 attached to each probe on the microarray were measured and plotted along the *E. coli* genome (Figure 1). A single peak of GgaR, binding was identified at the spacer between the *fbaB* gene and *yegTUV* operon, indicating that GgaR is a single-target TF. Downstream of *yegTUV*, *ggaR* is encoded on the reverse strand. Thus, the only target of the transcription factor, GgaR, is in close proximity to its own coding region. Detailed analysis of GgaR binding within the region corresponding to the single ChIP peak indicated the presence of two binding sites, one at a bidirectional transcription unit between the *fbaB* gene and *yegTUV* operon, and another within *yegT* open reading frame (ORF) (Figure 1).

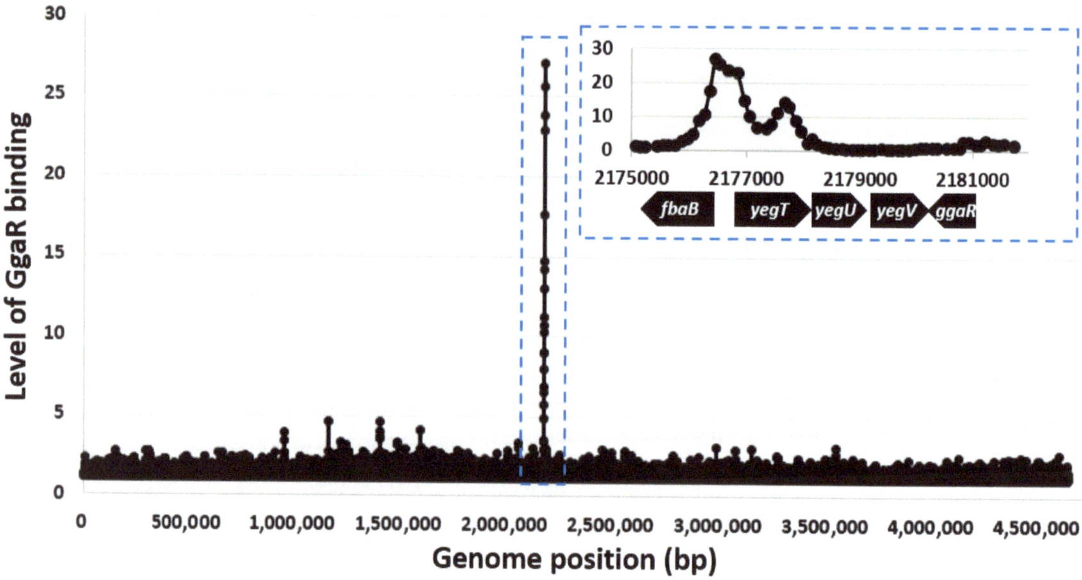

Figure 1. Distribution of GgaR binding across the *E. coli* K-12 chromosome. The figure shows an overview of results from the ChIP–chip experiments that measure the profile of GgaR binding across the *E. coli* K-12 chromosome during exponential growth. Binding signals (*y*-axis) are plotted against their location on the 4.64 Mbp *E. coli* chromosome (*x*-axis). A single GgaR-binding peak was identified and is indicated by the blue dotted box. The expanded region of the peak with neighboring genes is shown in the upper right blue dotted box. One peak is located within the spacer between the *fbaB* gene and the *yegTUV* operon and another is within *yegT* ORF.

3.2. Confirmation of GgaR-Binding to Its Targets

Using in vivo ChIP–chip screening, we identified two adjacent GgaR-binding sites in the *E. coli* K-12 genome (Figure 1). To experimentally confirm GgaR binding to these sites, we performed gel shift assays in vitro to detect GgaR-target DNA complexes. First, *fbaB/yegT* spacer DNA probe was tested. A gel shift assay indicated that the probe formed GgaR-DNA complexes in a GgaR concentration-dependent manner (Figure 2A).

Next, we analyzed the binding of GgaR to various segments of the *fbaB-yegT* region. We prepared six different DNA fragments and subjected these DNA probes to a gel shift assay (Figure 2B). Probes were prepared for the region from *fbaB* to *yegT* ORF, with several dozen base pairs overlapping each other (Figure S1), and their binding to GgaR was verified under the presence of different GgaR concentrations. In the presence of 5 pmol of GgaR, the formation of a complex with GgaR was observed only in probe-2, which contains 190 bp upstream of the *yegT* gene (Figure 2B). When the concentration of GgaR was increased to 10 pmol, the complex formation was also observed in probe-5, which is 571–869 bp inside the ORF of *yegT*. In contrast, probes 1, 3, 4, and 6, located upstream and downstream of each other, respectively, did not form complexes. The positions of the two probes to which GgaR was bound in vitro were consistent with the positions of the two peaks identified by the ChIP–chip screening, suggesting in vivo GgaR binding (Figure S1). The results of those in vivo and in vitro experiments were also consistent with the binding affinity with GgaR, which was higher between the *fbaB-yegT* spacer than the ORF of *yegT*. Taken together, we concluded that the target site for the initial binding of GgaR is the *fbaB/yegT* spacer.

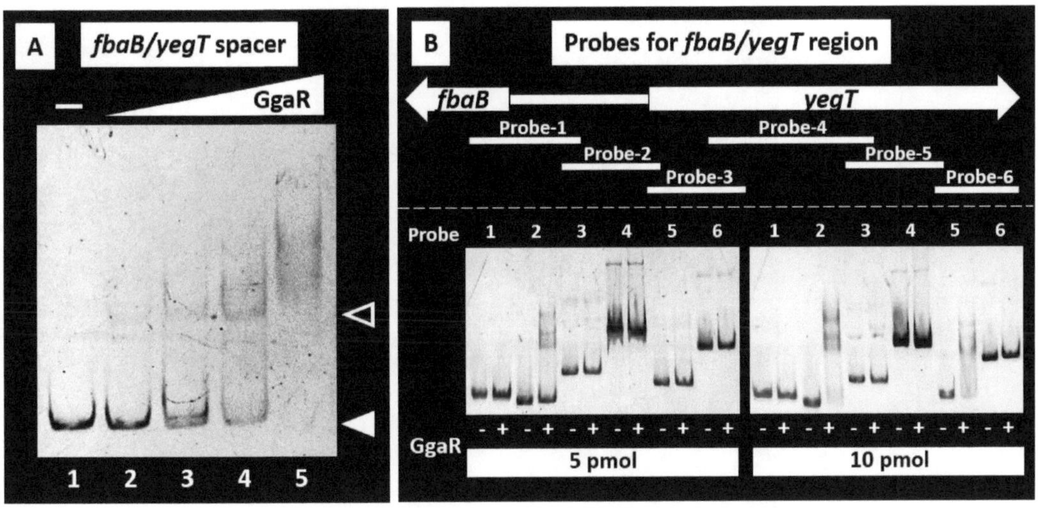

Figure 2. Gel shift assay of the GgaR-DNA complex formation. (**A**) Binding of GgaR to its predicted targets, the *fbaB/yegT* spacer, was examined. Then, 0.5 pmol of the target probe was mixed with purified GgaR. The concentration of GgaR added to lane 1, 2, 3, 4, and 5 was 0, 0.25, 0.5, 1, and 3 pmol, respectively. Filled triangles indicate the GgaR-DNA probe complex, whereas open triangles indicate free probes. (**B**) Mapping of the binding site of GgaR was carried using six different probe segments in *fbaB-yegT* region. The position of each probe in its region is shown at the top. An amount of 0.5 pmol of each probe was incubated with the absence or presence of 5 or 10 pmol of purified GgaR, and subjected to PAGE.

3.3. Regulatory Role of GgaR in the Expression of the Target Genes

Based on the in vivo ChIP–chip and in vitro gel shift assay results, the involvement of GgaR in the regulation of the *fbaB* and/or *yegTUV* operon was predicted. We then carried out Northern blot analysis to examine the possible influence of GgaR on these genes determined mRNA levels in vivo for each of the predicted GgaR target genes in the presence or absence of GgaR. Wild-type *E. coli* BW25113, *ggaR*-deleted mutants, and *ggaR*-deleted mutants harboring the GgaR expression vector strains were grown in LB medium, and total RNA was extracted in both the exponential and stationary phases. Based on the gene size, the predicted *yegTUV* mRNA length should be approximately 3.3 kb. In the *ggaR*-deleted mutant, the same transcripts of *yegTUV* could were detected using *yegT*- and *yegV*-specific probes (Figure 3). The size of this transcript was estimated to be 3.3 kb as compared with the size of the 2904 nucleotide 23S rRNA and was in agreement with the size predicted for *yegTUV* mRNA. The intensity of the *yegTUV* signal was somewhat reduced by complementation with the GgaR expression vector, was not detected in the wild-type strain, and showed a similar pattern during the exponential and stationary phases. In contrast, strong signals were observed in the stationary phase for the *fbaB* gene located divergently from *yegT*. This is in agreement with a previous report that the *fbaB* transcript was noticeably reduced in the log growth phase [37], but no clear effect of *ggaR* was observed. These results indicate that the *yegTUV* operon is transcribed into a single mRNA, and this transcription is repressed by GgaR, regardless of the growth phase.

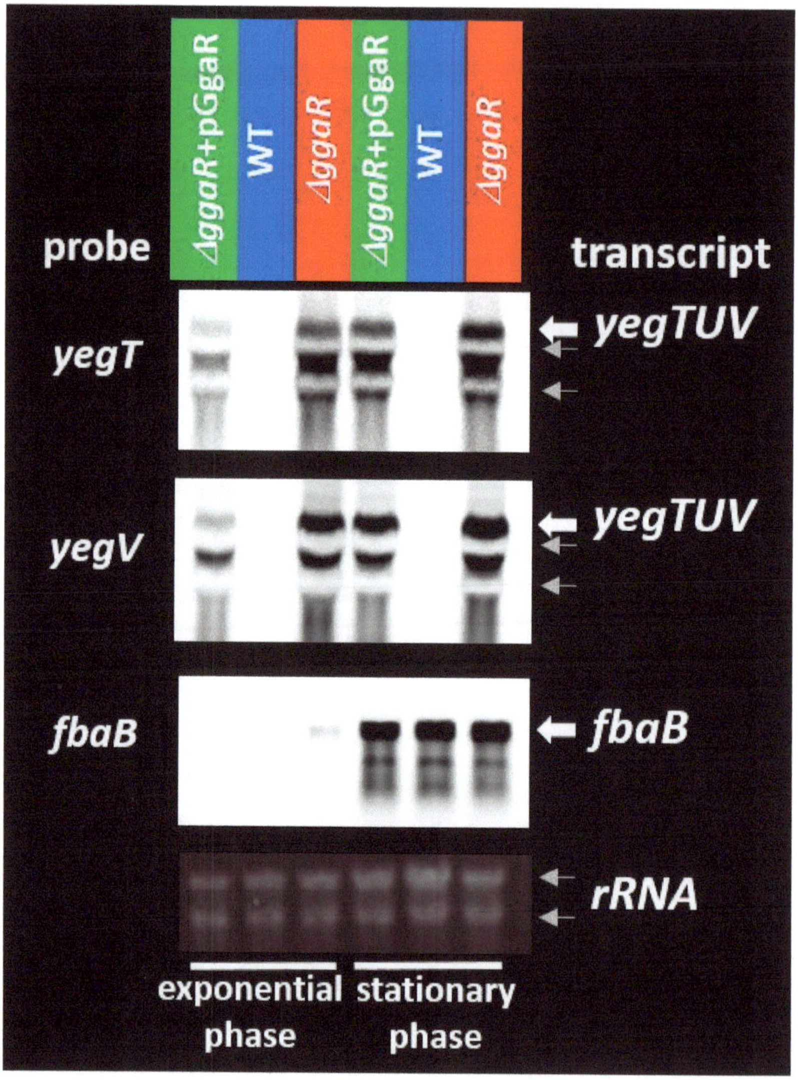

Figure 3. Northern blotting analysis of mRNAs from GgaR target genes. Wild-type *E. coli* K-12 BW25113, its *ggaR* mutant JW2088, and the *ggaR* mutant strain harboring the GgaR expression vector were grown in LB medium at 37 °C. Total RNA was prepared at log phase and stationary phase and subjected to Northern blotting analysis. DIG-labeled hybridization probes are shown on the left side of each panel, and the detected transcripts are shown on the right side. The positions of 23S rRNA (2904 bases) and 16S rRNA (1542 bases) were observed as a native shadow band and are indicated using gray arrows. The amounts of total RNA analyzed were examined by measuring the intensity of ribosomal RNAs.

3.4. Effect of GgaR on Glycogen Accumulation

To date, GgaR is a functionally unknown transcription factor; however, this study shows that it is a repressor of the *yegTUV* operon. Additionally, the functions of *yegT*, *yegU*, and *yegV* were unknown. Of these genes, *yegV*, the putative sugar kinase, was included in a list reported by Almagro et al. for genes whose deletion significantly reduced glycogen accumulation in KM [15]. Based on the protein sequence, YegT is predicted to belong to the Nucleoside:H+ symporter (NHS) family within the major facilitator superfamily (MFS) [38], and YegU is predicted to be an ADP-ribosylglycohydrolase [39]; any of these predicted enzymatic reactions could act on ADPG. Considering that in *E. coli*, glycogen is produced via ADPG converted from glucose as a substrate, the predicted functions of *yegTUV* genes were implicated in glycogen accumulation.

In this study, we investigated the effect of *ggaR* deficiency on cell growth and glycogen accumulation in KM, a rich medium that allows *E. coli* to enhance the accumulation of glycogen and the typically and commonly used medium in studies of glycogen metabolism in *E. coli* [29]. First, the growth curves of *E. coli* wild-type and *ggaR*-deficient strains in KM at various glucose concentrations were observed. The maximum turbidity increased with the addition of 0.25% to 1% glucose, but the deletion of *ggaR* did not affect growth (Figure 4A). Under the same conditions, the amount of intracellular glycogen was qualitatively determined by iodine staining of the collected cell pellets (Figure 4B). No significant glycogen accumulation was observed at any glucose concentration at the beginning of cultivation or without the addition of glucose. In contrast, after 6 h of incubation, glycogen accumulation was dependent on glucose concentration. Additionally, the *ggaR*-deficient strain stained more intensely than the wild-type strain at the same glucose concentration. The amount of glycogen accumulated during cell growth in medium with low concentrations of glucose, such as 0.25% and 0.5%, decreased after 24 h, which is consistent with previous reports [5,16]. In further experiments, the amount of glycogen that accumulated in the cells was quantitatively measured and compared (Figure 4C). At 6 h of growth in KM with 1% glucose, the amount of glycogen relative to total protein was measured to be approximately twice as much in the *ggaR*-deficient strain at 15 μg/mL compared to 8 μg/mL in the wild-type strain. The effect of *ggaR* deficiency was largely suppressed by complementation of *ggaR* with the harboring of a GgaR expression vector. After 24 h, glycogen accumulation increased in all strains, with only a slight increase in the *ggaR*-deficient strain compared to other strains. TEM was performed to directly observe the accumulation of intracellular glycogen granules (Figure 4D). The wild-type strain cells only had a few glycogen granules at the edge of cytoplasm at 6 h after inoculation in KM with 1% glucose, whereas large amounts of glycogen granules were observed in the *ggaR*-deficient strain. Similar observations after 4 h of cultivation revealed almost no glycogen granules in the wild-type strain, whereas some glycogen granules were observed in the *ggaR*-deficient strain (Figure S2). Taken together, these results suggest that GgaR downregulates glycogen accumulation in *E. coli* grown in KM medium in response to the addition of glucose.

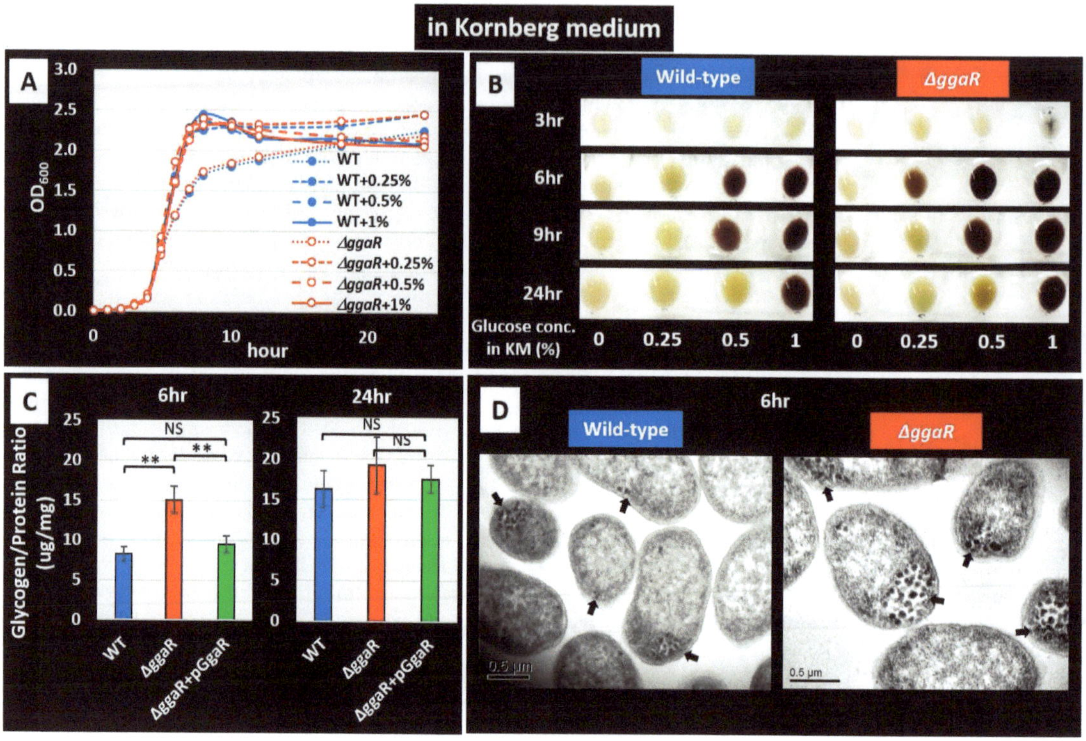

Figure 4. Influence of *ggaR* deficiency on glycogen accumulation in Kornberg medium. *E. coli* wild-type strain BW25113 and its *ggaR* mutant JW2088 were grown in KM with different concentrations of glucose (0, 0.25, 0.5, and 1.0%). (**A**) Cell growth was monitored by measuring the turbidity at 600 nm over time. (**B**) Iodine staining for measuring the glycogen accumulation of each wild-type strain and *ggaR* mutant cell pellets at different timepoints (3, 6, 9, 24 h). (**C**) Quantification of glycogen accumulation in *E. coli* strains (wild-type strain, *ggaR* mutant, and the *ggaR* mutant strain harboring the GgaR expression vector) were cultured in KM with 1.0% glucose and harvested at 6 h or 24 h after the start of inoculation. The glycogen amount represents the mean ± SD of three independent experiments. Statistical significance was analyzed using the Student's *t*-test with multiple comparisons' correction and were represented by asterisks (** $p < 0.01$). NS indicates not significant. (**D**) TEM analysis of glycogen granules. *E. coli* strains were cultured in KM with 1.0% glucose and harvested 6 h after the start of inoculation. Arrows indicate the position of glycogen granules.

3.5. The Influence of GgaR under Stress Conditions

Glycogen is an energy storage compound that plays an important role in bacterial environmental viability. Wang et al. [5] reported that *E. coli* strains lacking genes involved in glycogen synthesis or degradation promote the ability to form biofilms, and that strains that accumulate glycogen have an increased survival rate after starvation. Therefore, we also examined the effects of deficiency of the transcription factor GgaR, which represses glycogen accumulation. *E. coli* wild-type and *ggaR*-deficient strains were grown in plastic tubes in a static culture in KM medium supplemented with 1% glucose. After 24 h of incubation, the amount of biofilm formed on the inner tube surface was determined and quantified using crystal violet staining. The results indicated that biofilm formation was enhanced in the *ggaR*-deficient strain compared to that in the wild-type strain, with an approximately four-fold difference (Figure 5A). The survival rate after starvation was less than 60% for the wild-type strain after two days, whereas it was more than 80% for the *ggaR*-deficient strain (Figure 5B). The survival rate decreased with time, and this difference decreased as the number of days passed. After 12 days, there was almost no difference, a pattern similar to that reported for glycogen metabolism in gene-deficient strains [5]. These results indicate that the effects of the deletion of *ggaR* were generally consistent with the previously reported deletion of *glg* genes involved in glycogen metabolism in their effects on the ability of *E. coli* to adapt to the stress environment [5].

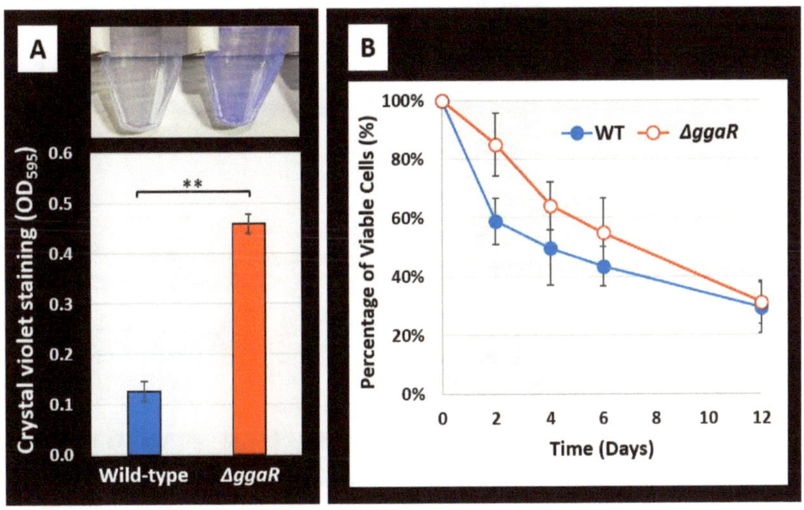

Figure 5. Influence of *ggaR* deficiency on survival rate under stress conditions. (**A**) Comparison of biofilm formation abilities of *E. coli* wild-type strain BW25113 and its *ggaR* mutant in KM with 1.0% glucose at 37 °C for 24 h in a plastic tube. The level of biofilm formation was measured using crystal violet staining. The upper panel shows the stained biofilm formed on the bottom of tube, and the lower panel shows the biofilm volume measured by elution with 70% ethanol. The relative level of biofilm represents the mean ± SD of three independent experiments. Statistical significance was analyzed using the Student's *t*-test with multiple comparisons' correction and were represented by asterisks (** $p < 0.01$). (**B**) Viabilities of *E. coli* wild-type strain BW25113 and its *ggaR* mutant under starvation conditions.

3.6. Search for Effectors Controlling GgaR Activity

In *E. coli*, glycogen is synthesized from glucose in the culture medium as a substrate, the imported Glucose 6-phosphate is converted to Glucose 1-phosphate and then to ADPG, which is the substrate for glycogen elongation. Considering that the transcription factor GgaR is involved in glycogen accumulation, as shown by the experiments of this study, and the predicted function of the genes constituting the *yegTUV* operon, the single target of GgaR on the *E. coli* genome is *yegT* for the putative nucleoside symporter, *yegU* for the putative ADP-ribosylglycohydrolase, and *yegV* for the putative sugar kinase. The effector of GgaR is thought to be ADPG; thus, we tested the possible influence of ADPG on the regulation of the *yegTUV* operon. The potential influence of ADPG on GgaR binding to its target was examined in vitro using a gel-shift assay (Figure 6). After adding increasing concentrations of ADPG, the complex of GgaR and the *fbaB*/*yegT* spacer probe was reduced, resulting in an increase in the number of free DNA probes. To examine the specificity of the inducer, glucose and ADP were also tested for their potential influence on the GgaR–probe DNA interactions. None of the compounds interfered with the binding of GgaR to *fbaB*/*yegT* spacer probes.

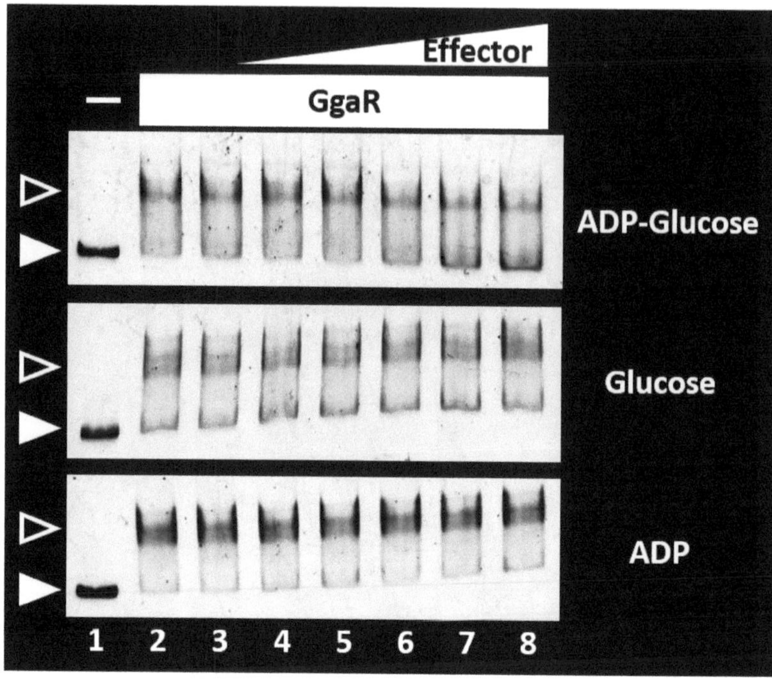

Figure 6. Search for GgaR ligands. The gel shift assay of GgaR-DNA complex formation and influence of each chemical. Purified GgaR was mixed with 0.5 pmol of each DNA probe corresponding to the *fbaB*/*yegT* spacer region. In the presence of 2.0 pmol GgaR (lanes 2–8), 0, 10, 20, 30, 40, 50, and 100 μM of ADPG, glucose, and ADP was added to lanes 1 and 2, lane 3, 4, 5, 6, 7, and 8, respectively. The filled triangles indicate the GgaR-DNA probe complex, whereas the open triangles indicate free probes.

To verify the effect of GgaR activity in vivo, the amount of GgaR target mRNA was measured using RT-qPCR. First, to confirm the validity of the experimental method, the target mRNA levels in the wild-type and *ggaR*-deficient strains grown in KM with 1.0% glucose were compared (Figure 7A). An approximately 30-fold increase in *yegTUV* mRNA levels was observed in the *ggaR*-deficient strain compared to that in the wild-type strain. Notably, *fbaB* mRNA, which is divergently located from the *yegTUV* operon, was not affected by the deletion of *ggaR*. These results are in agreement with the effects of GgaR on the single-target *yegTUV* in LB medium, as shown in Figure 3, suggesting that GgaR represses the *yegTUV* operon, similar to the effect observed in the KM medium. Next, we tested the effect of ADPG; however, it was difficult to obtain an ample sample of ADPG to add to the liquid medium. Therefore, we tested the effect of GgaR regulation on the concentration of glucose, the precursor of ADPG, added to KM and inoculated the wild-type strain. The results showed that the level of *yegTUV* mRNA increased approximately three-fold when 1% glucose was added to the medium, compared to when 0.25% glucose was added (Figure 7B). However, compared to 0.25% and 1% glucose concentrations, the *yegTUV* mRNA levels in the *ggaR*-deficient strain, in which GgaR was absent, did not change. These results suggest that the increase in the mRNA level of *yegTUV* upon the addition of higher concentrations of glucose to the medium in the wild-type strain was GgaR-dependent, implying that GgaR becomes inactive in the presence of ADPG and derepresses the *yegTUV* operon.

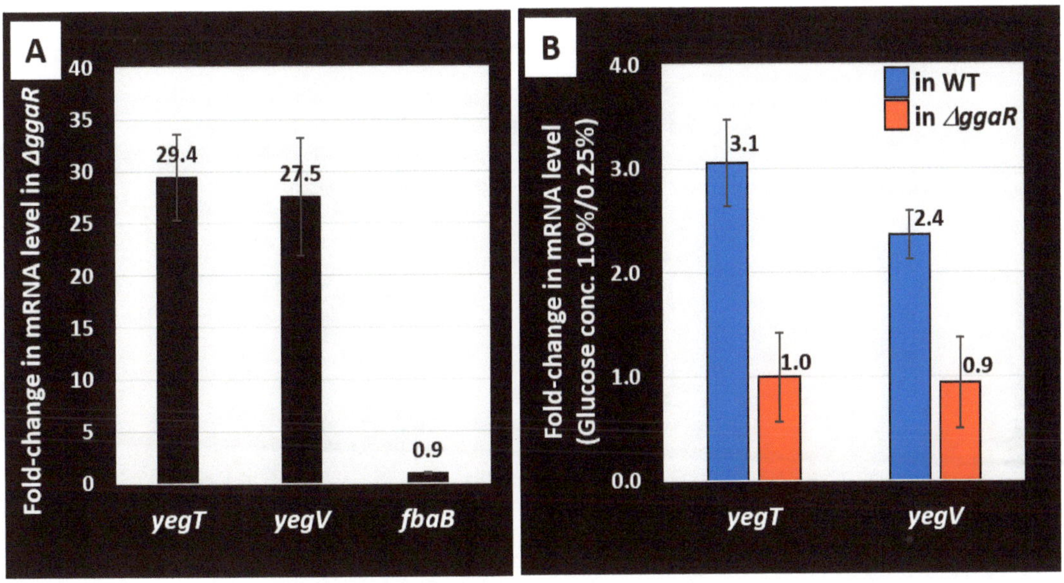

Figure 7. In vivo influence of *ggaR* deletion or glucose concentration in medium of GgaR targets using RT-qPCR. (**A**) *E. coli* BW25113 and its *ggaR*-deleted mutant were grown in KM with 1.0% glucose. (**B**) *E. coli* BW25113 wild-type strain and *ggaR*-deleted mutant were grown in KM supplemented with 0.25% or 1.0% glucose. Total RNA was extracted from exponential phase cells (OD_{600} = 0.5) and subjected to RT-qPCR analysis. The *y*-axis represents the relative level of mRNA of each GgaR target gene between the wild-type and *ggaR* mutant (**A**) as well the relative level between in KM with 0.25% glucose and with 1.0% glucose in the wild-type or *ggaR*-deleted mutant (**B**), respectively; the ratio of 16S rRNA is set as an internal control between the compared strains. Each experiment was repeated three times, and the average means are shown.

3.7. Physiological Role of GgaR on E. coli Growth and Glycogen Accumulation in Minimal Medium

In this study, the effects of GgaR on glycogen accumulation were tested using KM, a rich medium that promotes glycogen accumulation in *E. coli* and is typically used in research on glycogen metabolism. Deletion of *ggaR* results in an inability to repress the *yegTUV* operon and promote glycogen accumulation, but no effect on growth was observed in KM despite the storage of glucose uptake in the *ggaR*-deficient strain. Therefore, to further clarify the physiological role of GgaR, we verified the importance of the GgaR-mediated regulation of glucose consumption and accumulation as glycogen using minimal glucose medium. *E. coli* wild-type and *ggaR*-deficient strains were cultured in M9 minimal medium with different concentrations (0.2%, 0.4%, and 0.8%) of glucose, and their growth curves were obtained (Figure 8A). The results showed that both strains grew well in the presence of 0.4% glucose, a concentration commonly used in *E. coli* cultures, and grew slightly poorly in the presence of 0.8% glucose. As expected, a marked decrease in growth was observed in the *ggaR*-deficient strain at all glucose concentrations compared with the wild-type strain in minimal medium. Next, glycogen accumulation was measured in each strain during the growth phase under these conditions. The results showed that glycogen accumulation was approximately three-fold higher in the *ggaR*-deficient strain at all glucose concentrations (>80 μg/mg) compared to approximately 30 μg/mg in the wild-type strain (Figure 8B). These results indicated that in the *ggaR*-deficient strain, glycogen accumulation was increased by glucose utilization in exchange for impaired cell growth under growth conditions in which the carbon source was dependent on glucose. These results indicate that GgaR plays an important physiological role in the utilization of glucose, determining whether it is consumed for energy production or stored as glycogen.

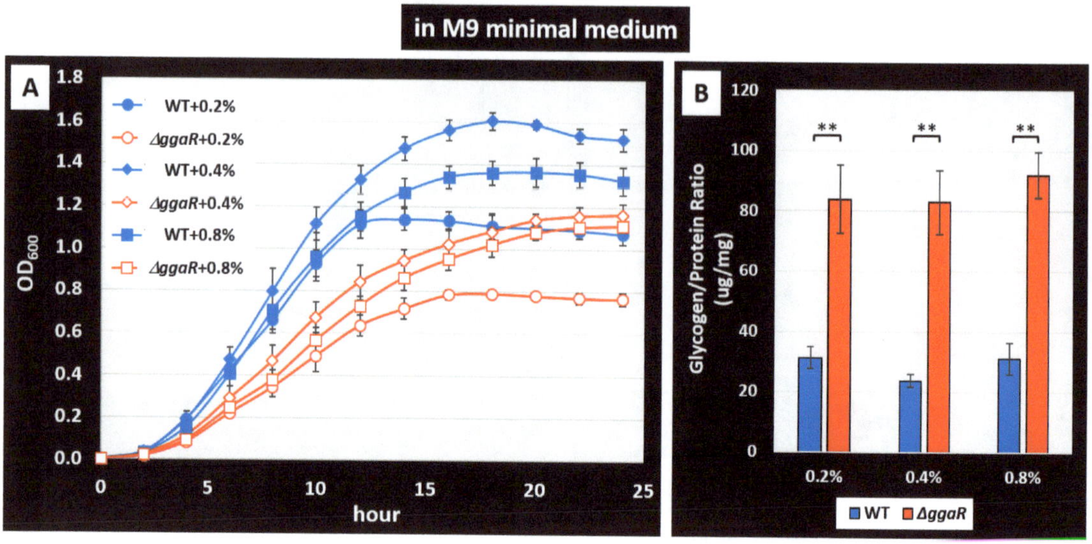

Figure 8. Influence of *ggaR* deficiency on cell growth and glycogen accumulation in glucose-supplemented M9 minimal medium. (**A**) *E. coli* BW25113 (blue closed symbols) and its *ggaR*-deleted mutant (red open symbols) were grown in M9 minimal medium with various concentrations of glucose: 0.25 (circle), 0.45 (diamond), and 0.85 (square). Cell density was monitored every 2 h and plotted along the time course. (**B**) Quantification of glycogen accumulation in *E. coli* wild-type and ggaR-deficient strains. The strains were cultured in M9 minimal medium supplemented with 0.25, 0.45, or 0.85 glucose and harvested at 10 h after the start of inoculation. The glycogen amount represents the mean ± SD of three independent experiments. Statistical significance was analyzed using the Student's *t*-test with multiple comparison correction and is represented by asterisks (** $p < 0.02$).

3.8. Protein Expression Level of GgaR Is Constant in Growth Phase

GgaR is a single-target TF that senses ADPG and represses the *yegTUV* operon, regardless of the growth phase. Therefore, to validate the intracellular expression of GgaR at the protein level, we measured its expression over time using antibodies against rich KM and poor M9 minimal media. The control protein was the α subunit of RNA polymerase, whose expression per cell has been reported to be 5000 molecules with virtually no fluctuation throughout cell growth [40,41]. As a result, the protein level of GgaR remained roughly constant in both KM and M9 minimal media with 1% glucose during the 24 h incubation of *E. coli* wild-type strains (Figure 9A). The negative control, the JW2088 *ggaR*-deficient strain, did not express GgaR. The expression level of GgaR per cell was calculated by comparing it to the expression level of RpoA and was estimated to be approximately 20–30 molecules per cell, regardless of the medium (Figure 9B).

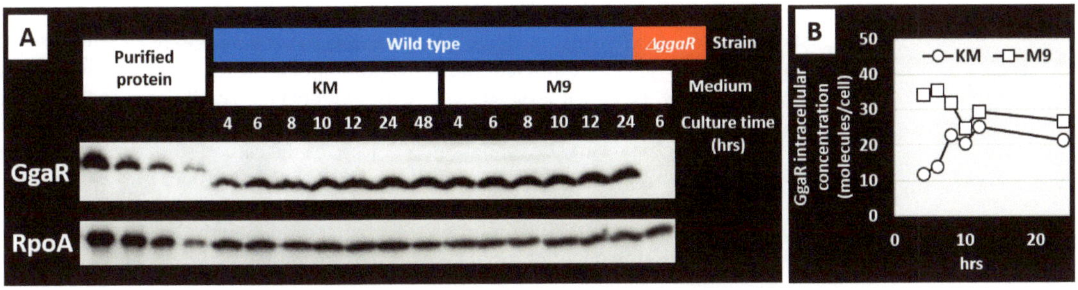

Figure 9. Intracellular levels of GgaR and RpoA. (**A**) *E. coli* BW25113 and its *ggaR*-deleted mutant were grown in KM with 1.05 glucose- or 0.45 glucose-supplemented M9 minimal medium and subjected to quantitative Western blotting analysis for determination of GgaR and RpoA intracellular levels. The left four lanes for the standards curve of purified proteins: His-tagged GgaR, 1, 0.5, 0.25, 0.1 ng; RpoA, 120, 60, 30, 10 ng. Each 3 μg of whole-cell extracts of cells was harvested at culture time. (**B**) The concentrations of GgaR monomer were calculated based on the concentration of RpoA monomer that stayed constant throughout the culture at the level of 5000 molecules per genome equivalent of DNA.

4. Discussion

Understanding the regulatory networks of genome transcription involving all seven sigma factors and all 300 TFs and identifying the association between each of these regulatory proteins and their direct targets is a major issue. We have previously elucidated the functions of more than 10 TFs of unknown function based on the identification of genomic transcriptional regulatory networks [42–44]. Here, we identified that the hitherto uncharacterized *E. coli* TF YegW (referred to here as GgaR) regulates a single target of the *yegTUV* operon in the *E. coli* genome and was therefore identified as a single-target regulator [42,45]. Most single-target TF genes in the *E. coli* genome are located in close proximity or adjacent to their regulatory target genes, forming a gene organization in which TFs and their regulatory targets exist as an adjacent set [45]. For GgaR, the *ggaR* gene is a typical single-target regulator located adjacent to the single-target *yegTUV* operon (Figure 1). Because single-target regulators have only a single target in the genome, their intracellular concentrations are considerably lower; for example, the expression of LacI, a single-target regulator known to regulate the lactose operon [42], has been reported to be less than 10 molecules in a cell [46]. Most transcription factors of the GntR family, to which GgaR belongs, form homodimers [25]. The intracellular concentration of GgaR with two binding sites in close proximity to the *yegT* promoter was approximately 20 molecules at constant concentration and approximately 10 molecules as a homodimer (Figures 2 and 9). This low intracellular concentration of GgaR is consistent with typical features of single-target regulators.

Although no clear function is known for any of the genes of the *yegTUV* operon, the single target of GgaR, the loss of *yegV* was screened to reduce glycogen accumulation [15], and the inferred function of any of the genes appeared to be related to ADPG, such as *yegT*, the putative transporter of nucleoside, *yegU*, the putative ADP-ribosylglycohydrolase, and *yegV*, the putative sugar kinase. In accordance with predictions based on this somewhat ambiguous information, we tested the effect of GgaR on glycogen accumulation in the KM, which has been typically used in previous studies on glycogen accumulation. As expected, the loss of *ggaR* increased glycogen accumulation (Figure 4), suggesting that ADPG inactivated GgaR as an effector (Figures 6 and 7). Although the expression of GgaR was constant in *E. coli* grown in KM (Figure 9), the effect of GgaR on glycogen accumulation was greater in the exponential phase and smaller in the stationary phase (Figure 4B,C). This may be because, as has long been thought, glycogen is largely produced by growth arrest or an excess of carbon sources due to the depletion of certain nutrients, such as nitrogen, and glycogen accumulates during the stationary phase by GlgBXCAP [1,3]. In the stationary phase in nutrient-rich medium conditions, GgaR played only a partial role. The regulatory role of GgaR is not specific to the growth phase, such as the growth arrest or stationary phase [1–3], but is dependent on the amount of intracellular carbon source by sensing ADPG. This is supported by the fact that under culture conditions in M9 minimal medium with glucose and dependence on glucose as the carbon source, the deletion of *ggaR* resulted in the accumulation of glycogen in exchange for poor growth (Figure 8). This suggests that the large amount of yeast extract added to the KM medium provided *E. coli* with sufficient nutrients for growth, such that even though GgaR was not present and glucose was stored inappropriately, glycogen accumulation did not affect the growth of *ggaR* mutant (Figure 4A). This indicates that GgaR plays an important physiological role in conditions in which the destination of glucose is important, such as whether it is consumed for energy production, necessary for cell proliferation, or stored as glycogen (Figure 10). Taken together, we propose that GgaR is a transcription factor that senses ADPG and represses glycogen accumulation in response to the amount of glucose available to the cell, and should thus be renamed as a repressor of glycogen accumulation.

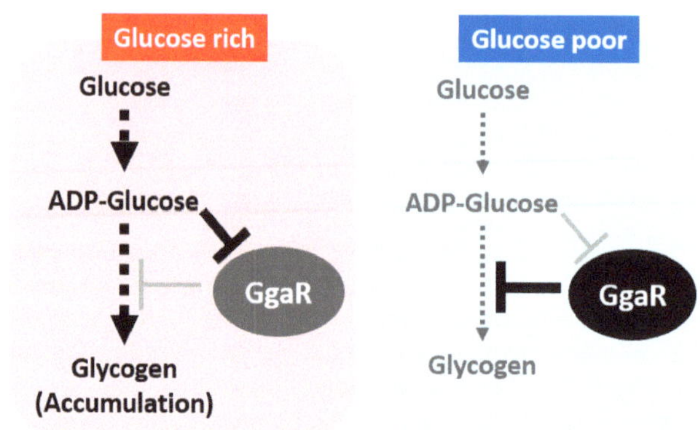

Figure 10. Model of regulatory network of ADPG-sensing glycogen accumulation repressor GgaR in *Escherichia coli* K-12. Under the presence of glucose, the concentration of ADPG increases, GgaR becomes inactive, glycogen accumulation is de-repressed, and glycogen accumulates. In contrast, in the absence of glucose, the concentration of ADPG decreases, GgaR becomes active, and glycogen accumulation is repressed.

Although screening for the genes involved in the glycogen metabolism has been conducted several times for all genes in *E. coli* [15–18], *ggaR* has not been identified. This was probably because they were all determined from steady-state glycogen accumulation in KM supplemented with glucose. This may be because they were all based on steady-state glycogen accumulation in KM, a nutrient-rich medium, and not under conditions requiring strict glucose utilization. The conservation of *ggaR* is high among Enterobacterales such as *Shigella*, *Citrobacter*, *Salmonella*, and *Vibrio*, suggesting that it plays an important role in adaptation to the gut environment. Recently, Gayán et al. [47] reported that the lack of *ggaR* caused *E. coli* to gain high hydrostatic pressure (HHP) resistance, which was RpoS-independent. Glycogen metabolism is important for the acquisition of various stress tolerances ([5]; see also Figure 5), and HHP resistance may be a part of this. Functional elucidation of the GgaR target genes, *yegT*, *yegU*, and *yegV*, will be helpful in understanding glycogen metabolism and awaits future studies.

Supplementary Materials: The following supporting information can be downloaded at: https://www.mdpi.com/article/10.3390/microorganisms12010115/s1, Table S1: Primers used in this study. Figure S1: Location of each probe in the *fbaB/yegT* region used to map the GgaR-binding site. Figure S2: TEM analysis of glycogen granules 4 h after inoculation.

Author Contributions: Conceptualization, T.S.; methodology, S.S., I.K., M.H. and T.S.; formal analysis, S.S., I.K., M.H. and T.S.; investigation, S.S., I.K., M.H. and T.S.; resources, E.U., A.S., S.I. and T.S.; writing—original draft preparation, T.S.; writing—review and editing, T.S., S.S., I.K., M.H., E.U., A.S. and S.I.; funding acquisition, T.S. All authors have read and agreed to the published version of the manuscript.

Funding: This research was funded by MEXT Grants-in-Aid for Scientific Research (C) (22K06184) to T.S.

Data Availability Statement: ChIP–chip data for GgaR have been deposited in the 'Transcription factor profiling of *Escherichia coli*' (TEC) database at the National Institute of Genetics (https://shigen.nig.ac.jp/ecoli/tec/ (accessed on 11 December 2023)).

Acknowledgments: We thank Akira Ishihama (Hosei University) for the research advice and providing GgaR anti-serum, and Michio Sato (Meiji University) for supporting the TEM analysis. We thank the National BioResource Project, National Institute of Genetics, Japan, for providing *E. coli* K-12 BW25113 and its single-gene deletion mutants.

Conflicts of Interest: The authors declare no conflict of interest.

References

1. Preiss, J.; Yung, S.; Baecker, P.A. Regulation of bacterial glycogen synthesis. *Mol. Cell Biochem.* **1983**, *57*, 61–80. [CrossRef] [PubMed]
2. Preiss, J. Bacterial glycogen synthesis and its regulation. *Annu. Rev. Microbiol.* **1984**, *38*, 419–458. [CrossRef] [PubMed]
3. Preiss, J.; Romeo, T. Physiology, biochemistry and genetics of bacterial glycogen synthesis. *Adv. Microb. Physiol.* **1989**, *30*, 183–238. [PubMed]
4. Yamamotoya, T.; Dose, H.; Tian, Z.; Fauré, A.; Toya, Y.; Honma, M.; Igarashi, K.; Nakahigashi, K.; Soga, T.; Mori, H.; et al. Glycogen is the primary source of glucose during the lag phase of *E. coli* proliferation. *Biochim. Biophys. Acta* **2012**, *1824*, 1442–1448. [CrossRef] [PubMed]
5. Wang, M.; Liu, Q.; Kang, X.; Zhu, Z.; Yang, H.; Xi, X.; Zhang, X.; Du, Y.; Guo, M.; Tang, D.; et al. Glycogen metabolism impairment via single gene mutation in the *glgBXCAP* operon alters the survival rate of *Escherichia coli* under various environmental stresses. *Front. Microbiol.* **2020**, *11*, 588099. [CrossRef] [PubMed]
6. Romeo, T.; Kumar, A.; Preiss, J. Analysis of the *Escherichia coli* glycogen gene cluster suggests that catabolic enzymes are encoded among the biosynthetic genes. *Gene* **1988**, *70*, 363–376. [CrossRef]
7. Preiss, J. Glycogen: Biosynthesis and regulation. *EcoSal Plus* **2014**, *6*, ESP-0015-2014. [CrossRef]
8. Greenberg, E.; Preiss, J. The occurrence of adenosine diphosphate glucose: Glycogen transglucosylase in bacteria. *J. Biol. Chem.* **1964**, *239*, 4314–4315. [CrossRef]
9. Dietzler, D.N.; Leckie, M.P.; Lais, C.J.; Magnani, J.L. Evidence for the allosteric regulation of bacterial glycogen synthesis in vivo. *Arch. Biochem. Biophys.* **1974**, *162*, 602–606. [CrossRef]
10. Ballicora, M.A.; Iglesias, A.A.; Preiss, J. ADP-glucose pyrophosphorylase: A regulatory enzyme for bacterial glycogen synthesis. *Microbiol. Mol. Biol. Rev.* **2003**, *67*, 213–225. [CrossRef]
11. Romeo, T.; Preiss, J. Genetic regulation of glycogen biosynthesis in *Escherichia coli*: In vitro effects of cyclic AMP and guanosine 5′-diphosphate 3′-diphosphate and analysis of in vivo transcripts. *J. Bacteriol.* **1989**, *171*, 2773–2782. [CrossRef] [PubMed]
12. Hengge-Aronis, R.; Fischer, D. Identification and molecular analysis of *glgS*, a novel growth-phase-regulated and *rpoS*-dependent gene involved in glycogen synthesis in *Escherichia coli*. *Mol. Microbiol.* **1992**, *6*, 1877–1886. [CrossRef] [PubMed]
13. Montero, M.; Almagro, G.; Eydallin, G.; Viale, A.M.; Munoz, F.J.; Bahaji, A.; Li, J.; Rahimpour, M.; Baroja-Fernandez, E.; Pozueta-Romero, J. *Escherichia coli* glycogen genes are organized in a single *glgBXCAP* transcriptional unit possessing an alternative suboperonic promoter within *glgC* that directs *glgAP* expression. *Biochem. J.* **2011**, *433*, 107–117. [CrossRef] [PubMed]
14. Romeo, T.; Gong, M.; Liu, M.Y.; Brun-Zinkernagel, A.M. Identification and molecular characterization of *csrA*, a pleiotropic gene from *Escherichia coli* that affects glycogen biosynthesis, gluconeogenesis, cell size, and surface properties. *J. Bacteriol.* **1993**, *175*, 4744–4755. [CrossRef]
15. Almagro, G.; Viale, A.M.; Montero, M.; Munoz, F.J.; Baroja-Fernandez, E.; Mori, H.; Pozueta-Romero, J. A cAMP/CRP-controlled mechanism for the incorporation of extracellular ADP-glucose in *Escherichia coli* involving NupC and NupG nucleoside transporters. *Sci. Rep.* **2018**, *8*, 15509. [CrossRef]
16. Eydallin, G.; Viale, A.M.; Morán-Zorzano, M.T.; Muñoz, F.J.; Montero, M.; Baroja-Fernández, E.; Pozueta-Romero, J. Genome-wide screening of genes affecting glycogen metabolism in *Escherichia coli* K-12. *FEBS Lett.* **2007**, *581*, 2947–2953. [CrossRef]
17. Eydallin, G.; Montero, M.; Almagro, G.; Sesma, M.T.; Viale, A.M.; Muñoz, F.J.; Rahimpour, M.; Baroja-Fernández, E.; Pozueta-Romero, J. Genome-wide screening of genes whose enhanced expression affects glycogen accumulation in *Escherichia coli*. *DNA Res.* **2010**, *17*, 61–71. [CrossRef]

18. Montero, M.; Eydallin, G.; Viale, A.M.; Almagro, G.; Muñoz, F.J.; Rahimpour, M.; Sesma, M.T.; Baroja-Fernández, E.; Pozueta-Romero, J. *Escherichia coli* glycogen metabolism is controlled by the PhoP-PhoQ regulatory system at submillimolar environmental Mg^{2+} concentrations, and is highly interconnected with a wide variety of cellular processes. *Biochem. J.* **2009**, *424*, 129–141 [CrossRef]
19. Ishihama, A.; Shimada, T.; Yamazaki, Y. Transcription profile of *Escherichia coli*: Genomic SELEX search for regulatory targets of transcription factors. *Nucleic Acids Res.* **2016**, *44*, 2058–2074. [CrossRef]
20. Ishihama, A.; Shimada, T. Hierarchy of transcription factor network in *Escherichia coli* K-12: H-NS-mediated silencing and Anti-silencing by global regulators. *FEMS Microbiol. Rev.* **2021**, *45*, fuab032. [CrossRef]
21. Karp, P.D.; Paley, S.; Caspi, R.; Kothari, A.; Krummenacker, M.; Midford, P.E.; Moore, L.R.; Subhraveti, P.; Gama-Castro, S.; Tierrafria, V.H.; et al. The EcoCyc Database. *EcoSal Plus* **2023**, *11*, eesp00022023. [CrossRef] [PubMed]
22. Tierrafría, V.H.; Rioualen, C.; Salgado, H.; Lara, P.; Gama-Castro, S.; Lally, P.; Gómez-Romero, L.; Peña-Loredo, P.; López-Almazo, A.G.; Alarcón-Carranza, G.; et al. RegulonDB 11.0: Comprehensive high-throughput datasets on transcriptional regulation in *Escherichia coli* K-12. *Microb. Genom.* **2022**, *8*, mgen000833. [CrossRef] [PubMed]
23. Grainger, D.C.; Hurd, D.; Harrison, M.; Holdstock, J.; Busby, S.J.W. Studies of the distribution of *Escherichia coli* cAMP-receptor protein and RNA polymerase along the *E. coli* chromosome. *Proc. Natl. Acad. Sci. USA* **2005**, *102*, 17693–17698. [CrossRef] [PubMed]
24. Shimada, T.; Ogasawara, H.; Ishihama, A. Genomic SELEX screening of regulatory targets of *Escherichia coli* transcription factors. *Methods Mol. Biol.* **2018**, *1837*, 49–69. [PubMed]
25. Jain, D. Allosteric control of transcription in GntR family of transcription regulators: A structural overview. *IUBMB Life* **2015**, *67*, 556–563. [CrossRef]
26. Shimada, T.; Ishihama, A.; Busby, S.J.W.; Granger, D.C. The *Escherichia coli* RutR transcription factor binds at targets within genes as well as intergenic regions. *Nucleic Acids Res.* **2008**, *36*, 3950–3955. [CrossRef]
27. Datsenko, K.A.; Wanner, B.L. One-step inactivation of chromosomal genes in *Escherichia coli* K-12 using PCR products. *Proc. Natl. Acad. Sci. USA* **2000**, *97*, 6640–6645. [CrossRef]
28. Baba, T.; Ara, T.; Hasegawa, M.; Takai, Y.; Okumura, Y.; Baba, M.; Datsenko, K.A.; Tomita, M.; Wanner, B.L.; Mori, H. Construction of *Escherichia coli* K-12 in-frame, single-gene knockout mutants: The Keio collection. *Mol. Syst. Biol.* **2006**, *2*, 2006.0008. [CrossRef]
29. Govons, S.; Vinopal, R.; Ingraham, J.; Preiss, J. Isolation of mutants of *Escherichia coli* B altered in their ability to synthesize glycogen. *J. Bacteriol.* **1969**, *97*, 970–972. [CrossRef]
30. Shimada, T.; Fujita, N.; Maeda, M.; Ishihama, A. Systematic search for the Cra-binding promoters using genomic SELEX system. *Genes. Cells* **2005**, *10*, 907–918. [CrossRef]
31. Anzai, T.; Imamura, S.; Ishihama, A.; Shimada, T. Expanded roles of pyruvate-sensing PdhR in transcription regulation of the *Escherichia coli* K-12 genome: Fatty acid catabolism and cell motility. *Microb. Genom.* **2020**, *6*, mgen000442. [CrossRef] [PubMed]
32. Anzai, T.; Kijima, K.; Fujimori, M.; Nakamoto, S.; Ishihama, A.; Shimada, T. Expanded roles of lactate-sensing LldR in transcription regulation of the *Escherichia coli* K-12 genome: Lactate utilisation and acid resistance. *Microb. Genom.* **2023**, *9*, mgen001015. [CrossRef] [PubMed]
33. Iijima, H.; Watanabe, A.; Sukigara, H.; Iwazumi, K.; Shirai, T.; Kondo, A.; Osanai, T. Four-carbon dicarboxylic acid production through the reductive branch of the open cyanobacterial tricarboxylic acid cycle in *Synechocystis* sp. PCC 6803. *Metab. Eng.* **2021**, *65*, 88–98. [CrossRef]
34. Shimada, T.; Katayama, Y.; Kawakita, S.; Ogasawara, H.; Nakano, M.; Yamamoto, K.; Ishihama, A. A novel regulator RcdA of the *csgD* gene encoding the master regulator of biofilm formation in *Escherichia coli*. *MicrobiologyOpen* **2012**, *1*, 381–394. [CrossRef] [PubMed]
35. Shimada, T.; Furuhata, S.; Ishihama, A. The whole set of constitutive promoters for RpoN sigma factor and the regulatory role of its enhancer protein NtrC in *Escherichia coli* K-12. *Microb. Genom.* **2021**, *7*, mgen000653. [CrossRef] [PubMed]
36. Shimada, T.; Bridier, A.; Briandet, R.; Ishihama, A. Novel roles of LeuO in transcription regulation of *E. coli* genome: Antagonistic interplay with the universal silencer H-NS. *Mol. Microbiol.* **2011**, *82*, 378–397. [CrossRef] [PubMed]
37. Lacour, S.; Landini, P. SigmaS-dependent gene expression at the onset of stationary phase in *Escherichia coli*: Function of sigmaS-dependent genes and identification of their promoter sequences. *J. Bacteriol.* **2004**, *186*, 7186–7195. [CrossRef]
38. Saier, M.H., Jr.; Reddy, V.S.; Tsu, B.V.; Ahmed, M.S.; Li, C.; Moreno-Hagelsieb, G. The Transporter Classification Database (TCDB): Recent advances. *Nucleic Acids Res.* **2016**, *44*, D372–D379. [CrossRef]
39. Finn, R.D.; Coggill, P.; Eberhardt, R.Y.; Eddy, S.R.; Mistry, J.; Mitchell, A.L.; Potter, S.C.; Punta, M.; Qureshi, M.; Sangrador-Vegas, A.; et al. The Pfam protein families database: Towards a more sustainable future. *Nucleic Acids Res.* **2016**, *44*, D279–D285. [CrossRef]
40. Ishihama, A. Molecular assembly and functional modulation of *Escherichia coli* RNA polymerase. *Adv. Biophys.* **1990**, *26*, 19–31. [CrossRef]
41. Jishage, M.; Iwata, A.; Ueda, S.; Ishihama, A. Regulation of RNA polymerase sigma subunit synthesis in *Escherichia coli*: Intracellular levels of four species of sigma subunit under various growth conditions. *J. Bacteriol.* **1996**, *178*, 5447–5451. [CrossRef]
42. Shimada, T.; Ogasawara, H.; Ishihama, A. Single-target regulators form a minor group of transcription factors in *Escherichia coli* K-12. *Nucleic Acids Res.* **2018**, *46*, 3921–3936. [CrossRef] [PubMed]

43. Shimada, T.; Murayama, R.; Mashima, T.; Kawano, N.; Ishihama, A. Regulatory role of CsuR (YiaU) in determination of cell surface properties of *Escherichia coli* K-12. *Microbiology* **2022**, *168*, 001166. [CrossRef] [PubMed]
44. Kobayashi, I.; Mochizuki, K.; Teramoto, J.; Imamura, S.; Takaya, K.; Ishihama, A.; Shimada, T. Transcription factors SrsR (YgfI) is a novel regulator for the stress-response genes in stationary phase in *Escherichia coli* K-12. *Int. J. Mol. Sci.* **2022**, *23*, 6055. [CrossRef] [PubMed]
45. Shimada, T.; Ogasawara, H.; Kobayashi, I.; Kobayashi, N.; Ishihama, A. Single-target regulators constitute the minority group of transcription factors in *Escherichia coli* K-12. *Front. Microbiol.* **2021**, *12*, 697803. [CrossRef]
46. Choi, P.; Cai, L.; Frieda, K.; Xie, X.S. A stochastic single-molecule event triggers phenotype switching of a bacterial cell. *Science* **2008**, *322*, 442–446. [CrossRef]
47. Gayán, E.; Wang, Z.; Salvador, M.; Gänzle, M.G.; Aertsen, A. Dynamics of high hydrostatic pressure resistance development in RpoS-deficient *Escherichia coli*. *Food Res. Int.* **2023**, *164*, 112280. [CrossRef]

microorganisms MDPI

Article

A Novel Regulator PepR Regulates the Expression of Dipeptidase Gene *pepV* in *Bacillus thuringiensis*

Xin Zhang [1,†], Hengjie Wang [1,†], Tinglu Yan [1], Yuhan Chen [1,2], Qi Peng [1] and Fuping Song [1,*]

[1] State Key Laboratory for Biology of Plant Diseases and Insect Pests, Institute of Plant Protection, Chinese Academy of Agricultural Sciences, Beijing 100193, China; zhangxin_live@outlook.com (X.Z.); whj18911651571@163.com (H.W.); yantinglu9969@163.com (T.Y.); chenyuhan0909@163.com (Y.C.); qpeng@ippcaas.cn (Q.P.)

[2] College of Life Science, Northeast Agricultural University, Harbin 150030, China

* Correspondence: fpsong@ippcaas.cn

† These authors contributed equally to this work.

Abstract: *Bacillus thuringiensis* produces insecticidal crystal proteins encoded by *cry* or *cyt* genes and targets a variety of insect pests. We previously found that a strong promoter of a DeoR family transcriptional regulator (HD73_5014) can efficiently drive *cry1Ac* expression in *B. thuringiensis* HD73. Here, we investigated the regulation of neighbor genes by HD73_5014. The HD73_5014 homologs are widely distributed in Gram-positive bacterial species. Its neighbor genes include *pepV*, *rsuA*, and *ytgP*, which encode dipeptidase, rRNA pseudouridine synthase and polysaccharide biosynthesis protein, respectively. The four open reading frames (ORFs) are organized to be a *pepR* gene cluster in HD73. RT-PCR analysis revealed that the *rsuA* and *ytgP* genes formed a transcriptional unit (*rsuA-ytgP* operon), while *pepV* formed a transcriptional unit in HD73. Promoter-*lacZ* fusion assays showed that the *pepV* and *rsuA-ytgP* promoters are regulated by HD73_5014. EMSA experiments showed that HD73_5014 directly binds to the *pepV* promoter region but not to the *rusA-ytgP* promoter region. Thus, the HD73_5014 transcriptional regulator, which controls the expression of the dipeptidase *pepV*, was named PepR (dipeptidase regulator). We also confirmed the direct regulation between PepR and PepV by the increased sensitivity to vancomycin in Δ*pepV* and Δ*pepR* mutants compared to HD73.

Keywords: transcriptional regulator PepR; dipeptidase gene *pepV*; resistant vancomycin resistance; *Bacillus thuringiensis*

check for updates

Citation: Zhang, X.; Wang, H.; Yan, T.; Chen, Y.; Peng, Q.; Song, F. A Novel Regulator PepR Regulates the Expression of Dipeptidase Gene *pepV* in *Bacillus thuringiensis*. *Microorganisms* **2024**, *12*, 579. https://doi.org/10.3390/microorganisms12030579

Academic Editor: Tomohiro Shimada

Received: 23 February 2024
Revised: 11 March 2024
Accepted: 12 March 2024
Published: 14 March 2024

1. Introduction

Bacillus thuringiensis (Bt) is a very important microorganism in the biological control of plant pests and diseases [1]. During sporulation, it can produce crystals which are mainly formed by the insecticidal crystal proteins and are toxic to many pests, including more than 500 species in nine orders such as *Lepidoptera*, *Hymenoptera*, *Diptera*, *Coleoptera*, *Trichoptera*, *Orthoptera*, and so on [2]. As of an update dated 22 February 2024, approximately 858 *cry* and *cyt* genes encoding crystal proteins have been discovered (http://www.lifesci.sussex.ac.uk/home/Neil_Crickmore/Bt/). The *cry* genes express during the stationary phase, and crystal production constitutes 20 to 30% of the dry weight of sporulating cells [3].

Several *cry* and non-*cry* gene promoters have been identified to effectively direct *cry* gene expression. For example, the *cry1Ac* promoter can direct the expression of a variety of *cry*-like genes [4], such as *cry1Ac*, *cry1Ab*, *cry1Ac5*, *cry1Ac-av3*, *cry2Ab27*, and *cry8*. Similarly, the expression of Cry1AbMod/Cry1AcMod [5], Cry1Ac [6], Cry1c [7], Cry3A [8], Cry8Ga [9], and Cry69Aa1 [10] can be directed by the *cry3A* promoter. Recent reports have shown that some non-*cry* gene promoters with high-level activity were utilized for the expression of *cry* genes [11,12]. P*exsY* is a strong promoter of the exosporium basal layer structural gene *exsY* in late sporulation which has been used to express *cry1Ac* genes in Bt [11]. P*5014*, a non-*cry* gene strong promoter controlled by Sigma E, was used to strongly direct *cry1Ac* expression in Bt HD73. The *HD73_5014* gene with promoter P*5014* was annotated as the DeoR family transcriptional regulator [12]. However, the function and the target gene of HD73_5014 as transcriptional regulator are still unknown.

In the HD73 genome, the upstream of the *HD73_5014* gene was the *pepV* gene, which encodes a dipeptidase. Dipeptidases are involved in the final breakdown of protein degradation fragments produced by other peptidases (e.g., aminopeptidasesare PepN, PepC, PepP, PepX, PepA [13] and endopeptidasesare PepO1, PepO2, PepF1 and PepF2 [14,15]). In *Lactococcus lactis* MG1363, PepV has been found to be involved in resistance to vancomycin. The transcription of *pepV* in *L. helveticus* was regulated by a CodY-like regulator and the BCAA-responsive transcriptional regulator which binds adjacent to the *pepV* promoter region [16,17]. However, whether PepV had a similar function and its transcription regulated by PepR (HD73_5014, encoding dipeptidase regulator) required further investigation in *B. thuringiensis*.

2. Materials and Methods

2.1. Bacterial Strains, Plasmids, and Growth Conditions

Bacillus thuringiensis HD73 and its derivatives were cultured at 30 °C in Luria–Bertani (LB) medium (1% tryptone, 0.5% yeast extract, and 0.5% NaCl) or on solid LB medium supplemented with 1.5% agar. Schaeffer's sporulation medium [18] (SSM; 0.8% nutrient broth, 0.012% $MgSO_4$, 0.1% KCl, 0.5 mM NaOH, 1 mM $Ca(NO_3)_2$, 0.01 μM $MnCl_2$, and 1 μM $FeSO_4$) was used to observe the development of bacterial cells. *Escherichia coli* TG1 was used for molecular cloning experiments, and *E. coli* ET 12567 was used for producing non-methylated plasmid DNA for *B. thuringiensis* transformations [19,20]. These strains were cultured at 37 °C in LB medium. When required, antibiotics were added at the following concentrations for growth of *B. thuringiensis*: 5 μg/mL erythromycin, 50 μg/mL kanamycin, and 6 ng/mL vancomycin. For growth of *E. coli*, 100 μg/mL ampicillin was added. The bacterial strains and plasmids used in this study are summarized in Table 1.

Table 1. Strains and plasmids used in this study.

Strain or Plasmid	Relevant Details [a]	Reference or Source
	E. coli strains	
E. coli TG1	Δ(*lac-proAB*) *supE thi hsd-5* (*F′ traD36 proA*⁺ *proB*⁺ *lacI*^{q} *lacZ*ΔM15), general purpose cloning host	Laboratory collection
E. coli ET 12567	F^- *dam-13*::Tn *9 dcm-6 hsdM hsdR recF143 zjj-202*::Tn10 *galK2 galT22 ara14 pacY1 xyl-5 leuB6 thi-1*, for generation of unmethylated DNA	Laboratory collection
BL21 (DE3)	F^- *dcmopmT hsds* (r_B^- m_B^-) galλ(DE3)	[21]
BL21 (pET*pepR*)	BL21(DE3) with pET*pepR* plasmid	This study
	B. thuringiensis strains	
HD73	Wild-type strain containing plasmid pHT73 carrying *cry1Ac* gene	Laboratory collection
Δ*pepR*	HD73 mutant, *pepR* gene was deleted by homologs recombination	This study
Δ*pepV*	HD73 mutant, *pepV* gene was deleted by homologs recombination	This study
HD (P*pepV*)	HD73 strain containing plasmid pHT304-P*pepV*-18Z	This study
Δ*pepR* (P*pepV*)	*pepR* mutant containing plasmid pHT304-P*pepV*-18Z	This study
HD (P*rsuA-ytgP*)	HD73 strain containing plasmid pHT304- P*rsuA-ytgP*-18Z	This study
Δ*pepR* (P*rsuA-ytgP*)	*pepR* mutant containing plasmid pHT304- P*rsuA-ytgP*-18Z	This study
Plasmids		
pHT304-18Z	*E. coli*-Bt shuttle vector with promoter-less *lacZ* reporter, Amp^R, Erm^R	[22]
pET-21b	Expression vector; Amp^r 5.4 kb	Laboratory collection
pMAD	Amp^R, Erm^R, temperature-sensitive *E. coli-B. thuringiensis* shuttle vector	[23]
pMADΩD*spo0A*	pMAD with *spo0A* deletion fragment, Amp^R, Erm^R, Kan^R	[24]
pMADΩD*pepR*	pMAD with *pepR* deletion fragment, Amp^R, Erm^R, Kan^R	This study
pMADΩD*pepV*	pMAD with *pepV* deletion fragment, Amp^R, Erm^R, Kan^R	This study
pET*pepR*	pET-21b containing *pepR* gene; Amp^r	This study
304P*pepV*	pHT304-18Z carrying P*pepV*, Amp^R, Erm^R	This study
304P*rsuA-ytgP*	pHT304-18Z carrying P*rsuA-ytgP*, Amp^R, Erm^R	This study

[a] Antibiotic resistance cassettes are indicated as follows: Erm^R, erythromycin resistance; Kan^R, kanamycin resistance; Amp^R, ampicillin resistance.

2.2. DNA Manipulation and Transformation

Plasmid DNA was extracted from *E. coli* cells with a Plasmid Miniprep Kit (Axygen, Beijing, China). Restriction enzymes and T4 DNA ligase (Takara Biotechnology Corporation, Dalian, China) were used according to the manufacturer's instructions. PCR was performed with high-fidelity PrimeSTAR HS DNA polymerase (Takara Biotechnology Corporation, Dalian, China) or *Taq* DNA polymerase (BioMed, Beijing, China). DNA fragments were purified from 1% agarose gels using an AxyPrep DNA Gel Extraction Kit (Axygen) Standard procedures were followed for *E. coli* transformation [25], and Bt HD73 cells were transformed by electroporation, as previously described [26].

2.3. RNA Isolation and Reverse Transcription PCR (RT-PCR) of pepR Neighbor Genes

The HD73 strain was cultured in SSM at 30 °C and harvested at T_6. Total RNA was extracted using the RNAprep Pure Bacteria Kit (Aidlab, Beijing, China). The RNA (500 ng) was used for reverse transcription using the HiScript III 1st Strand cDNA Synthesis Kit (Vazyme, Nanjing, China). The primers were used to detect the expression of *pepR* and neighbor genes locus, as shown in Table 2. To confirm the absence of DNA contamination, all RNA samples were routinely subjected to 16S rRNA gene PCR using 16S rRNA-F/16S rRNA-R primers.

Table 2. Oligonucleotide primers used in this study.

Primer Name	Sequence (5′–3′) [a]	Restriction Site
pET*pepR*-F	CGGGATCCGTTGAAACCTACAACTACTCG	BamHI
pET*pepR*-R	ACGCGTCGACTGAGGTCATTCTCACTTTC	SalI
pepV-AF	GTACCCGGGAGCTCGAATTCCGAAATGTCCGACTTGTTCCATACG	EcoRI
pepV-AR	TCACCTCAAATGGTTCGCTGGAATTAGCGAAGTAATGGATATATAAATAATGTCCGCTC	
pepV-KF	GAGCGGACATTATTTATATATCCATTACTTCGCTAATTCCAGCGAACCATTTGAGGTGA	
pepV-KR	GAAGGATGGATGCGTGATGTCAGCAATTAATTGGAAATTCCTCGTAGGCGCTCG	
pepV-BF	CGAGCGCCTACGAGGAATTTCCAATTAATTGCTGACATCACGCATCCATCCTTC	
pepV-BR	CGTCGGGCGATATCGGATCCGCACACGTTGCAGGAGTAGTAACAGAAG	BamHI
pepV-WF	GATACACAGCACCTAAATCTGTACCTTCG	
pepV-WR	CCCAGTTGGACGACTTGATATCGATACGGAAGG	
pepR-AF	GTACCCGGGAGCTCGAATTCAATCAAATGAAACAAGTTCAT	EcoRI
pepR-AR	TCAAATGGTTCGCTGAATGCGTGTTAGCATACGAG	
pepR-KF	ATGCTAACACGCATTCAGCGAACCATTTGAGGTGA	
pepR-KR	TTATGAGGTCATTCTAAATTCCTCGTAGGCGCTCG	
pepR-BF	GCCTACGAGGAATTTAGAATGACCTCATAATGAAA	
pepR-BR	CGTCGGGCGATATCGGATCCCAATTTGTGCAGGTATTGGC	BamHI
pepR-WF	CTTCTCCTGTAATAACCGCTTCTGC	
pepR-WR	CTAACACTTATAATGGTCGTTGCTG	
pMAD-F	CCAAATTTCCTCTGGCCATT	
pMAD-R	CCTATACCTTGTCTGCCTCC	
P*pepV*-F	CGCGGATCCCATCACGCATCCATCCTTCACT	BamHI
P*pepV*-R	AACTGCAGGAGAAAACCACTCCTCTAC	PstI
P*rsuA-ytgP*-F	AACTGCAGCTGCAGCACCAATTGCAGCCATTAG	PstI
P*rsuA-ytgP*-R	CGCGGATCCGTAACAATAAGCGTTCCGCGCAG	BamHI
RT-*thpR*-F	ACGAGCAACGGTAATA	
RT-*thpR*-R	GGGCGAAGGTAAGTGA	
RT-*pepV*-F	TGATGTGGGCGAGAAT	
RT-*pepV*-R	ACGGTGGTAACTTAGGG	
RT-*thpR-pepV*-F	GTACATCTTCTCCTTCTTTC	
RT-*thpR-pepV*-R	GGCCCGCTATTCCCTGG	
RT-*rsuA*-F	CATCCTCTTCTGTTACTACTCC	
RT-*rsuA*-R	CAGCGACGGAAGATGATAATC	
RT-*ytgP*-F	GCTCCTACTGTACCGAAATAACG	
RT-*ytgP*-R	GATTCAGATCCACTAGGTGGAC	
RT-*rsuA-ytgP*-F	GTTTTTCACCTTCTTTATTTATC	
RT-*rsuA-ytgP*-R	CATGATTTCGCCTGATGGCCG	
RT-*16S*-F	TCGCATTAGCTAGTTGGTGAG	
RT-*16S*-R	TCTTCCCTAACAACAGAGTTT	

[a] Restriction enzyme sites are underlined.

2.4. β-Galactosidase Activity Assays

The *pepV* (P*pepV*, 360 bp) and *rsuA-ytgP* (P*rsuA-ytgP*, 402 bp) promoters were amplified from HD73 genomic DNA using primers P*pepV*-F/P*pepV*-R and P*rsuA-ytgP*-F/P*rsuA-ytgP*-R, respectively. P*pepV* and P*rsuA-ytgP* were digested with PstI and BamHI sites, followed by ligation into the linearized pHT304-18Z plasmid, which harbors a promoterless *lacZ* [22], to obtain the recombinant plasmid, 304P*pepV* and 304P*rsuA-ytgP*. The 304P*pepV* and 304P*rsuA-ytgP* plasmid were introduced into HD73 and Δ*pepR*, resulting in HD (P*pepV*), Δ*pepR* (P*pepV*), HD (P*rsuA-ytgP*), and Δ*pepR* (P*rsuA-ytgP*), respectively. The HD (P*pepV*), Δ*pepR* (P*pepV*), HD (P*rsuA-ytgP*), and Δ*pepR* (P*rsuA-ytgP*) strains were validated by erythromycin and PCR.

To detect the transcriptional activity of the P*pepV* and P*rsuA-ytgP* promoters in HD73 and Δ*pepR* strains, HD (P*pepV*), Δ*pepR* (P*pepV*), HD (P*rsuA-ytgP*), and Δ*pepR* (P*rsuA-ytgP*) were cultured in SSM medium at 30 °C with shaking 220 rpm. Two milliliters of culture were collected at 1 h intervals from T_0 to T_7 (T_0 indicates the end of the exponential growth phase and T_n indicates n hours after time T_0). The cells were centrifuged (12,000× *g*, 1 min), and the pellets were stored at −20 °C until use. β-galactosidase activities were measured as previously described [24]. Values are reported as the mean and standard error of at least three independent assays.

2.5. Expression and Purification of PepR

The pET*pepR* plasmid containing *pepR* from HD73 was constructed by amplifying *pepR* with primers pET*pepR*-F/pET*pepR*-R and cloning into BamHI/SalI-digested pET21b. The pET*pepR* was transferred into the *E. coli* BL21(DE3), and the positive transformants were BL21 (pET*pepR*). The transformants were grown to the OD_{600nm} of approximately 1.0 in LB medium supplemented with ampicillin; then, they were incubated with isopropyl-β-D-thiogalactoside (IPTG) at a final concentration of 1 mM in an 18 °C shaking incubator for 12 h. Total cell proteins from each optimization experiment were analyzed by sodium dodecyl sulfate–polyacrylamide gel electrophoresis (SDS-PAGE). The expression and purification of the PepR-His protein was performed as previously described [27].

2.6. Electrophoresis Mobility Shift Assays

The *pepV* fragment was PCR amplified from HD73 genomic DNA by specific primers labeled with a 5′-end FAM modification and confirmed by *pepV* sequencing. Electrophoresis mobility shift assays (EMSA) [28] were performed as described to analyze the binding of purified PepR.

2.7. Vancomycin Sensitivity Assay

The Δ*pepR*, Δ*pepV*, and HD73 strains were grown in LB medium until they reached an optical density of approximately 1.0 at 600 nm (OD 600 nm). Then, each strain suspension was divided into triplicates and transferred to 100 mL of SSM medium with a vancomycin concentration of 6 ng/mL. The OD 600 nm was measured every hour at 30 °C.

3. Results

3.1. PepR Homologs Are Widely Distributed in Various Gram-Positive Bacteria

In a previous study, we found that the *pepR* gene promoter can efficiently direct *cry1Ac* expression in HD73 [12]. However, the function of the *pepR* gene and the target gene of PepR as a transcriptional regulator are still unknown. In this study, we analyzed PepR homologs through a phylogenetic evolutionary tree. PepR homologs with amino acid similarity greater than 70% and 100% amino acid coverage from the NCBI database were screened (Table S1). The PepR homologs were searched with an E-value lower than 2.07×10^{-45}. The phylogenetic tree showed that the PepR homologs were present in 143 bacterial species, all of which were Gram-positive and mainly included *Anoxybacillus* spp., *Bacillus* spp., *Caldibacillus* spp., *Cytobacillus* spp., *Domibacillus* spp., *Ectobacillus* spp., *Heyndrickxia* spp., *Jeotgalibacillus* spp., *Kurthia* spp., *Lysinibacillus* spp., *Metabacillus* spp.,

Neobacillus spp., *Paenisporosarcina* spp., *Peribacillus* spp., *Planococcus* spp., *Psychrobacillus* spp., *Sporosarcina* spp., and *Sutcliffiella* spp., etc. (Figure 1). This finding suggests that PepR homologs are highly conserved across a wide range of Gram-positive bacteria.

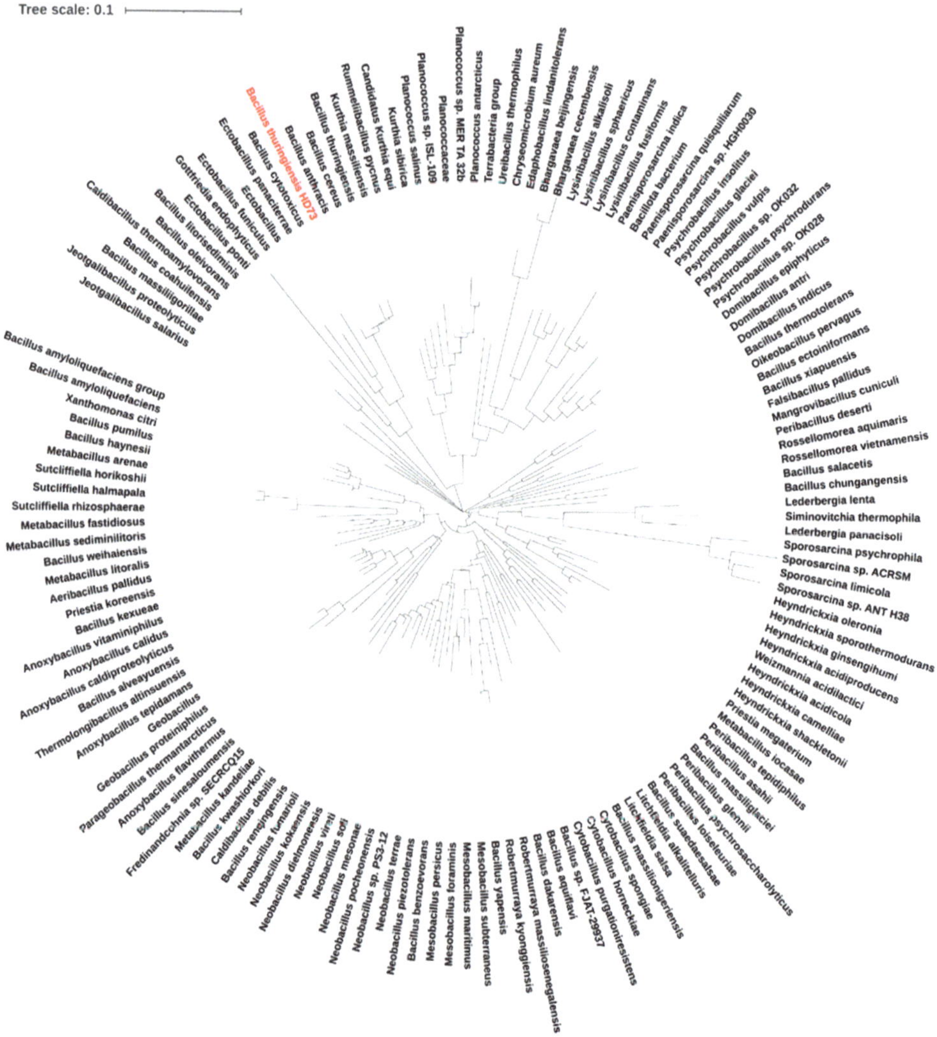

Figure 1. Evolution of PepR homologs in bacteria. Evolutionary analyses were conducted in MEGA7 using the neighbor-joining method. A total of 143 amino acid sequences were analyzed and an optimal tree was shown. Red text indicates the *B. thuringiensis* HD73 strain from which the reference sequence of the PepR homologs originated.

3.2. Characterization of the Transcription Units in the pepR Gene Cluster

The nucleotide sequence of the *pepR* gene cluster (4489 bp) of HD73 is comprised of four ORFs, which are annotated as dipeptidase (*pepV*, *HD73_5013*), DeoR family transcriptional regulator (*pepR*, *HD73_5014*), rRNA pseudouridine synthase (*rsuA*, *HD73_5015*), and polysaccharide biosynthesis protein (*ytgP*, *HD73_5016*) (Figure 2A). To determine the transcription units of the *pepR* cluster, a series of primers were designed. The total RNA was extracted at T_6 from *B. thuringiensis* HD73 cultures grown in SSM. RT-PCR results showed mRNA overlapping the *rsuA* and *ytgP* genes (Figure 2B). However, no mRNA overlapping *thpR* and *pepV* was transcribed. These results indicate that *rsuA* and *ytgP* genes form a transcriptional unit (*rsuA-ytgP* operon), while *thpR* and *pepV* each form a transcriptional unit independently.

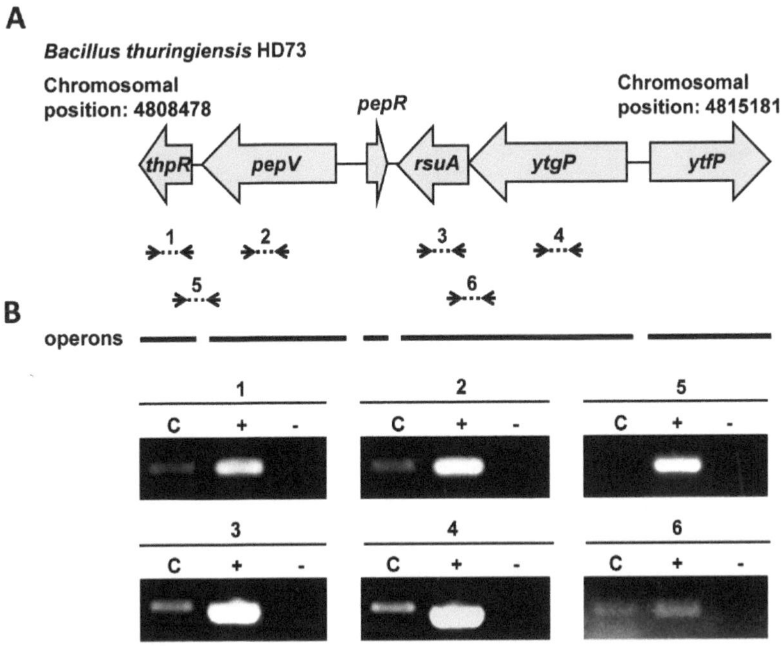

Figure 2. RT-PCR identified the *pepV* and *rsuA-ytgP* transcription units at the *pepR* gene locus in *B. thuringiensis* HD73. (**A**) Genetic organization of the *pepR* locus. Open reading frames (ORFs) are represented by grey arrows. The RT-PCR amplicons (see lanes in panel (**B**)) correspond to the dashed lines with small black arrows. The solid lines below the ORFs represent the operons. (**B**) RT-PCR analyzed the transcription units of *pepR* neighboring genes. RNA samples were prepared at T_6 of bacterial culture in SSM medium. The RT-PCR labelled "C" was performed with 500 ng RNA. Positive controls are labelled "+" (PCR with 100 ng of genomic DNA). Negative controls are labelled "−" (RT-PCR with 500 ng RNA using heat-inactivated reverse transcriptase). Numbers represent different RT-PCR amplicons: numbers 1–4 represent *thpR*, *pepV*, *rsuA*, and *ytgP*; numbers 5 and 6 represent *thpR-pepV* and *rsuA-ytgP*, respectively.

3.3. *PepR Positively Regulates pepV and rsuA-ytgP*

To verify whether PepR regulates the transcription of *pepV* and *rsuA-ytgP*, we first constructed a *pepR* deletion mutant. The upstream (*pepR*-A, 1065 bp) and downstream (*pepR*-B, 1057 bp) homologous fragments of *pepR* were amplified from the HD73 genome DNA. Then, the *pepR*-A, *pepR*-B, and kanamycin resistance gene (*pepR*-K, 1506 bp) fragments were ligated together by overlapping PCR. The resulting 3563-bp fragment was inserted into the temperature-sensitive pMAD plasmid, leading to pMADΩD*pepR* plasmid. The pMADΩD*pepR* plasmid was electroporated into HD73. This transformant was then screened at 37 °C to identify the *pepR* deletion mutant, which lacked erythromycin resistance but was resistant to kanamycin. The diagram shows that the kanamycin resistance gene on the recombinant plasmid was exchanged by homolog recombination with *pepR* on the HD73 chromosome (Figure 3A). HD73 and Δ*pepR* strains were confirmed by PCR using primers of *pepR*-WF/*pepR*-WR. PCR products with the size of 3677 bp from Δ*pepR* and 2372 bp from HD73 were detected by agarose gel electrophoresis (Figure 3B). The Δ*pepR* DNA as template was amplified by PCR using primers (pMAD-F/pMAD-R) for detecting the pMAD plasmid, resulting in no PCR products (Figure 3B). It proved that the Δ*pepR* mutant was obtained.

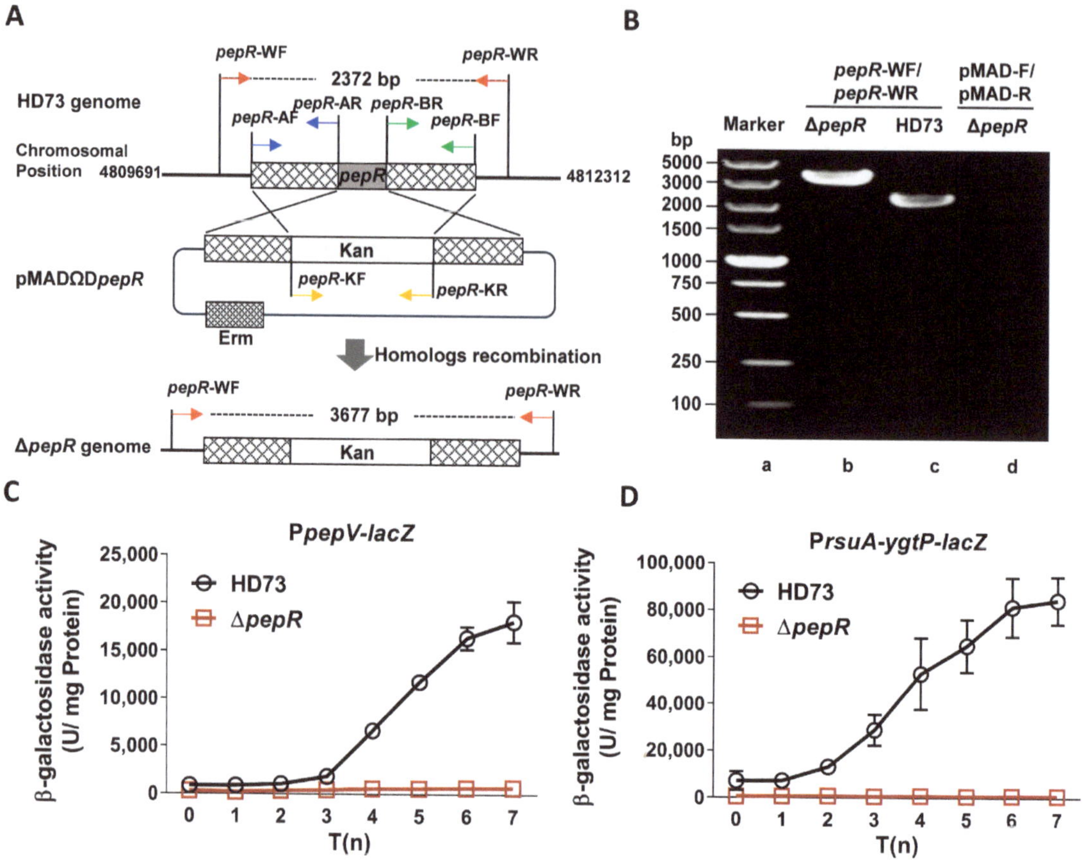

Figure 3. Transcription of *pepV* and *rsuA-ygtP* depended on *pepR*. (**A**) Construction of an in-frame deletion *pepR* mutant in HD73. The *pepR* gene was substituted by a kanamycin-resistant gene through double homolog recombination events (**B**) Identification of Δ*pepR* knockout mutant by PCR. PCR products were amplified from Δ*pepR* (Lane b) and HD73 (Lane c) with primer pair *pepR*-WF/*pepR*-WR. To confirm the presence of the pMADΩD*pepR* plasmid in the Δ*pepR* genome, PCR was performed using pMAD plasmid universal primers (pMAD-F/pMAD-R) (Lane d). The numbers indicate the size of the DNA standards in kilobase pairs (Lane a). (**C**) β-galactosidase activity of P*pepV-lacZ* in HD73 and Δ*pepR*. (**D**) β-galactosidase activity of P*rsuA-ygtP-lacZ* in HD73 and Δ*pepR*. The promoters of *pepV* and *rsuA-ygtP* were fused with the *lacZ* reporter and transformed into HD73 and Δ*pepR*, respectively. The β-galactosidase activities of three clones were determined at the indicated times after growing the cells in SSM at 30 °C. Each value represents the mean and standard error of at least three independent replicates.

Then, P*pepV* and P*rsuA-ytgP* promoter activity was measured in HD73 and Δ*pepR*. The P*pepV* and P*rsuA-ytgP* promoters were fused with the *lacZ* reporter gene and transformed into HD73 and Δ*pepR*, respectively. β-galactosidase activity showed that the activity of P*pepV* increased from T_0 to T_7 and reached the highest level at T_7 in HD73. However, the transcriptional activity of P*pepV* was significantly reduced from T_0 to T_7 in Δ*pepR* (Figure 3C). The result revealed that the P*pepV* promoter is regulated by PepR. Similarly, the transcriptional activity of P*rsuA-ytgP* in the *pepR* mutant strain was also significantly decreased compared to that in HD73 from T_0 to T_7 (Figure 3D), which suggested that the *rsuA-ytgP* operon was also regulated by PepR.

3.4. PepR Binds to the pepV Promoter

To determine whether the PepR protein directly binds to the *pepV* or *rsuA-ytgP* promoter regions, the electrophoretic mobility shift assay (EMSA) was performed. We constructed the recombinant plasmid, pET*pepR*, which was able to express the His-tagged PepR protein and introduced it into the *E. coli* BL21(DE3). The PepR-His protein with a molecular weight of approximately 10.06 kDa was expressed in *E. coli* BL21(pET*pepR*) and purified by Ni^{2+}-affinity chromatography (Figure 4A). Then, we tested the binding of the PepR protein to the *pepV* or *rsuA-ytgP* promoter regions. A 0.21 nM P*pepV*-labeled probe was used to bind to PepR (low concentrations of 34.79 μM). The P*pepV*-labeled probe at 0.21 nM was strongly shifted with a 44.73 nM PepR (Figure 4B). Notably, a 200-fold excess of unlabeled probe competed with the labeled probe, confirming the specific binding (Figure 4B). These results demonstrated that PepR directly binds to the *pepV* promoter region. However, when 0.19 nM of the P*rsuA-ytfP* probe was exposed to PepR concentrations ranging from 14.91 μM to 44.73 μM, no bind-shift could be detected, indicating that PepR cannot interact with the *rusA-ytfP* promoter region (Figure 4C). These results strongly support the theory that PepR directly regulates the expression of *pepV* and indirectly regulates *rusA-ytfP* expression.

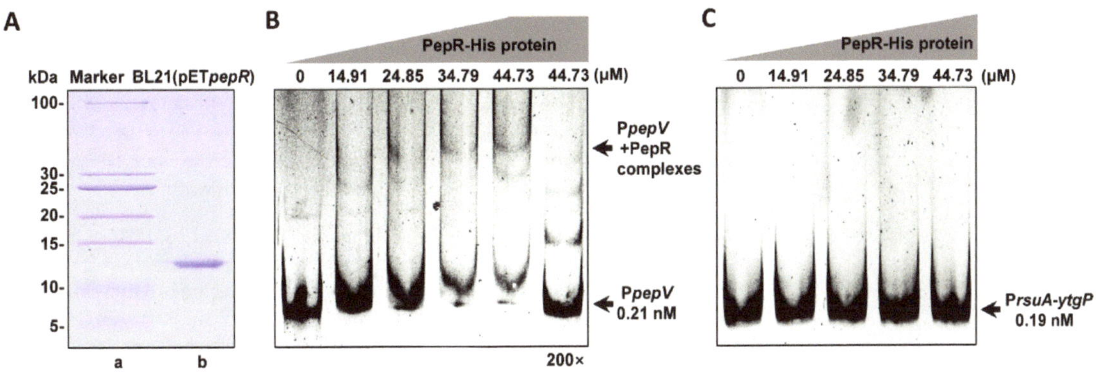

Figure 4. PepR directly bound *pepV* but not the *rsuA-ygtP* promoter region. (**A**) SDS-PAGE confirmed the expression and purification of the PepR-His protein by Ni^{2+}-affinity chromatography. Lane a, standard proteins (PageRuler Prestained Protein Ladder 26632, Thermo, Rockford, IL, USA). Lane b, purified PepR from the BL21(pET*pepR*) strain. (**B**) EMSA detecting protein–DNA interactions using FAM-labeled P*pepV* and increasing concentrations of recombinant PepR. The lanes contained 0, 14.91, 24.85, 34.79, and 44.73 μM of the PepR protein. The last lane was employed for the 200-fold unlabeled probe. Protein–DNA complexes were separated by native-PAGE and FAM-labeled bands scanned using Typhoon9410 (Cytiva, Marlborough, MA, USA). (**C**) EMSA detecting protein–DNA interactions using FAM-labeled P*rusA-ytgP* and increasing concentrations of recombinant PepR.

3.5. ΔpepR and ΔpepV Mutants Are More Sensitive to Vancomycin Than HD73

To characterize the function of the *pepV* gene, we constructed a *pepV* deletion mutant. The homolog arms on both sides of *pepV* were *pepV*-A (993 bp) and *pepV*-B (1023 bp) amplified from HD73 using primers *pepV*-AF/*pepV*-AR and *pepV*-BF/*pepV*-BR, respectively. The *pepV*-K fragment was obtained using PCR with *pepV*-KF/*pepV*-KR. The overlapping PCR products of *pepV*-A, *pepV*-B, and *pepV*-K were amplified with *pepV*-AF/*pepV*-BR primers, resulting in a 3452 bp fragment. The fragment was digested with BamHI and EcoRI sites and ligated to the temperature-sensitive vector of pMAD to obtain recombinant plasmid pMADΔ*pepV*. The recombinant plasmid pMADΔ*pepV* was electroporated into HD73. Transformants were selected for anti-sensitivity to erythromycin and kanamycin

(Figure 5A). Positive transformants were verified at 37 °C. Colonies lacking erythromycin resistance and containing kanamycin resistance were selected for Δ*pepV*.

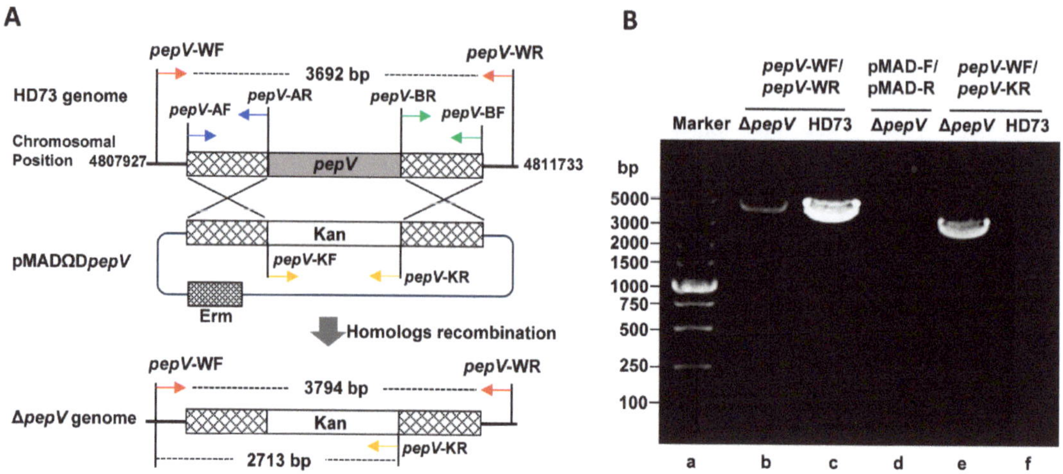

Figure 5. Construction of an in-frame deletion *pepV* mutant in HD73. (**A**) Construction of an in-frame deletion *pepV* mutant. The *pepV* gene was substituted by a kanamycin-resistant gene through double homolog recombination events. (**B**) Identification of the Δ*pepR* knockout mutant by PCR. PCR products were amplified from HD73 and Δ*pepV* strains using the primer pairs of *pepV*-WF/*pepV*-WR (Lanes b and c) and *pepV*-WF/*pepV*-KR (Lanes e and f). To confirm the presence of the pMADΩD*pepV* plasmid in the Δ*pepV* genome, PCR was performed using pMAD plasmid universal primers (pMAD-F/pMAD-R) (Lane d). The numbers indicate the size of the DNA standards in kilobase pairs (Lane a).

The *pepV* deletion mutant and HD73 strains were confirmed by PCR using *pepV*-WF/*pepV*-WR primers. PCR was performed with HD73 and Δ*pepV* chromosomal DNA as template. The product from Δ*pepV* was a 3794 bp fragment, while the product from HD73 was a 3692 bp fragment (Figure 5B, Lanes b and c). The Δ*pepV* mutant was confirmed using PCR with pMAD plasmid primers of pMAD-F/pMAD-R, which did not produce any PCR product (Figure 5B, Lane d). In addition, PCR products were generated using HD73 and Δ*pepV* chromosomal DNA as templates with primers (*pepV*-WF/*pepV*-KR). There was a 2713 bp fragment from Δ*pepV*, while no product from HD73 was obtained (Figure 5B, Lanes f and e). The result confirms that the Δ*pepV* mutant was successfully obtained.

It has been demonstrated that PepV is related to vancomycin resistance in *L. lactis* MG1363 [29]. A vancomycin resistance test was performed by inoculating Δ*pepR*, Δ*pepV*, and HD73 in SSM medium with 6 ng/mL concentrations of vancomycin. The result showed that the absence of vancomycin had no effect on the growth of Δ*pepR*, Δ*pepV*, and HD73 strains. However, the addition of vancomycin slowed HD73 growth. Δ*pepV* growth stagnated while Δ*pepR* lysed and died after vancomycin was added (Figure 6). The result suggests that Δ*pepR* and Δ*pepV* were more sensitive to vancomycin than HD73, and Δ*pepR* was more sensitive to vancomycin than Δ*pepV*.

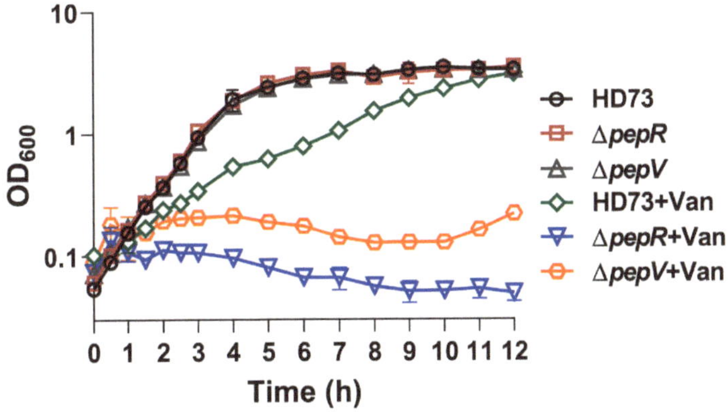

Figure 6. Δ*pepR* and Δ*pepV* mutant strains were more sensitive to vancomycin compared to HD73. Bacteria growth was monitored every hour after inoculation to culture medium with vancomycin at 6 ng/mL. Growth curves represent means of three clones.

4. Discussion

In our previous study, we validated the strong promoter activity from the *pepR* gene which encodes a DeoR family transcriptional regulator [12]. The *pepR* promoter is activated by Sigma E and can direct *cry1Ac* expression efficiently [12]. However, the functions of the *pepR* gene and its targets were unknown. In this study, we found that PepR homologs are widely distributed in various Gram-positive bacteria, and the amino acid similarities are highly conserved in different bacteria (Figure 1). We selected 18 strains that have been extensively studied for exhaustive analyses of the *pepR* genetic loci. We found that *pepV*, *pepR* and *rusA-ytgP* were closely linked in some strains, such as *B. thuringiensis* HD73, *B. cereus* ATCC 14579, *B. cytotoxicus*, *B. anthracis* Ames, *Planococcus antarcticus*, *Ectobacillus* sp. JY-23, *Sporosarcina psychrophile*, and *Sutcliffiella horikoshii*, while in some other strains they were not closely linked, such as in *Anoxybacillus caldiproteolyticus*, *A. flavithermus*, *Priestia koreensis*, *Mangrovibacillus cuniculi*, *P. megaterium* QM B1551, *B. pumilus*, *Lysinibacillus fusiformis*, *Aeribacillus pallidus* KCTC 3564, *Metabacillus sediminilitoris*, and *Cytobacillus spongiae* (Figure S1). In HD73, *pepV* and *rusA-ytgP* were two independent operons, and their promoters failed to transcript in the *pepR* knockout strain (Figures 2 and 3), suggesting these genes are all controlled by transcriptional regulator PepR. We further found that PepR directly binds to the *pepV* promoter but not to the *rusA-ytgP* promoter (Figures 4 and 5), revealing that *pepV* acts as a downstream target of *pepR*. The *pepV* promoter regions were 98.95% similar in different *B. thuringiensis* strains, and we determined that the expression of *pepV* is regulated by PepR. Whether PepR homologs also regulate *pepV* transcription in other bacteria remains to be confirmed. In addition, other targets of PepR should be investigated using RNA-seq and ChIP-seq methods in following studies.

PepV is highly conserved within the dipeptidase M20 family in the metallopeptidases [30]. It has been reported that the active sites and conserved domains of all PepVs are similar [30–32]. We analyzed the amino acid sequence of the PepV homologs and found that the identity varies in *B. subtilis* (57.36%), *L. delbrueckii* (34.62%), *L. helveticus* (33.12%), *S. gordonii* (40.86%) and *L. lactis* (38.92%) compared to HD73 (Figure S2). This result suggests that PepV in HD73 may have a similar function in catalyzing substrates to that reported in other bacteria. It is worth mentioning that the deletion of *pepV* in HD73 increased the susceptibility to vancomycin, whereas the deletion of *pepV* in *L. lactis* MG1363 decreased the susceptibility of *L. lactis* to vancomycin [29]. In addition to BCARR regulating *pepV* expression in the presence of BCAAs in *L. helveticus* [17] and a CodY-like regulatory system controlling the expression of *pepV* in *L. helveticus* CM4 [16], *pepV* might have multiple functions in responding to different signals in different bacteria strains.

Supplementary Materials: The following supporting information can be downloaded at: https://www.mdpi.com/article/10.3390/microorganisms12030579/s1, Figure S1: The *pepR* gene cluster is a conserved in bacteria; Figure S2: Comparison of the amino acid sequences revealed a conserved PepV homologs in bacteria; Table S1: List of species in Figure 1.

Author Contributions: F.S. and X.Z. designed the experiments. X.Z., H.W., Y.C. and T.Y. performed the experiments. X.Z., Q.P. and F.S. analyzed the results. X.Z. wrote the manuscript. F.S. revised the manuscript. All authors have read and agreed to the published version of the manuscript.

Funding: This work was supported by grants from the National Key R&D Program of China (Grant No. 2022YFE0116500) and the National Natural Science Foundation of China (General Program, grants No. 32072499 and No. 32372623).

Data Availability Statement: Data are contained within the article.

Conflicts of Interest: The authors declare no conflict of interest.

References

1. Argôlo-Filho, R.C.; Loguercio, L.L. *Bacillus thuringiensis* Is an Environmental Pathogen and Host-Specificity Has Developed as an Adaptation to Human-Generated Ecological Niches. *Insects* **2013**, *5*, 62–91. [CrossRef] [PubMed]
2. Palma, L.; Muñoz, D.; Berry, C.; Murillo, J.; Caballero, P. *Bacillus thuringiensis* toxins: An overview of their biocidal activity. *Toxins* **2014**, *6*, 3296–3325. [CrossRef] [PubMed]
3. Agaisse, H.; Lereclus, D. How does *Bacillus thuringiensis* produce so much insecticidal crystal protein? *J. Bacteriol.* **1995**, *177*, 6027–6032. [CrossRef] [PubMed]
4. Peng, Q.; Yu, Q.; Song, F. Expression of *cry* genes in *Bacillus thuringiensis* biotechnology. *Appl. Microbiol. Biot.* **2019**, *103*, 1617–1626. [CrossRef]
5. García-Gómez, B.I.; Sánchez, J.; Martínez de Castro, D.L.; Ibarra, J.E.; Bravo, A.; Soberón, M. Efficient production of *Bacillus thuringiensis* Cry1AMod toxins under regulation of *cry3Aa* promoter and single cysteine mutations in the protoxin region. *Appl. Environ. Microb.* **2013**, *79*, 6969–6973. [CrossRef] [PubMed]
6. Chaoyin, Y.; Wei, S.; Sun, M.; Lin, L.; Faju, C.; Zhengquan, H.; Ziniu, Y. Comparative study on effect of different promoters on expression of *cry1Ac* in *Bacillus thuringiensis* chromosome. *J. Appl. Microbiol.* **2007**, *103*, 454–461. [CrossRef]
7. Sanchis, V.; Agaisse, H.; Chaufaux, J.; Lereclus, D. Construction of new insecticidal *Bacillus thuringiensis* recombinant strains by using the sporulation non-dependent expression system of *cryIIIA* and a site specific recombination vector. *J. Bacteriol.* **1996**, *48*, 81–96. [CrossRef]
8. de Souza, M.T.; Lecadet, M.M.; Lereclus, D. Full expression of the *cryIIIA* toxin gene of *Bacillus thuringiensis* requires a distant upstream DNA sequence affecting transcription. *J. Bacteriol.* **1993**, *175*, 2952–2960. [CrossRef]
9. Jia, Y.; Zhao, C.; Wang, Q.; Shu, C.; Feng, X.; Song, F.; Zhang, J. A genetically modified broad-spectrum strain of *Bacillus thuringiensis* toxic against *Holotrichia parallela, Anomala corpulenta* and *Holotrichia oblita*. *World J. Microb. Biot.* **2014**, *30*, 595–603. [CrossRef]
10. Guan, P.; Dai, X.; Zhu, J.; Li, Q.; Li, S.; Wang, S.; Li, P.; Zheng, A. Bacillus thuringiensis subsp. *sichuansis* strain MC28 produces a novel crystal protein with activity against *Culex quinquefasciatus* larvae. *World J. Microb. Biot.* **2014**, *30*, 1417–1421. [CrossRef]
11. Zheng, Q.; Wang, G.; Zhang, Z.; Qu, N.; Zhang, Q.; Peng, Q.; Zhang, J.; Gao, J.; Song, F. Expression of *cry1Ac* gene directed by P*exsY* promoter of the *exsY* gene encoding component protein of exosporium basal layer in *Bacillus thuringiensis*. *Acta Microbiol. Sin.* **2014**, *54*, 1138–1145.
12. Zhang, X.; Gao, T.; Peng, Q.; Song, L.; Zhang, J.; Chai, Y.; Sun, D.; Song, F. A strong promoter of a non-*cry* gene directs expression of the *cry1Ac* gene in *Bacillus thuringiensis*. *Appl. Microbiol. Biot.* **2018**, *102*, 3687–3699. [CrossRef]
13. Guédon, E.; Renault, P.; Ehrlich, S.D.; Delorme, C. Transcriptional pattern of genes coding for the proteolytic system of *Lactococcus lactis* and evidence for coordinated regulation of key enzymes by peptide supply. *J. Bacteriol.* **2001**, *183*, 3614–3622. [CrossRef]
14. Mierau, I.; Tan, P.S.; Haandrikman, A.J.; Mayo, B.; Kok, J.; Leenhouts, K.J.; Konings, W.N.; Venema, G. Cloning and sequencing of the gene for a lactococcal endopeptidase, an enzyme with sequence similarity to mammalian enkephalinase. *J. Bacteriol.* **1993**, *175*, 2087–2096. [CrossRef]
15. Nardi, M.; Renault, P.; Monnet, V. Duplication of the *pepF* gene and shuffling of DNA fragments on the lactose plasmid of *Lactococcus lactis*. *J. Bacteriol.* **1997**, *179*, 4164–4171. [CrossRef]
16. Wakai, T.; Yamaguchi, N.; Hatanaka, M.; Nakamura, Y.; Yamamoto, N. Repressive processing of antihypertensive peptides, Val-Pro-Pro and Ile-Pro-Pro, in *Lactobacillus helveticus* fermented milk by added peptides. *J. Biosci. Bioeng.* **2012**, *114*, 133–137. [CrossRef]
17. Wakai, T.; Yamamoto, N. A novel branched chain amino acids responsive transcriptional regulator, BCARR, negatively acts on the proteolytic system in *Lactobacillus helveticus*. *PLoS ONE* **2013**, *8*, e75976. [CrossRef] [PubMed]
18. Schaeffer, P.; Millet, J.; Aubert, J.P. Catabolic repression of bacterial sporulation. *Proc. Natl. Acad. Sci. USA* **1965**, *54*, 704–711. [CrossRef]

19. Macaluso, A.; Mettus, A.M. Efficient transformation of *Bacillus thuringiensis* requires nonmethylated plasmid DNA. *J. Bacteriol.* **1991**, *173*, 1353–1356. [CrossRef] [PubMed]
20. Wang, G.; Zhang, J.; Song, F.; Wu, J.; Feng, S.; Huang, D. Engineered *Bacillus thuringiensis* GO33A with broad insecticidal activity against *lepidopteran* and *coleopteran* pests. *Appl. Microbiol. Biot.* **2006**, *72*, 924–930. [CrossRef] [PubMed]
21. Munson, R.S., Jr.; Sasaki, K. Protein D, a putative immunoglobulin D-binding protein produced by *Haemophilus influenzae*, is glycerophosphodiester phosphodiesterase. *J. Bacteriol.* **1993**, *175*, 4569–4571. [CrossRef]
22. Agaisse, H.; Lereclus, D. Structural and functional analysis of the promoter region involved in full expression of the *cryIIIA* toxin gene of *Bacillus thuringiensis*. *Mol. Microbiol.* **1994**, *13*, 97–107. [CrossRef]
23. Arnaud, M.; Chastanet, A.; Débarbouillé, M. New vector for efficient allelic replacement in naturally nontransformable, low-GC-content, gram-positive bacteria. *Appl. Environ. Microb.* **2004**, *70*, 6887–6891. [CrossRef]
24. Yang, H.; Wang, P.; Peng, Q.; Rong, R.; Liu, C.; Lereclus, D.; Zhang, J.; Song, F.; Huang, D. Weak transcription of the *cry1Ac* gene in nonsporulating *Bacillus thuringiensis* cells. *Appl. Environ. Microb.* **2012**, *78*, 6466–6474. [CrossRef]
25. Sambrook, B.J.; Russell, D.W. *Molecular Cloning: A Laboratory Manual*; Cold Spring Harbor Laboratory Press: Cold Spring Harbor, NY, USA, 2015.
26. Lereclus, D.; Arantes, O.; Chaufaux, J.; Lecadet, M. Transformation and expression of a cloned delta-endotoxin gene in *Bacillus thuringiensis*. *FEMS Microbiol. Lett.* **1989**, *51*, 211–217.
27. Peng, Q.; Yang, M.; Wang, W.; Han, L.; Wang, G.; Wang, P.; Zhang, J.; Song, F. Activation of *gab* cluster transcription in *Bacillus thuringiensis* by γ-aminobutyric acid or succinic semialdehyde is mediated by the Sigma 54-dependent transcriptional activator GabR. *BMC Microbiol.* **2014**, *14*, 306. [CrossRef]
28. Li, R.; Liu, G.; Xie, Z.; He, X.; Chen, W.; Deng, Z.; Tan, H. PolY, a transcriptional regulator with ATPase activity, directly activates transcription of *polR* in polyoxin biosynthesis in *Streptomyces cacaoi*. *Mol. Microbiol.* **2010**, *75*, 349–364. [CrossRef]
29. Huang, C.; Hernandez-Valdes, J.A.; Kuipers, O.P.; Kok, J. Lysis of a *Lactococcus lactis* Dipeptidase Mutant and Rescue by Mutation in the Pleiotropic Regulator CodY. *Appl. Environ. Microb.* **2020**, *86*, e02937-19. [CrossRef]
30. Goldstein, J.M.; Kordula, T.; Moon, J.L.; Mayo, J.A.; Travis, J. Characterization of an extracellular dipeptidase from *Streptococcus gordonii* FSS2. *Infect. Immun.* **2005**, *73*, 1256–1259. [CrossRef]
31. Jozic, D.; Bourenkow, G.; Bartunik, H.; Scholze, H.; Dive, V.; Henrich, B.; Huber, R.; Bode, W.; Maskos, K. Crystal structure of the dinuclear zinc aminopeptidase PepV from *Lactobacillus delbrueckii* unravels its preference for dipeptides. *Structure* **2002**, *10*, 1097–1106. [CrossRef]
32. Mori, S.; Miyamoto, M.; Kaneko, S.; Nirasawa, S.; Komba, S.; Kasumi, T. Characterization and kinetic analysis of enzyme-substrate recognition by three recombinant *lactococcal* PepVs. *Arch. Biochem. Biophys.* **2006**, *454*, 137–145. [CrossRef] [PubMed]

microorganisms

Article

AraC Functional Suppressors of Mutations in the C-Terminal Domain of the RpoA Subunit of the *Escherichia coli* RNA Polymerase

Dominique Belin [1,*], Jordan Costafrolaz [2] and Filo Silva [1,†]

[1] Department of Pathology and Immunology, Faculty of Medicine, University of Geneva, 1 Rue Michel-Servet, CH-1211 Geneva, Switzerland; filomena.silva@unige.ch
[2] Department of Microbiology and Molecular Medicine, Faculty of Medicine, University of Geneva, 1 Rue Michel-Servet, CH-1211 Geneva, Switzerland; jordan.costafrolaz@unige.ch
* Correspondence: dominique.belin@unige.ch; Tel.: +41-79-664-43-46
† Current address: Department of Microbiology and Molecular Medicine, Faculty of Medicine, University of Geneva, 1 Rue Michel-Servet, CH-1211 Geneva, Switzerland.

Abstract: In *E. coli*, transcriptional activation is often mediated by the C-terminal domain of RpoA, the α subunit of RNA polymerase. Random mutations that prevent activation of the arabinose P_{BAD} promoter are clustered in the RpoA C-terminal domain (α-CTD). We have isolated functional suppressors of *rpoA* α-CTD mutations that map to *araC*, the main transcriptional regulator of *ara* genes, or to the P_{BAD} promoter. No mutation was found in the DNA regulatory region between *araC* and P_{BAD}. Most suppressors that improve P_{BAD} transcription are localized to the N-terminal domain of AraC. One class of *araC* mutations generates substitutions in the core of the N-terminal domain, suggesting that they affect its conformation. Other suppressors localize to the flexible N-terminal arm of AraC. Some, but not all, suppressors confer an arabinose constitutive phenotype. Suppression by both classes of *araC* mutations requires the α-CTD to stimulate expression from P_{BAD}. Surprisingly, in *rpoA*$^+$ strains lacking Crp, the cAMP receptor protein, these *araC* mutations largely restore arabinose gene expression and can essentially bypass Crp activation. Thus, the N-terminal domain of AraC exhibits at least three distinct activities: dimerization, arabinose binding, and transcriptional activation. Finally, one mutation maps to the AraC C-terminal domain and can synergize with AraC mutations in the N-terminal domain.

Keywords: arabinose operon; genetic suppression; transcription activation; P_{BAD} promoter

Citation: Belin, D.; Costafrolaz, J.; Silva, F. AraC Functional Suppressors of Mutations in the C-Terminal Domain of the RpoA Subunit of the *Escherichia coli* RNA Polymerase. *Microorganisms* **2024**, *12*, 1928. https://doi.org/10.3390/microorganisms12091928

Academic Editor: Tomohiro Shimada

Received: 29 August 2024
Revised: 18 September 2024
Accepted: 20 September 2024
Published: 23 September 2024

1. Introduction

The arabinose regulon was the first system in which positive control of gene expression was genetically demonstrated [1]. AraC, the product of the activator gene, binds to arabinose and stimulates transcription of the catabolic genes *araB*, *araA*, and *araD* (*araBAD* operon), as well as those of the arabinose import systems, *araE* and *araFGH*. In the absence of arabinose, AraC represses transcription from the P_{BAD} promoter. In both instances, one AraC monomer is bound to the *I1* site (Figure 1). The transition between the repressor and activator states of AraC involves a large conformational movement, named the light switch, that displaces one subunit of the AraC dimer from the distant *O2* operator site to the promoter proximal *I2* activator site. During the switch, the N-terminal arm of AraC covers the arabinose-binding pocket of the N-terminal dimerization domain (AraC-NTD) [2–4]. The P_{BAD} promoter is intrinsically very weak, and AraC bound to the *I2* site enhances RNA polymerase binding and open complex formation [5]. In addition, Crp, the cAMP receptor protein, is required for full activity of the P_{BAD} promoter [6,7]. Crp activates many bacterial promoters, and several mechanisms of activation have been described [8]. The properties of the arabinose system were exploited to construct a widely used set of expression plasmids [9].

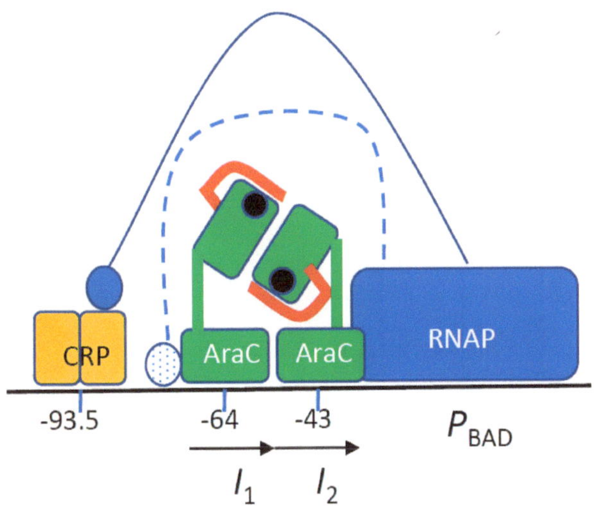

Figure 1. Transcription activation at the arabinose P_{BAD} promoter (adapted from [3]). The sequence represented extends from −120 to +10; the *araC* promoter and *O2* site are not included. RNA polymerase (blue) binding to the promoter is helped by the proximal AraC monomer bound to the *I2* site (position −35 to −51 from the start site) that overlaps the promoter −35 site. Each AraC monomer (green) consists of an N-terminal dimerization domain that binds arabinose (black circle) attached by a linker to the C-terminal DNA-binding domain. In the presence of arabinose, the AraC N-terminal arm (red) covers the sugar-binding pocket. The Crp dimer (yellow) binds to a distal site (position −83 to −104). One α-CTD domain of RpoA contacts the Crp domain (filled blue circle, solid line). A potential contact site for the other α-CTD domain (dotted blue circle, stippled blue line) is indicated.

The α subunit of RNA polymerase, encoded by the *rpoA* gene, plays at least two roles in transcription. Its N-terminal domain (α-NTD) is involved in the assembly of the enzyme [10,11]. In contrast, the C-terminal domain (α-CTD), which is attached by a flexible linker to the α-NTD, plays multiple roles in transcription activation [8,12–14]. It activates transcription by interacting with activators in different ways and/or by binding to UP DNA sites upstream of promoters [15,16]. Since RNA polymerase contains two α subunits, each subunit can interact independently with two activating regions [13]. In addition to α-CTD structures [17,18], several structures of α-CTD-containing complexes have been resolved, including that of α-CTD bound to Fis at the *proP* P2 promoter [19], that of the α-CTD–Crp–DNA complex [14], and that of the RNA polymerase holoenzyme bound to a synthetic promoter in the presence of Crp [20].

At the arabinose P_{BAD} promoter, one α-CTD interacts with Crp [7]. The potential target of the second α-CTD has not yet been identified. However, by analogy with other Crp-dependent promoters [8], it has been proposed to bind the AraC DNA-binding domain and possibly a DNA motif localized between the Crp and AraC binding sites [3,21] (Figure 1). Several random *rpoA* α-CTD mutations that interfere with P_{BAD} activity have been characterized.

We have used a genetic approach to search for functional suppressors that compensate for the detrimental effects of mutations in the RpoA α-CTD. Random mutagenesis of *araC* and of the intergenic regulatory region between *araC* and *araB* led to the isolation of *araC* suppressors. Most of these random mutations are scattered at several positions in the AraC-NTD, and their effect requires the RpoA α-CTD. We have also characterized a mutation in the DNA-binding C-terminal domain of AraC that exerts a synergistic effect with mutations in the AraC-NTD. However, we did not identify an AraC contact site with the RpoA α-CTD. Rather, all the mutations may promote or stabilize a conformation of

AraC that more efficiently promotes transcription, thereby functionally compensating for the deleterious effects of the *rpoA* mutations.

2. Experimental Procedures

Strains. MC4100 is an F-*araD*139 Δ(*argF-lac*)U169 *flhD*5301 *fruA*25 *relA*1 *rpsL*150 *rbsR*22 Δ(*fimB-fimE*)632 *deoC*1 *thi* strain [22]. DB504 is MC4100 *ara*Δ*714 malE18* [23]. JW5702 contains the Δ*crp-765*::*kan* allele [24] and was obtained from the Coli Genetic Stock Center (CGSC strain 11596). HfrH was obtained from the Coli Genetic Stock Center (CGSC strain 5081).

The *rpoA* mutations were introduced into DB504 by bacteriophage P1-mediated co-transduction with a *gspA*::Tn*10* marker (90–95% co-transduction) [25]. The presence of the *rpoA* mutations was verified by suppression of the toxic phenotype caused by induction of the appropriate proteins, by growth on M63 minimal media with or without cysteine and methionine, and by fermentation on McConkey-melibiose indicator plates [26]. The Δ*crp-765*::*kan* allele was introduced by P1-mediated transduction in strains carrying the *rpoA* alleles; the transductants were plated on streptomycin (100 μg/mL) containing plates to select for recombination between the *rpsL150* marker of DB504 and *crp*, and retention of the *gspA*::Tn*10* marker. Strains containing the Δ*crp-765*::*kan* allele were grown in LB containing 0.1% glucose [25]. Information on the *E. coli* genes and regulatory sequences was obtained from https://ecocyc.org/ (accessed on 25 August 2024).

Plasmids. Plasmids derived from pBAD24 [9] contain the *araC* gene, the intergenic region between *araC* and P_{BAD}, the P_{BAD} promoter, and the first 13 nt of the *araBAD* transcript. pBAD18-vs.1 expresses a lysozyme encoded by bacteriophage T4 [27]. pBAD72K expresses a chimeric protein containing the signal sequence of murine PAI2 fused to the mature portion of PhoA [28]. pDB2114 expresses gene *55.2* of bacteriophage T4 [29]. These three plasmids confer a toxic phenotype that prevents growth in the presence of arabinose.

Maspin (SERPINB5, P70124) is a member of the ovalbumin-related protease inhibitors family [30] whose N-terminal region is devoid of signal sequence activity [31]. psM74c expresses a Maspin::PhoA chimera: the initial Met is fused to residues 23–46 of Maspin, with a P32L substitution to allow export, followed by a synthetic ALA-A c-region cleaved by signal peptidase, and residues 23–471 of PhoA (P00634). The export of this chimeric protein is slightly more efficient than that of wild-type PhoA; hence psM74c was used to isolate and characterize suppressors of the *rpoA* mutations.

The *rpoA*$^+$ gene was PCR amplified from chromosomal DNA with the primers RPOAUP (5′cgggatcccacctgatcgtcg) and RPOADON (5′ccggtaccttaacctgtgatccggttactcg), and cloned between the BamHI and KpnI restriction sites of pUC19 to generate pUC-*rpoA*. The deletion of the α-CTD was performed by PCR amplification of pUC-*rpoA* with the primers RPOAUP and RPOAdel (5′cggggtaccttacacttcaggctgacg). The pUC-*rpoA*del plasmid expresses amino acids 1 to 242 of RpoA. The expression level of the proteins from the P_{lac} promoter is estimated to be ~50-fold higher than that from the chromosomal *rpoA* gene in *rpoA*$^+$ *crp*$^+$ strains; the level should be ~5-fold in Δ*crp* strains. The pUC-*rpoA*del plasmid confers a Mel$^-$ phenotype in an *rpoA*$^+$ strain.

The original pBAD24-derived plasmids (AraC$^+$, AraC D7N, and AraC I71V) were digested with NsiI and XbaI. The *araC*-Maspin::PhoA fragments were cloned into the cognate sites of pBAD33 whose origin is compatible with that of the *rpoA* plasmids [9].

A complete deletion of the *araC* coding region was performed by removing the two MfeI-digested fragments of psM74c.

Enzyme assays. PhoA activity was qualitatively detected on LB plates supplemented with 0.2% arabinose and 40 μg/mL XP (5-bromo-4-chloro-3 indolyl phosphate). PhoA enzymatic activity was determined by measuring the rate of *p*-nitro-phenyl-phosphate hydrolysis [32]. Cultures were assayed after a 1 h induction with 0.2% arabinose. LacZ activity was determined by measuring the rate of *o*-nitro-phenyl-galactoside hydrolysis [25]. For each strain, triplicate independent cultures were assayed twice. The values were analysed with a Mann–Whitney test or an unpaired *t*-test.

Random mutagenesis. Hydroxylamine mutagenesis of psM74c plasmid DNA was performed as described [31]. Upon transformation in DB504, approximately 6% of the colonies carried inactivating mutations in *araC*, the regulatory region, or the maspin::*phoA* gene and formed white colonies on PhoA indicator plates containing arabinose. Plasmids allowing the formation of light blue colonies in *rpoA* Δ*crp* strains occurred at a frequency of 1–2%; in these strains, the parental plasmid formed white colonies. More than half of the mutant plasmids exhibited a higher plasmid copy number than the original plasmid. The complete *araC* gene was recloned in the parental plasmid. In a second experiment, a complete *araC* fragment (NsiI-MluI, 1130 bp long) was randomly mutagenized by PCR [28] and cloned in psM74c. Ten percent of the resulting plasmids transformed into an *rpoA*$^+$ *crp*$^+$ strain formed white colonies on PhoA indicator plates containing arabinose. After transformation in the *rpoA* A272E Δ*crp* strain, candidate suppressors were sequenced. Three additional AraC residues were isolated: N16S, I51T, and I71V; three mutants had different substitutions at previously identified residues (S14P, A152T, and A152P), and the A152V substitution was isolated again. The *araC* mutations described here have been independently isolated once (D7N, N16S, I51T, I71V, A152T, and A152P), or 4–6 times (S14L, S14P, G22D, and A152V). To isolate suppressors of *rpoA* N268T, the mutagenized *araC* PCR fragment was digested with NsiI and EcoRV and cloned in psM74c. After transformation in the *rpoA* N268T *crp*$^+$ strain, three suppressors were identified: R61H was isolated six times, while R261C and C205Y were isolated twice.

3. Results

3.1. RpoA Mutations in the α-CTD Region That Interfere with P_{BAD} Expression

The chimeric protein PAI2::PhoA contains the signal sequence of murine PAI-2 fused to the mature portion of PhoA. Expression of this chimeric protein from the arabinose P_{BAD} promoter is lethal because it interferes with protein export to the periplasm [31]. Suppressors of this toxicity were selected as colony-forming mutants in the presence of arabinose. While many chromosomal suppressors map to *sec* genes involved in protein export to the periplasm thus preventing the toxicity of the chimera [28,33], we also isolated fifteen *rpoA* mutants (Table 1). To modulate the severity of the toxicity in our selective system, we used several genes of bacteriophage T4 that also confer a toxic phenotype when expressed from the arabinose P_{BAD} promoter. The expression of one of these genes, *vs.1*, encoding an exported lysozyme whose toxicity is weaker than that of PAI2::PhoA [27] led to the isolation of six *rpoA* mutants. Finally, one *rpoA* mutant was isolated as a suppressor of phage T4 gene *55.2*, a topoisomerase inhibitor that exhibit the strongest toxicity of the three proteins [29]. A total of twenty-two independent *rpoA* mutations have been isolated (Table 1).

Table 1. *ropA* mutations that decrease expression from the P_{BAD} promoter.

	rpoA **Substitutions**						
Toxic Protein	**N268T**	**L270F**	**K271E**	**K271I**	**K271V**	**K271T**	**A272E**
T4 *vs.1*		2	2			2	
PAI2::PhoA			7	3	2	1	2
T4 *55.2*	1						
growth phenotypes							
MC-melibiose	W	R	W	K	K	K	W
min-glucose	no	yes	no	no	no	no	no
+cys-Met	no	yes	yes	yes	yes	yes	yes

The randomly selected *rpoA* mutations all map to amino acids 268–272 of the RpoA C-terminal domain (α-CTD) (Figure 2A). The RpoA K271E substitution (*rpoA341* [26,34]) was the most frequent, but three other new substitutions (V, T, and I) were also identified at this position (Table 1). The four substitutions of K271 exert similar effects on P_{BAD} promoter expression, although the K271E substitution has a slightly stronger effect (Figure 2B, Table 1). These mutations affect CysB regulatory function and prevent growth on M63 minimal medium in the absence of cysteine or methionine (Cym phenotype) [26]. The L270F substitution appears to be the weakest of the *rpoA* alleles since it allows growth on minimal media (Table 1) and thus does not detectably interfere with activation of the *cysA* operon by CysB [35]. The N268T substitution is the strongest *rpoA* mutation observed (Figure 2B), as it prevents growth on minimal media, even in the presence of casamino acids. Thus, this mutation must affect one or more promoter(s) in addition to P_{BAD} and P_{cysA}.

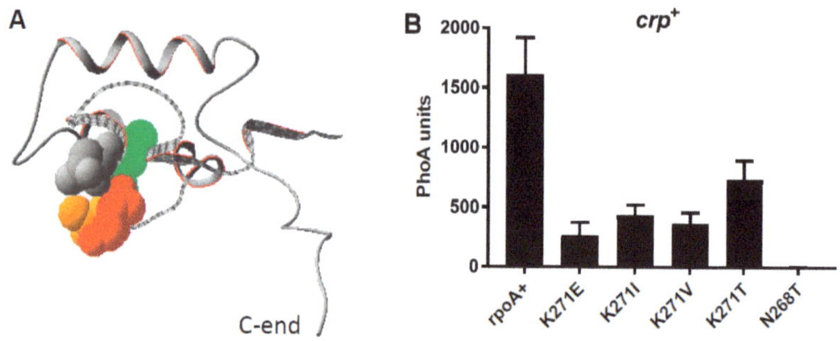

Figure 2. (**A**) The structure of the RpoA C-terminal domain as determined by NMR (PDB 1C00) [17]. The image was generated with Swiss-PdbViewer v4.1.0. The domain is represented as a ribbon, and the side chains of relevant residues are shown in space filling (residues 268–272). N268 is shown in grey, L270 in green, K271 in red and A272 in orange. (**B**) *crp*$^+$ strains with the indicated *rpoA* alleles contain the psM74c plasmid expressing wild-type AraC and a Maspin::PhoA chimera. Cultures were induced for 1 h with arabinose.

The *rpoA* mutations were selected on arabinose-containing LB plates, under conditions where the expression of the indicated toxic protein prevents growth. For each substitution, the number of independent isolates is shown. The N268T allele suppresses the toxicity of all three proteins. The L270F allele suppresses the toxicity of T4 *vs.1* and partially that of PAI2::PhoA. The other alleles suppress the toxicity of T4 *vs.1* and PAI2::PhoA but not that of T4 *55.2*.

Colonies grown on McConkey-melibiose indicator plates were scored as R (red), identical to those of *rpoA** strains; K (pink), reflecting an intermediate melibiose fermentation; and W (white), reflecting no detectable melibiose fermentation. On M63-minimal-glucose plates, only the weak L270F allele allows growth. Upon addition of Cys and Met, most *rpoA* mutants grew. The N268T allele must prevent the expression of one or more auxotrophic genes.

3.2. Suppressors of the rpoA Mutations

Both AraC and Crp promote transcription from the P_{BAD} promoter in the presence of arabinose [3,4]. While AraC is absolutely required, a 20- to 40-fold reduction in expression can be detected in an *rpoA*$^+$ strain in the absence of Crp (Figure 3A). Quantification of *phoA* expression from the P_{BAD} promoter in strains deleted for *crp* and carrying five alleles of *rpoA* demonstrated a strong reduction of promoter activity with the four substitutions of K271, while the N268T mutation totally abolishes expression (Figure 3B). These findings imply that the α-CTD is required for P_{BAD} transcription in the absence of Crp.

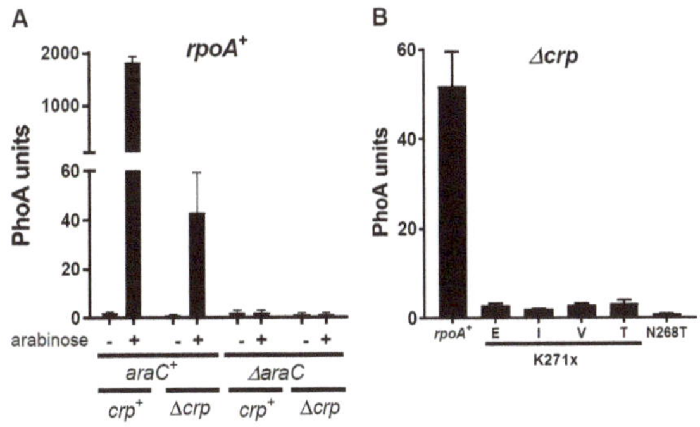

Figure 3. Residual activity of the P_{BAD} promoter in the absence of Crp. (**A**) Plasmid psM74c or its derivative with a complete deletion of *araC* was introduced in *rpoA*$^+$ *crp*$^+$ and *rpoA*$^+$ Δ*crp* strains. Cultures were assayed with or without arabinose induction. (**B**) The psM74c plasmid expressing wild-type AraC was introduced in Δ*crp* strains carrying the indicated *rpoA* alleles. Cultures were induced for 1 h with arabinose.

To genetically define whether the interactions defective in the RpoA α-CTD mutants can be functionally suppressed, we carried out a suppressor screen. We used Δ*crp* strains containing either the K271E or K271V RpoA substitution. We also employed a Maspin::PhoA fusion that contains a very efficient signal sequence. This chimeric protein provides an extremely sensitive reporter, compensating for the reduced expression from the P_{BAD} promoter in the absence of Crp. Cells were transformed with mutagenized plasmids harboring the *araC* gene and the P_{BAD} regulatory region and plated on LB plates containing arabinose and XP (a color indicator of PhoA activity). Colonies in which AraC-dependent expression of maspin::*phoA* is restored are blue. Plasmids isolated from such strains were sequenced.

We isolated four plasmids that contain mutations in the P_{BAD} promoter (Figure 4). One of these, at position −36, was located in the −35 hexamer, between the *cip*-5 site [6,36] and the I^c mutations [37,38]. The *cip*-5 and I^c promoters allow AraC-independent constitutive expression, and the −36 mutation could also facilitate RNA polymerase binding. The second one, at position −23, has a weaker stimulatory effect. The last two, located at position −10, are identical to the X^c mutation, which has been described as creating a strong AraC-independent promoter [38,39]; a P_{BAD} promoter carrying the X^c mutation is not affected by the *rpoA341* K271E mutation [26]. No mutation was isolated in the regulatory region between the *araC* coding sequence and the P_{BAD} promoter (Figure 1).

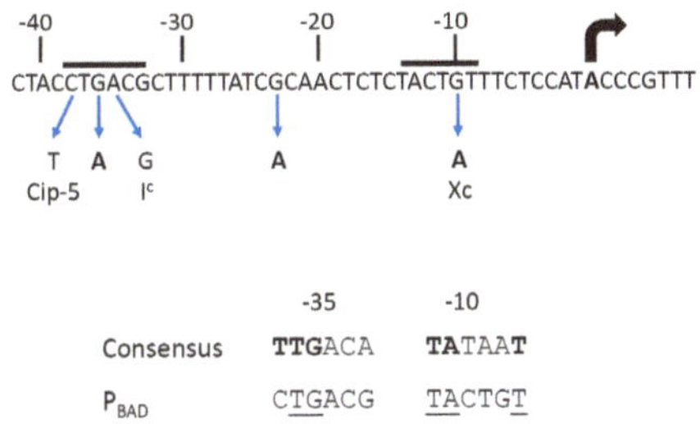

Figure 4. Nucleotide sequence of the P_{BAD} promoter. The positions of the −35 and −10 hexamers and the transcription start site are indicated. The three G to A mutations isolated in this study are shown in bold. Previously described promoter mutations are discussed in the text. The sequence consensus of the −35 and −10 hexamers [40] is compared to that of the P_{BAD} promoter. The most conserved residues in the consensus are in bold, and the conserved residues in the P_{BAD} promoter are underlined.

We isolated four mutations in *araC*. These *araC* suppressors all localize to the N-terminal domain of AraC (AraC-NTD). Since most *araC* constitutive mutations map to this region [41,42], we measured expression from the P_{BAD} promoter in the presence and absence of arabinose with these AraC mutant proteins. In a *rpoA*⁺ Δ*crp* strain, the A152V substitution confers a nearly complete constitutive phenotype (Figure 5A). In contrast, G22D and S14L confer an expression level of only 12–27% in the absence of the inducer (Figure 5A,C). Constitutive expression with A152V has been observed in a *crp*⁺ strain [42]. Finally, the D7N substitution is entirely devoid of constitutive activity (Figure 5C). In the *rpoA* K271E strain, only A152V confers a partial constitutive phenotype (34% of the induced level) (Figure 5B,D); this suggests that the RpoA α-CTD may stabilize the activator form of the constitutive AraC proteins. However, an arabinose constitutive phenotype is not necessary for functional suppression of the *rpoA* mutations.

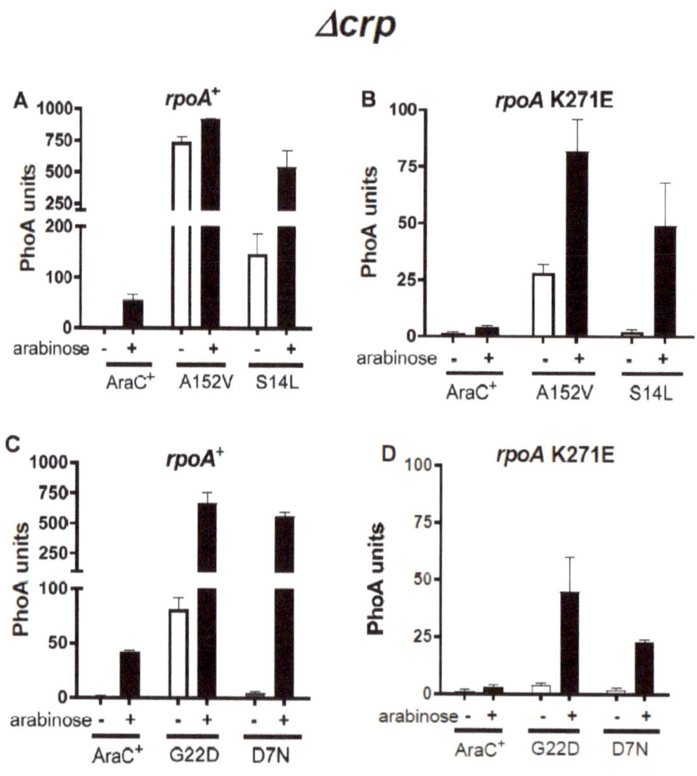

Figure 5. Constitutive P_{BAD} activity mediated by mutant AraC proteins in the absence of Crp. Plasmids expressing AraC⁺ or the indicated AraC suppressor proteins were introduced in *rpoA*⁺ Δ*crp* (**A**,**C**) or *rpoA* K271E Δ*crp* (**B**,**D**) strains. Cultures grown in the absence of arabinose were assayed to detect the constitutive activity of each AraC protein. Cultures induced for 1 h with arabinose were used to determine the full activity of each AraC protein.

While induced expression mediated by wild-type AraC is strongly decreased in an *rpoA*⁺ Δ*crp* strain, the *araC* mutations confer robust expression in this strain, at levels of 30–50% of that measured in an *araC*⁺ *rpoA*⁺ *crp*⁺ strain. Thus, they can partially compensate for the absence of Crp in *rpoA*⁺ strains (Figure 5A,C).

These *araC* mutants were isolated in strains carrying either the K271E or K271V RpoA substitutions. We determined the capacity of the *araC* mutations to suppress the four different K271X substitutions, as well as the N268T one. The results shown in Figure 6 were obtained with AraC D7N and A152V; similar results were obtained with S14L and G22D (Supplementary Figure S1). In induced *rpoA*$^+$ *crp*$^+$ strains, expression from the P_{BAD} promoter was slightly increased (1.5- to 3-fold) by the *araC* mutations (Figure 6A,C). In almost all cases, the expression observed with the *rpoA* mutants did not reach the levels observed in *rpoA*$^+$ strains, indicating that the AraC suppressors do not fully compensate for the *rpoA* defect. However, the AraC D7N substitution restored the Ara^S phenotype conferred by PAI2::PhoA in an *rpoA* K271T *crp*$^+$ strain; an effect of the other *araC* mutations could not be tested because of their constitutive activity. In Δ*crp* strains, stimulation by the *araC* mutations was much higher (Figure 6B,D). In *rpoA*$^+$ Δ*crp* strains, the suppressors nearly compensated for the absence of Crp, as shown above in Figure 5. With the four *rpoA* K271 substitutions, the *araC* mutations increased expression 8- to 30-fold when compared to *araC*$^+$. In no case was expression as high as that detected in *rpoA*$^+$ strains, confirming that the *araC* mutations only partially compensate for the *rpoA* defects. Finally, the RpoA N268T substitution, which exerts the most drastic effect on P_{BAD} expression, was not suppressed by any of these *araC* mutations, both in the presence and absence of Crp, suggesting two independent mechanisms for α-CTD activation.

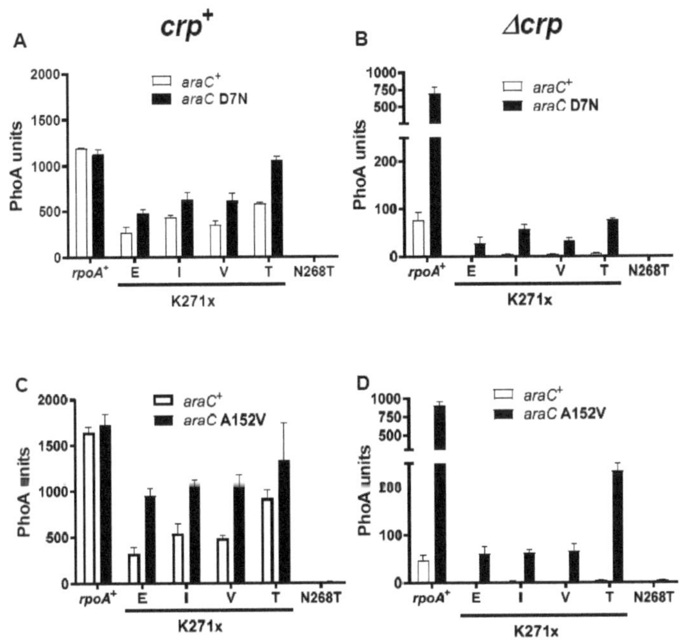

Figure 6. P_{BAD} activity mediated by wild-type AraC$^+$ and D7N (**A**,**B**), or A152V (**C**,**D**) mutant AraC proteins. Plasmids expressing AraC$^+$ or the indicated AraC mutant proteins were introduced in *crp*$^+$ (**A**,**C**) or Δ*crp* (**B**,**D**) strains with the indicated *rpoA* alleles. Cultures were assayed after a 1 h induction with arabinose.

To determine whether any specificity could be observed between the AraC suppressors and other α-CTD mutated residues, we compared the K271E substitution with substitutions at the two flanking residues, L270 and A272 (Figure 7). The three mutant AraC proteins exerted similar effects in all cases, although small quantitative differences were detected. Expression was highest with the *rpoA* L270F mutant, the weakest *rpoA* allele, and lowest with the *rpoA* A272E one. When compared to *araC*$^+$, increased expression with the *araC* mutants was found to be stronger in Δ*crp* (Figure 7) than in *crp*$^+$ strains (Supplementary Figure S2). In conclusion, we did not succeed in isolating a single *araC* mutation that selectively suppressed, in an allele-specific manner, any of the RpoA mutated residues tested.

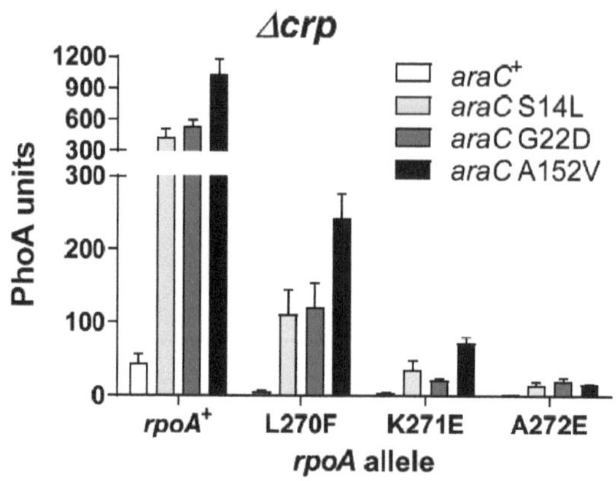

Figure 7. Effects of the AraC suppressors on the two residues that flank K271 of the RpoA α-CTD. Plasmids expressing AraC$^+$ or the indicated AraC mutant proteins were introduced in Δ*crp* strains carrying the *rpoA*$^+$ or the three indicated *rpoA* mutant alleles.

In a different approach, using a mutagenic PCR protocol on a plasmid harboring the wild-type *araC* gene, we identified three additional AraC residues (N16S, I51T, and I71V) whose substitutions resulted in increased P_{BAD} expression in an *rpoA* A272E Δ*crp* strain. In addition, different substitutions were found at two previously identified residues (S14P and A152T). In the *rpoA*$^+$ Δ*crp* strain, all compensated to a varying extent for the absence of Crp (Figure 8). The effect was maximal for the S14P mutant and minimal for the I51T mutant; the A152T mutant was less efficient than the A152V one (see Figure 6D). Furthermore, while the S14P and N16S mutants were partially constitutive, the other three mutants remained fully inducible; the A152T mutant, therefore, behaves clearly differently from the A152V mutant (Figure 8). These results confirm that a constitutive phenotype is not required for the suppression of the *rpoA* defects. The AraC I71V substitution restored the Ara^S phenotype conferred by PAI2::PhoA in the *rpoA* K271T *crp*$^+$ strain, confirming that increased expression from the P_{BAD} promoter can also be observed with this mutant in the presence of Crp. In Δ*crp* mutant strains carrying different *rpoA* alleles, increased expression was also detected (Supplementary Figure S3). Here again, we did not observe any allele specificity.

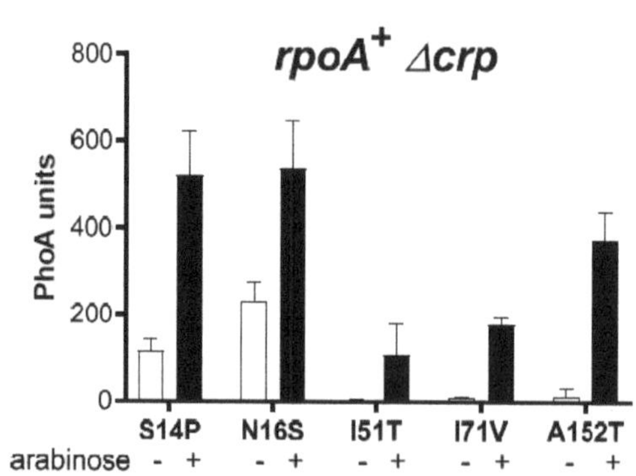

Figure 8. P_{BAD} activity mediated by *araC* suppressors of RpoA A272E. Plasmids expressing AraC$^+$ or the indicated AraC suppressor proteins were introduced in the *rpoA*$^+$ Δ*crp* strain. Cultures grown in the absence of arabinose were assayed to detect the constitutive activity of each AraC protein. Cultures induced for 1 h with arabinose were used to determine the full activity of each AraC protein.

3.3. The Activity of the AraC Mutant Proteins Requires the α-CTD of RpoA

The observed functional suppression of RpoA α-CTD mutations and the lack of allele specificity of the AraC mutations are compatible with at least two models. Firstly, the suppressors could create or improve an interaction between the AraC-NTD and the RNA polymerase core. Alternatively, RpoA α-CTD could contact a specific site of AraC that remains to be identified. In this first scenario, functional suppression would not depend on the RpoA α-CTD. To discriminate between these two possibilities, we expressed either wild-type RpoA or an RpoA deletion that lacks the α-CTD (residues 243–329). Both proteins were expressed from the P_{lac} promoter on a plasmid, with levels predicted to be ~5-fold higher than that of RpoA derived from the chromosomal *rpoA*$^+$ gene in the absence of Crp.

In *rpoA*$^+$ *Δcrp* strains, the activity of AraC$^+$ decreased about 2-fold in the presence of the RpoA deletion (Figure 9A). A similar decrease was observed with the two AraC suppressors, namely the D7N mutant located in the AraC N-terminal arm and the I71V mutant located in the core of the AraC-NTD (Figure 9B,C). This result is more evident when the activity of the three AraC proteins in cells carrying the RpoA$^+$ (Figure 9E) or the RpoA deletion plasmid (Figure 9F) are directly compared. As a control, expression from the *lacZ* promoter, which is activated only by an interaction of the RpoA α-CTD with Crp [8], was decreased to the same extent in the presence of the RpoA deletion (Figure 9D). When RpoA and the RpoA deletion were expressed from the strong bacteriophage T4 gene *32* promoter [43], we observed an almost complete inhibition of P_{BAD} expression (Supplementary Figure S4). Taken together, these results demonstrate that the RpoA α-CTD is required for transcriptional activation at the P_{BAD} promoter with both wild-type and mutant AraC in the absence of Crp, suggesting that the AraC mutants functionally compensate for defects in the RpoA α-CTD by increasing AraC activation of RNA polymerase.

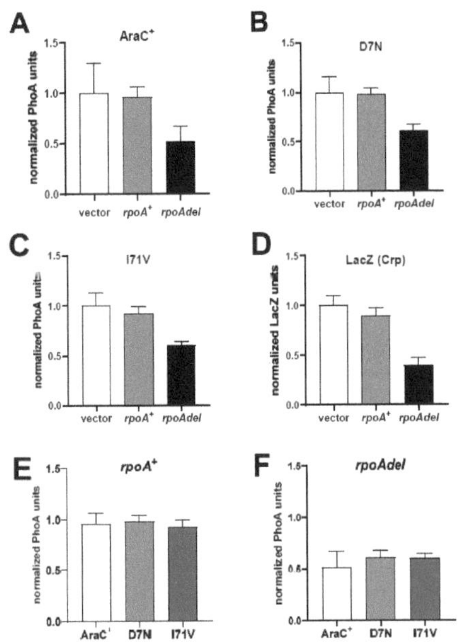

Figure 9. A deletion of the RpoA α-CTD decreases expression from the *araBAD* and *lac* promoters. (**A–D**) Empty bars: pUC19 vector; grey bars: pUC-*rpoA*$^+$; black bars: pUC-rpoAdelCTD. The PhoA and LacZ activities were normalized to those measured with the empty vector. PhoA expression was measured in *rpoA*$^+$ *Δcrp* strains. PhoA units with the empty vector: 4.5 ± 1 (AraC$^+$); 390 ± 60 (D7N); 188 ± 30 (I71V). Expression of *lacZ* was measured in the HfrH strain: LacZ units with the empty vector: 3530 ± 340. The normalized activities of the three AraC proteins in the presence of the pUC-*rpoA*$^+$ (**E**) or the pUC-rpoAdelCTD (**F**) plasmid are directly compared.

3.4. A Suppressor of RpoA N268T Synergizes with Mutations in the NTD of AraC

Since none of the AraC mutants described above suppressed the *rpoA* N268T mutant even in the presence of Crp, we repeated the mutant screen in the *rpoA* N268T *crp*$^+$ strain and isolated three mutants in the C-terminal DNA-binding domain of AraC. The largest effect was observed with R251H, while a 3-fold lower effect was observed with R251C and C205Y. The AraC R251H mutant compensated to the same extent for the effect of the *rpoA* N268T mutant in a *crp*$^+$ strain (Figure 10A) or in the absence of Crp in a *rpoA*$^+$ strain (Figure 10B); in *rpoA* K271X *crp*$^+$ strains, expression was >50% of that measured in an *araC*$^+$ *rpoA*$^+$ *crp*$^+$ strain. In contrast, this AraC mutant was totally inactive in the *rpoA* N268T Δ*crp* strain (Figure 10C).

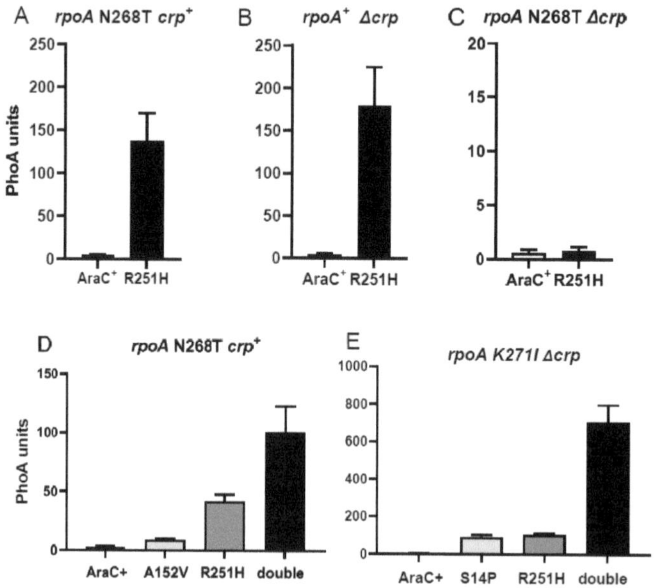

Figure 10. P_{BAD} activity mediated by AraC R251H and AraC double mutants. Plasmids expressing AraC$^+$ or the AraC R251H mutant were introduced into the *rpoA* N268T *crp*$^+$ (**A**), *rpoA*$^+$ Δ*crp* (**B**), or *rpoA* N268T Δ*crp* (**C**) strains. Plasmids expressing AraC$^+$, AraC A152V, AraC R251H, or the double mutant A152V-R251H were introduced into the *rpoA* N268T *crp*$^+$ strain (**D**). Plasmids expressing AraC$^+$, AraC S14P, AraC R251H or the double mutant S14P-R251H were introduced into the *rpoA* K271I Δ*crp* strain (**E**). Cultures were assayed after a 1 h induction with arabinose.

Since R251H also strongly increased expression in the *rpoA* K271E *crp*$^+$ strain, we investigated the effect of combining AraC mutations in the CTD (R251H) and in the NTD (S14L, in the N-terminal arm, or A152V, in the NTD core). We found that in the *rpoA* N268T *crp*$^+$ strain, the combined effect of both AraC mutations was synergistic and not additive (Figure 10D,E). This synergistic effect was most striking in an *rpoA* K271I Δ*crp* strain, reaching about 40% of the expression observed in the *araC*$^+$ *rpoA*$^+$ *crp*$^+$ strain. These results suggest that the R251H mutation in the DNA-binding domain of AraC exerts its effect at a different step of transcriptional activation than mutations in the AraC-NTD, underlying a regulatory mechanism not directly related to AraC-NTD activation.

In conclusion, we have characterized mutations in AraC that functionally compensate for the deleterious effect of *rpoA* mutations. Mutations in the AraC-NTD can almost fully compensate for the absence of Crp. Finally, combinations of mutations in the AraC-NTD and in the AraC-CTD synergistically activate transcription at the P_{BAD} promoter.

4. Discussion

The RpoA α-CTD domain is a pleiotropic hub required for transcription activation in multiple systems [8,12]. The residues that functionally interact with each activator are often different. For instance, the E261K/G substitutions affect a subset of Crp-dependent promoters, while P322S affects OmpR [44], and D250 and R310 affect TyrR [45]. The α-CTD has also been shown to play a role in genome-wide transcription regulation in *B. subtilis* [46].

Considering the number of *rpoA* mutations detected through our genetic selections, a large fraction of the RpoA residues necessary for P_{BAD} activation have probably now been identified. However, truncations of the α-CTD and several substitutions, including R265A and N268A, are not viable in the absence of a functional RpoA and must be studied in strains containing an *rpoA*$^+$ gene on the chromosome [15,47,48]. Thus, our experimental setup does not allow the detection of some mutations, including nonsense ones that would truncate RpoA.

The RpoA K271 substitutions affect CysB and MelR activity in addition to AraC [26]. However, several RpoA C-terminal truncations allow MelR transcriptional activity without any effect on AraC or CysB [11]. More recently, K271 was shown to mediate transcription activation by Fis at the *proP* P2 and *rrnB* P1 promoters, and the Fis residues that interact with the α-CTD were identified through structure-based models [19,49]. The K271E substitution also impaired phage λ PRE activation by *cII* [50,51]. Finally, we observed a 3- to 5-fold decrease in MalE synthesis in strains carrying the *rpoA* K271E substitution; it had no effect on *malPQ* transcription, which is independent of Crp [26]. The A272E change described here has similar effects to the K271E substitution. In contrast, the A272T substitution decreased *ompF* transcription mediated by OmpR, but had no detectable effect on activation by AraC, CysB, and MelR [52]. The replacement of the wild-type A residue by a larger, charged one likely confers a more drastic effect on α-CTD-mediated transcription.

The properties of the *rpoA* mutants provided the basis for the isolation of suppressor mutations that could compensate for the RpoA defects. A similar approach identified mutations in the phage P2 *Ogr* activator gene that suppress the *rpoA109* (L290H) mutation [53]. Furthermore, extensive genetic interactions between TyrR and RpoA have recently been described [45]. In our initial experiments, we used Δ*crp* strains, and only two elements were targets of mutagenesis: the *araC* coding region and the regulatory sequence between *araC* and P_{BAD}. Since expression is strongly decreased in the absence of Crp (Figure 3), we used a highly sensitive reporter for screening on indicator plates. In our study, suppressor mutations in genes expressed from the chromosome, such as potential compensatory mutations in *rpoA* or *crp*, could not be identified. The collection of suppressors described here is limited for two reasons. Firstly, the number of characterized mutants is not large, although most *araC* mutations were isolated several times. Secondly, the suppressors were isolated after hydroxylamine treatment of plasmid DNA, using a mutagen that induces only GC to AT transitions, or after mutagenic PCR, which mostly induces both transitions.

We first isolated two types of suppressors. Mutations in the P_{BAD} promoter are believed to improve intrinsic promoter efficiency, thus reducing the need for activation by AraC (Figure 4). This compensates indirectly for the negative effect of the *rpoA* mutations. The other suppressors carry a mutation in either the N-terminal arabinose-binding and dimerization domain of AraC or the C-terminal DNA-binding domain (Figure 11. Our results suggest that the AraC-NTD is a major target for functional suppression of the RpoA α-CTD mutations and contributes to transcriptional activation of the P_{BAD} promoter, particularly in the absence of Crp.

The *araC*-NTD mutations largely compensate for the activation defect caused by the absence of Crp in *rpoA*$^+$ strains (Figures 5–8), thus strengthening activation by AraC. If one assumes that one α-CTD plays a crucial role by interacting with the DNA and possibly with the AraC monomer bound to *I1*, the second α-CTD could bind to another AraC monomer (Figure 1). In the presence of Crp, only one α-CTD would be available to interact with AraC and/or the DNA. The toxic PAI2::PhoA protein confers an Ara^S phenotype in an *rpoA*$^+$ *crp*$^+$

strain but an Ara^R phenotype in strains carrying an *rpoA* α-CTD mutation; several *araC* mutations restore the Ara^S phenotype in an *rpoA crp*$^+$ strain. Thus, a functional interaction of one α-CTD with AraC could also occur in the presence of Crp.

The positions of the *araC* mutations on the structure of the AraC N-terminal domain [54] are shown in Figure 11A. None of the affected residues are directly involved in arabinose binding. Arabinose constitutive mutations have been isolated in the N-terminal domain of *araC* [41,42,55]. The S14L, N16S, G22D, and A152V mutations confer a constitutive phenotype, with A152V being the most efficient. In contrast, D7N and I71V are devoid of activity in the absence of arabinose but allow efficient induction (Figures 5 and 8). Thus, an arabinose constitutive phenotype is not required for *rpoA* suppression.

Several residues identified here must affect the conformation of the AraC-NTD. I51 and I71 are in the core of the N-terminal domain, on the side of the β-sheet that faces the two α-helices promoting dimerization, with A152 being part of one of these helices. The other residues (D7, S14, and N16) are in the arm of the AraC-NTD, and their lateral chains face the outside of the N-terminal domain. Although it is possible that these residues contact the α-CTD, the major argument against a direct interaction of the flexible arm with the α-CTD lies in the absence of allele specificity between pairs of *araC* and *rpoA* mutant alleles. Therefore, it appears more likely that the *araC* mutations exert an indirect effect and generate a conformation that improves the functional interaction of the α-CTD with the AraC-NTD.

Alternatively, the AraC-NTD could interact with the RNA polymerase core. If this were the case, the activity of the AraC mutant proteins would be maintained in the presence of an RpoA protein lacking its C-terminal domain. Since this deletion is not viable, we expressed the RpoA deletion protein in strains carrying the *rpoA*$^+$ gene on the chromosome. Under these conditions, the activity of wild-type and mutant AraC proteins was inhibited to the same extent (Figure 9 and Supplementary Figure S4). These results indicate that a putative interaction of the AraC-NTD with the RNA polymerase core does not participate in the suppression of the RpoA defects by the mutant AraC proteins. In agreement with this notion, we did not isolate RNA polymerase mutants, outside of the α-CTD, that selectively decrease P_{BAD} transcription mediated by mutant AraC.

The suppressors that localize to the AraC-NTD suggest a new function for this domain. In the absence of arabinose, the arm domain binds to the DNA-binding domain to ensure DNA looping and P_{BAD} repression. In the presence of arabinose, the arm domain folds on top of the inducer binding pocket, and one of the arm residues, F15, contacts arabinose [41]. In a third function, the AraC-NTD could functionally interact with the RpoA α-CTD to promote efficient transcription at the P_{BAD} promoter, particularly in the absence of Crp. This genetic approach may be easily extended to other activators that interact with the RpoA C-terminal domain, a hub for transcriptional activation.

The first evidence that the α-CTD domain binds DNA was observed with the UP element, upstream of the *rrnB* promoter [15]. R265 is a critical residue for DNA binding, along with several other residues, including N268 [14,20,48]. A crystal structure of a DNA-Crp-α-CTD complex showed that R265 and N268, together with other RpoA residues, bind to the DNA backbone and to the minor groove [14]. The RpoA N268D substitution decreased *lacZ* expression [56], while the N268A, which cannot complement an *rpoA ts* allele [48], and W substitutions abolished activation by an UP element [57]. The R265A and N268A substitutions decreased *rhaBAD* activation by RhaS [58]. The N268T substitution described here abolishes P_{BAD} promoter activity, both in the presence and absence of Crp (Figures 2B and 3B). Unlike the K271 and A272 substitutions, this viable mutation affects at least one more gene than *cysA*, as it prevented growth on minimal media in the presence of Cys and Met.

Transcription activation at P_{BAD} has been proposed to involve the interaction of one of the α-CTD with Crp, and it has been suggested that the other α-CTD could bind to the DNA-binding domain of AraC and to a DNA sequence element between the Crp and AraC binding sites [3] (Figure 1). We did not identify any mutation in the regulatory region that would have defined a DNA site responsible for α-CTD binding, possibly because the target size is small, because a single base pair change would not compensate for the drastic effect of the RpoA N268T change or because it could interfere with AraC binding to I1. Nevertheless, the effect of this change and the role of this residue in DNA binding at other promoters indicate that an α-CTD–DNA interaction could be critical for transcription at the P_{BAD} promoter. The isolation of mutants in the C-terminal DNA-binding domain of AraC that suppress the RpoA N268T allele provides the first genetic evidence to supports this model (Figure 11B). The activity of the R251H mutant in a *rpoA*$^{+}$ Δ*crp* strain strongly suggests that the AraC-CTD could functionally interact with one α-CTD. The synergistic effect of mutations in the NTD and CTD of AraC is compatible with the notion that each domain could function independently to activate transcription.

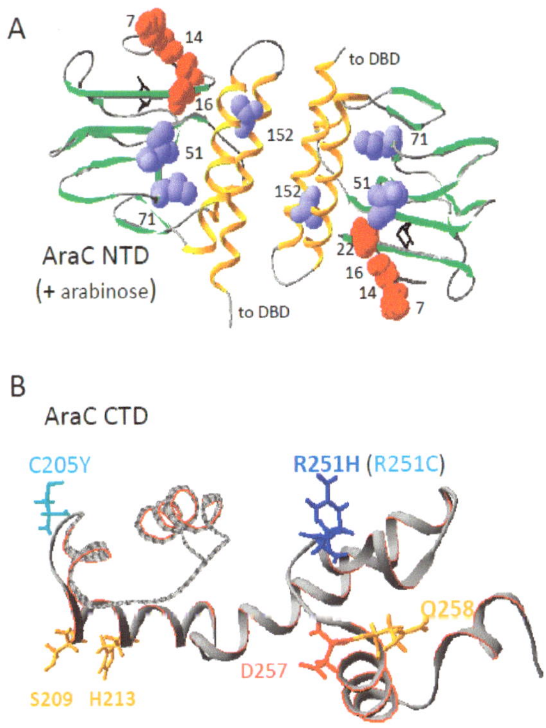

Figure 11. Distribution of the AraC suppressors on the structures of the AraC N-terminal domain dimer (PDB 2ARC) [54] and of the C-terminal DNA-binding domain (PDB 2K9S) [59]. The images were generated with Swiss-PdbViewer v4.1.0. The domains are represented as ribbons. (**A**) The two large α helices that promote NTD dimerization are shown in orange and the β strands are in green. N-terminal arm D7, S14, N16, and G22 residues are shown in red and space-filling. The side chains of the internal residues I51, I71, and A152 are shown in blue and space-filling. The bound arabinose inducers are shown in black. (**B**) The two residues mutated in suppressors of *rpoA* N268T are shown in blue; R251H has the strongest phenotype and is shown in dark blue. Mutations of D257 (red) abolish DNA binding while mutations of the three other residues (orange) are defective in DNA binding.

Supplementary Materials: The following supporting information can be downloaded at: https://www.mdpi.com/article/10.3390/microorganisms12091928/s1, Figure S1: P_{BAD} activity mediated by wild type AraC^{+} and S14L (A,B) or G22D (C,D). Plasmids expressing AraC^{+} or the indicated AraC suppressor proteins were introduced in *crp*$^{+}$ (A,C) or Δ*crp* (B,D) strains with the indicated *rpoA* alleles.

Cultures were assayed after a 1h induction with arabinose; Figure S2: Effects of the AraC suppressors on the two residues that flank K271 of the RpoA α-CTD. Plasmids expressing AraC$^+$ or the indicated AraC suppressor proteins were introduced in *crp*$^+$ strains carrying the *rpoA*$^+$ or the three indicated *rpoA* alleles. Cultures were assayed after a 1h induction with arabinose; Figure S3: P_{BAD} activity mediated by *araC* suppressors of RpoA A272E. Plasmids expressing AraC$^+$ or the indicated AraC suppressor proteins were introduced in Δ*crp* strains expressing either the RpoA K271E or the K271V proteins. Cultures were assayed after a 1h induction with arabinose; Figure S4: High expression of the *rpoA* α-CTD deletion strongly decreased expression from the *araBAD* promoter. RpoA and the RpoA α-CTD deletion were expressed from the bacteriophage T4 *32* promoter of pRDB8 [43]. Black bars: pRDB8-*rpoA*$^+$; white bars: pRDB8-*rpoA*delCTD. PhoA expression was measured in a *rpoA*$^+$ δ*crp* strain. With AraC$^+$, the expression of PhoA was barely detectable with both plasmids. The expression of *lacZ* was measured in the HfrH strain.

Author Contributions: Conceptualization, D.B. and J.C.; formal analysis, D.B.; investigations, F.S and D.B.; writing, D.B. All authors have read and agreed to the published version of the manuscript.

Funding: The work was supported by the Swiss National Foundation, the Canton de Genève, and the CPG Research Foundation, grant number 3100-068157.02.

Data Availability Statement: The original contributions presented in the study are included in the article/Supplementary Materials, further inquiries can be directed to the corresponding author/s.

Acknowledgments: We thank Sandrine Bost, Corinne Chauffat, Giuseppe Plaia and Nessim Kaufman for the isolation of *rpoA* mutants, and Sylvie Trichard and Halil Bagci for performing some of the PhoA assays. We also thank Joseph Curran, Costa Georgopoulos, Gael Panis, Robert Schleif, David Shore, Michel Strubin, and Benjamin Weiss for discussions and critical, constructive reading of the manuscript. D.B. thanks Joseph Curran for his kind hospitality.

Conflicts of Interest: The authors declare no conflict of interest.

Abbreviations

α-CTD	C-terminal domain of RpoA
α–NTD	N-terminal domain of RpoA
AraC-NTD	N-terminal domain of AraC

References

1. Englesberg, E.; Irr, J.; Power, J.; Lee, N. Positive control of enzyme synthesis by gene C in the L-arabinose system. *J. Bacteriol.* **1965**, *90*, 946–957. [CrossRef] [PubMed]
2. Cole, S.D.; Schleif, R. A new and unexpected domain-domain interaction in the AraC protein. *Proteins* **2012**, *80*, 1465–1475. [CrossRef] [PubMed]
3. Schleif, R. Regulation of the L-arabinose operon of *Escherichia coli*. *Trends Genet.* **2000**, *16*, 559–565. [CrossRef] [PubMed]
4. Schleif, R. AraC protein: A love-hate relationship. *BioEssays News Rev. Mol. Cell. Dev. Biol.* **2003**, *25*, 274–282. [CrossRef] [PubMed]
5. Zhang, X.; Reeder, T.; Schleif, R. Transcription activation parameters at *ara* pBAD. *J. Mol. Biol.* **1996**, *258*, 14–24. [CrossRef] [PubMed]
6. Hahn, S.; Dunn, T.; Schleif, R. Upstream repression and CRP stimulation of the *Escherichia coli* L-arabinose operon. *J. Mol. Biol.* **1984**, *180*, 61–72. [CrossRef] [PubMed]
7. Zhang, X.; Schleif, R. Catabolite gene activator protein mutations affecting activity of the *araBAD* promoter. *J. Bacteriol.* **1998**, *180*, 195–200. [CrossRef] [PubMed]
8. Busby, S.; Ebright, R.H. Transcription activation by catabolite activator protein (CAP). *J. Mol. Biol.* **1999**, *293*, 199–213. [CrossRef] [PubMed]
9. Guzman, L.M.; Belin, D.; Carson, M.J.; Beckwith, J. Tight regulation, modulation, and high-level expression by vectors containing the arabinose P_{BAD} promoter. *J. Bacteriol.* **1995**, *177*, 4121–4130. [CrossRef]
10. Kimura, M.; Fujita, N.; Ishihama, A. Functional map of the alpha subunit of *Escherichia coli* RNA polymerase. Deletion analysis of the amino-terminal assembly domain. *J. Mol. Biol.* **1994**, *242*, 107–115. [CrossRef]
11. Zou, C.; Thomas, M.S.; Keen, J.; Glass, R.E. A nested set of C-terminal deletions of the alpha subunit of *Escherichia coli* RNA polymerase define regions concerned with assembly, proteolysis, stabilization and transcriptional activation in vivo. *Genes Cells* **1997**, *2*, 81–94. [CrossRef] [PubMed]
12. Russo, F.D.; Silhavy, T.J. Alpha: The Cinderella subunit of RNA polymerase. *J. Biol. Chem.* **1992**, *267*, 14515–14518. [CrossRef] [PubMed]

13. Lloyd, G.S.; Niu, W.; Tebbutt, J.; Ebright, R.H.; Busby, S.J. Requirement for two copies of RNA polymerase alpha subunit C-terminal domain for synergistic transcription activation at complex bacterial promoters. *Genes Dev.* **2002**, *16*, 2557–2565. [CrossRef] [PubMed]
14. Benoff, B.; Yang, H.; Lawson, C.L.; Parkinson, G.; Liu, J.; Blatter, E.; Ebright, Y.W.; Berman, H.M.; Ebright, R. Structural basis of transcription activation: The CAP-alpha CTD-DNA complex. *Science* **2002**, *297*, 1562–1566. [CrossRef] [PubMed]
15. Ross, W.; Gosink, K.K.; Salomon, J.; Igarashi, K.; Zou, C.; Ishihama, A.; Severinov, K.; Gourse, R.L. A third recognition element in bacterial promoters: DNA binding by the alpha subunit of RNA polymerase. *Science* **1993**, *262*, 1407–1413. [CrossRef] [PubMed]
16. Gourse, R.L.; Ross, W.; Gaal, T. UPs and downs in bacterial transcription initiation: The role of the alpha subunit of RNA polymerase in promoter recognition. *Mol. Microbiol.* **2000**, *37*, 687–695. [CrossRef] [PubMed]
17. Jeon, Y.H.; Negishi, T.; Shirakawa, M.; Yamazaki, T.; Fujita, N.; Ishihama, A.; Kyogoku, Y. Solution structure of the activator contact domain of the RNA polymerase alpha subunit. *Science* **1995**, *270*, 1495–1497. [CrossRef] [PubMed]
18. Lara-Gonzalez, S.; Birktoft, J.J.; Lawson, C.L. Structure of the *Escherichia coli* RNA polymerase alpha subunit C-terminal domain. *Acta Crystallogr. D Biol. Crystallogr.* **2010**, *66*, 806–812. [CrossRef] [PubMed]
19. McLeod, S.M.; Aiyar, S.E.; Gourse, R.L.; Johnson, R.C. The C-terminal domains of the RNA polymerase alpha subunits: Contact site with Fis and localization during co-activation with CRP at the *Escherichia coli* proP P2 promoter. *J. Mol. Biol.* **2002**, *316*, 517–529. [CrossRef] [PubMed]
20. Liu, B.; Hong, C.; Huang, R.K.; Yu, Z.; Steitz, T.A. Structural basis of bacterial transcription activation. *Science* **2017**, *358*, 947–951. [CrossRef] [PubMed]
21. Dhiman, A.; Schleif, R. Recognition of overlapping nucleotides by AraC and the sigma subunit of RNA polymerase. *J. Bacteriol.* **2000**, *182*, 5076–5081. [CrossRef] [PubMed]
22. Casadaban, M.J. Transposition and fusion of the *lac* genes to selected promoters in *Escherichia coli* using bacteriophage lambda and Mu. *J. Mol. Biol.* **1976**, *104*, 541–555. [CrossRef] [PubMed]
23. Belin, D.; Plaia, G.; Boulfekhar, Y.; Silva, F. *Escherichia coli* SecG is required for residual export mediated by mutant signal sequences and for SecY-SecE complex stability. *J. Bacteriol.* **2015**, *197*, 542–552. [CrossRef] [PubMed]
24. Baba, T.; Ara, T.; Hasegawa, M.; Takai, Y.; Okumura, Y.; Baba, M.; Datsenko, K.A.; Tomita, M.; Wanner, B.L.; Mori, H. Construction of *Escherichia coli* K-12 in-frame, single-gene knockout mutants: The Keio collection. *Mol. Syst. Biol.* **2006**, *2*, 2006.0008. [CrossRef] [PubMed]
25. Miller, J.H. *A Short Course in Bacterial Genetics*; Cold Spring Harbor Laboratory Press: Cold Spring Harbor, NY, USA, 1992.
26. Giffard, P.M.; Booth, I.R. The *rpoA341* allele of *Escherichia coli* specifically impairs the transcription of a group of positively-regulated operons. *Mol. Gen. Genet.* **1988**, *214*, 148–152. [CrossRef] [PubMed]
27. Plaia, G. Ruolo di SecG nel Trasporto Proteico in *Escherichia coli* e Studio di una Proteina Sconosciuta del Fago T4. Ph.D. Thesis, University of Geneva, Geneva, Switzerland, 2009.
28. Bost, S.; Belin, D. A new genetic selection identifies essential residues in SecG, a component of the *Escherichia coli* protein export machinery. *EMBO J.* **1995**, *14*, 4412–4421. [CrossRef]
29. Mattenberger, Y.; Silva, F.; Belin, D. *55.2*, a phage T4 ORFan gene, encodes an inhibitor of Escherichia coli topoisomerase I and increases phage fitness. *PLoS ONE* **2015**, *10*, e0124309. [CrossRef]
30. Benarafa, C.; Remold-O'Donnell, E. The ovalbumin serpins revisited: Perspective from the chicken genome of clade B serpin evolution in vertebrates. *Proc. Natl. Acad. Sci. USA* **2005**, *102*, 11367–11372. [CrossRef] [PubMed]
31. Belin, D.; Guzman, L.M.; Bost, S.; Konakova, M.; Silva, F.; Beckwith, J. Functional activity of eukaryotic signal sequences in *Escherichia coli*: The ovalbumin family of serine protease inhibitors. *J. Mol. Biol.* **2004**, *335*, 437–453. [CrossRef] [PubMed]
32. Belin, D. In vivo analysis of protein translocation to the *Escherichia coli* periplasm. *Methods Mol. Biol.* **2010**, *619*, 103–116. [CrossRef] [PubMed]
33. Khatib, K.; Belin, D. A novel class of *secA* alleles that exert a signal-sequence-dependent effect on protein export in *Escherichia coli*. *Genetics* **2002**, *162*, 1031–1043. [CrossRef] [PubMed]
34. Thomas, M.S.; Glass, R.E. *Escherichia coli rpoA* mutation which impairs transcription of positively regulated systems. *Mol. Microbiol.* **1991**, *5*, 2719–2725. [CrossRef] [PubMed]
35. Lochowska, A.; Iwanicka-Nowicka, R.; Zaim, J.; Witkowska-Zimny, M.; Bolewska, K.; Hryniewicz, M.M. Identification of activating region (AR) of Escherichia coli LysR-type transcription factor CysB and CysB contact site on RNA polymerase alpha subunit at the *cysP* promoter. *Mol. Microbiol.* **2004**, *53*, 791–806. [CrossRef] [PubMed]
36. Martin, K.; Huo, L.; Schleif, R.F. The DNA loop model for ara repression: AraC protein occupies the proposed loop sites in vivo and repression-negative mutations lie in these same sites. *Proc. Natl. Acad. Sci. USA* **1986**, *83*, 3654–3658. [CrossRef] [PubMed]
37. Englesberg, E.; Sheppard, D.; Squires, C.; Meronk, F., Jr. An analysis of "revertants" of a deletion mutant in the C gene of the L-arabinose gene complex in *Escherichia coli* B-r: Isolation of initiator constitutive mutants (Ic). *J. Mol. Biol.* **1969**, *43*, 281–298. [CrossRef] [PubMed]
38. Cass, L.G.; Horwitz, A.H.; Wilcox, G. The araI^c mutation in *Escherichia coli* B/r. *J. Bacteriol.* **1981**, *146*, 1098–1105. [CrossRef] [PubMed]
39. Colome, J.; Wilcox, G.; Englesberg, E. Constitutive mutations in the controlling site region of the *araBAD* operon of *Escherichia coli* B/r that decrease sensitivity to catabolite repression. *J. Bacteriol.* **1977**, *129*, 948–958. [CrossRef] [PubMed]

40. Shultzaberger, R.K.; Chen, Z.; Lewis, K.A.; Schneider, T.D. Anatomy of *Escherichia coli* sigma70 promoters. *Nucleic Acids Res.* **2007**, *35*, 771–788. [CrossRef] [PubMed]
41. Ross, J.J.; Gryczynski, U.; Schleif, R. Mutational analysis of residue roles in AraC function. *J. Mol. Biol.* **2003**, *328*, 85–93. [CrossRef] [PubMed]
42. Dirla, S.; Chien, J.Y.; Schleif, R. Constitutive mutations in the *Escherichia coli* AraC protein. *J. Bacteriol.* **2009**, *191*, 2668–2674. [CrossRef] [PubMed]
43. Duvoisin, R.M.; Belin, D.; Krisch, H.M. A plasmid expression vector that permits stabilization of both mRNAs and proteins encoded by the cloned genes. *Gene* **1986**, *45*, 193–201. [CrossRef] [PubMed]
44. Slauch, J.M.; Russo, F.D.; Silhavy, T.J. Suppressor mutations in *rpoA* suggest that OmpR controls transcription by direct interaction with the alpha subunit of RNA polymerase. *J. Bacteriol.* **1991**, *173*, 7501–7510. [CrossRef] [PubMed]
45. Camakaris, H.; Yang, J.; Fujii, T.; Pittard, J. Activation by TyrR in *Escherichia coli* K-12 by Interaction between TyrR and the alpha Subunit of RNA Polymerase. *J. Bacteriol.* **2021**, *203*, e0025221. [CrossRef] [PubMed]
46. Murayama, S.; Ishikawa, S.; Chumsakul, O.; Ogasawara, N.; Oshima, T. The Role of alpha-CTD in the Genome-Wide Transcriptional Regulation of the *Bacillus subtilis* Cells. *PLoS ONE* **2015**, *10*, e0131588. [CrossRef] [PubMed]
47. Hayward, R.S.; Igarashi, K.; Ishihama, A. Functional specialization within the alpha-subunit of *Escherichia coli* RNA polymerase. *J. Mol. Biol.* **1991**, *221*, 23–29. [CrossRef] [PubMed]
48. Gaal, T.; Ross, W.; Blatter, E.E.; Tang, H.; Jia, X.; Krishnan, V.V.; Assa-Munt, N.; Ebright, R.H.; Gourse, R.L. DNA-binding determinants of the alpha subunit of RNA polymerase: Novel DNA-binding domain architecture. *Genes Dev.* **1996**, *10*, 16–26. [CrossRef] [PubMed]
49. Aiyar, S.E.; McLeod, S.M.; Ross, W.; Hirvonen, C.A.; Thomas, M.S.; Johnson, R.C.; Gourse, R.L. Architecture of Fis-activated transcription complexes at the *Escherichia coli rrnB* P1 and *rrnE* P1 promoters. *J. Mol. Biol.* **2002**, *316*, 501–516. [CrossRef] [PubMed]
50. Kedzierska, B.; Lee, D.J.; Wegrzyn, G.; Busby, S.J.; Thomas, M.S. Role of the RNA polymerase alpha subunits in CII-dependent activation of the bacteriophage lambda pE promoter: Identification of important residues and positioning of the alpha C-terminal domains. *Nucleic Acids Res.* **2004**, *32*, 834–841. [CrossRef] [PubMed]
51. Kedzierska, B.; Szambowska, A.; Herman-Antosiewicz, A.; Lee, D.J.; Busby, S.J.; Wegrzyn, G.; Thomas, M. The C-terminal domain of the *Escherichia coli* RNA polymerase alpha subunit plays a role in the CI-dependent activation of the bacteriophage lambda pM promoter. *Nucleic Acids Res.* **2007**, *35*, 2311–2320. [CrossRef] [PubMed]
52. Sharif, T.R.; Igo, M.M. Mutations in the alpha subunit of RNA polymerase that affect the regulation of porin gene transcription in *Escherichia coli* K-12. *J. Bacteriol.* **1993**, *175*, 5460–5468. [CrossRef] [PubMed]
53. King, R.A.; Anders, D.L.; Christie, G.E. Site-directed mutagenesis of an amino acid residue in the bacteriophage P2 ogr protein implicated in interaction with Escherichia coli RNA polymerase. *Mol. Microbiol.* **1992**, *6*, 3313–3320. [CrossRef] [PubMed]
54. Soisson, S.M.; MacDougall-Shackleton, B.; Schleif, R.; Wolberger, C. Structural basis for ligand-regulated oligomerization of AraC. *Science* **1997**, *276*, 421–425. [CrossRef] [PubMed]
55. Saviola, B.; Seabold, R.; Schleif, R.F. Arm-domain interactions in AraC. *J. Mol. Biol.* **1998**, *278*, 539–548. [CrossRef] [PubMed]
56. Zou, C.; Fujita, N.; Igarashi, K.; Ishihama, A. Mapping the cAMP receptor protein contact site on the alpha subunit of *Escherichia coli* RNA polymerase. *Mol. Microbiol.* **1992**, *6*, 2599–2605. [CrossRef] [PubMed]
57. Murakami, K.; Fujita, N.; Ishihama, A. Transcription factor recognition surface on the RNA polymerase alpha subunit is involved in contact with the DNA enhancer element. *EMBO J.* **1996**, *15*, 4358–4367. [CrossRef] [PubMed]
58. Holcroft, C.C.; Egan, S.M. Roles of cyclic AMP receptor protein and the carboxyl-terminal domain of the alpha subunit in transcription activation of the *Escherichia coli rhaBAD* operon. *J. Bacteriol.* **2000**, *182*, 3529–3535. [CrossRef] [PubMed]
59. Rodgers, M.E.; Schleif, R. Solution structure of the DNA binding domain of AraC protein. *Proteins* **2009**, *77*, 202–208. [CrossRef] [PubMed]

microorganisms

MDPI

Article

Regulatory Functions of PurR in *Yersinia pestis*: Orchestrating Diverse Biological Activities

Liting Xiao [1,2,†], Junyan Jin [2,†], Kai Song [2], Xiuwei Qian [1,2], Yarong Wu [2], Zhulin Sun [2], Ziyao Xiong [2], Yanbing Li [2], Yanting Zhao [2], Leiming Shen [2], Yiming Cui [2], Wenwu Yao [2], Yujun Cui [1,2,*] and Yajun Song [1,2,*]

1 School of Basic Medical Sciences, Anhui Medical University, Hefei 230032, China; xlt15576480220@163.com (L.X.); qxw_2020@163.com (X.Q.)
2 State Key Laboratory of Pathogen and Biosecurity, Beijing Institute of Microbiology and Epidemiology, Beijing 100071, China; jinjunyan97@163.com (J.J.); wuyarong525@126.com (Y.W.); sunzhulin0902@126.com (Z.S.); xiaongzy@163.com (Z.X.); liyanbing@csu.edu.cn (Y.L.); zhaoyanting2121@163.com (Y.Z.); 1581182094@126.com (L.S.); m15762561593@163.com (Y.C.); wwyao@cdc.zj.cn (W.Y.)
* Correspondence: cuiyujun.new@gmail.com (Y.C.); songyj@bmi.ac.cn (Y.S.)
† These authors contributed equally to this work.

Abstract: The bacterium *Yersinia pestis* has developed various strategies to sense and respond to the complex stresses encountered during its transmission and pathogenic processes. PurR is a common transcriptional regulator of purine biosynthesis among microorganisms, and it modulates the transcription level of the *pur* operon to suppress the production of hypoxanthine nucleotide (IMP). This study aims to understand the functions and regulatory mechanisms of *purR* in *Y. pestis*. Firstly, we constructed a *purR* knockout mutant of *Y. pestis* strain 201 and compared certain phenotypes of the null mutant (201-Δ*purR*) and the wild-type strain (201-WT). The results show that deleting *purR* has no significant impact on the biofilm formation, growth rate, or viability of *Y. pestis* under different stress conditions (heat and cold shock, high salinity, and hyperosmotic pressure). Although the cytotoxicity of the *purR* knockout mutant on HeLa and 293 cells is reduced, the animal-challenging test found no difference of the virulence in mice between 201-Δ*purR* and 201-WT. Furthermore, RNA-seq and EMSA analyses demonstrate that PurR binds to the promoter regions of at least 15 genes in *Y. pestis* strain 201, primarily involved in purine biosynthesis, along with others not previously observed in other bacteria. Additionally, RNA-seq results suggest the presence of 11 potential operons, including a newly identified co-transcriptional T6SS cluster. Thus, aside from its role as a regulator of purine biosynthesis, *purR* in *Y. pestis* may have additional regulatory functions.

Keywords: *Yersinia pestis*; *purR*; transcriptional regulation; purine biosynthesis

Citation: Xiao, L.; Jin, J.; Song, K.; Qian, X.; Wu, Y.; Sun, Z.; Xiong, Z.; Li, Y.; Zhao, Y.; Shen, L.; et al. Regulatory Functions of PurR in *Yersinia pestis*: Orchestrating Diverse Biological Activities. *Microorganisms* **2023**, *11*, 2801. https://doi.org/10.3390/microorganisms11112801

Academic Editors: Thomas Proft and Tomohiro Shimada

Received: 7 October 2023
Revised: 11 November 2023
Accepted: 15 November 2023
Published: 17 November 2023

1. Introduction

Yersinia pestis is the causative agent of the plague, a highly infectious disease that has caused three global pandemics throughout human history [1,2]. This bacterium possesses the ability to adapt to both flea (26 °C) and mammalian (37 °C) body temperatures, as transmission between these hosts is crucial for its natural life cycle [3]. However, *Y. pestis* can undergo physiological changes in anabolism when exposed to different environments. As a relatively young bacterium that diverged from *Yersinia pseudotuberculosis* approximately 7000 years ago [4,5], *Y. pestis* displays a moderate degree of sequence diversity.

Purine plays a vital role in the survival of microorganisms, and most bacteria rely on de novo synthesis for its production. The *purR* gene is widely present in bacterial genomes and functions as a transcriptional repressor, regulating purine biosynthesis by controlling the expression of the *pur* operon [6–8]. The deletion of *purR* had been proven to enhance the metabolic flow of the purine pathway and improve the production of riboflavin in *Escherichia coli*, *Bacillus subtilis*, and *Ashbya gossypii* [9–11]. However, in certain bacteria, *purR* serves additional roles beyond its involvement in purine regulation. For instance in *Escherichia coli*, *purR* contributes to the bacterium's tolerance to organic solvents and enhances its viability when exposed to them [12]. Moreover, mutations in *purR* have been shown to increase the virulence of *Staphylococcus aureus* [13]. Nevertheless, the precise functions of the *purR* gene in *Y. pestis* remains incompletely understood.

To investigate the functions of the *purR* gene in *Y. pestis*, we compared the adaptabilities of *Y. pestis* strain 201 (201-WT) and its *purR* knockout mutant (201-Δ*purR*). We observed that the deletion of *purR* had minimal impact on the environmental adaptability of *Y. pestis* strain 201. To further elucidate the role of *purR* as a transcriptional regulator, we conducted RNA-seq and assessed the expression of the PurR protein. RNA-seq results revealed significant alterations in the transcription levels of numerous genes across the entire genome upon deletion of *purR*, indicating specific associations between certain genes and *purR* in *Y. pestis* Moreover, Electrophoretic Mobility Shift Assay (EMSA) results demonstrated that PurR could directly bind to the promoter regions of multiple genes within the *Y. pestis* genome, suggesting its direct regulation of their transcription and expression. Collectively, our findings offer valuable insights into the regulatory function of *purR* in governing gene expression in *Y. pestis*.

2. Materials and Methods

2.1. Bacterial Strains and Culture Conditions

The strains and plasmids used in this study are presented in Table 1, with the primer sequences listed in Supplementary Table S1. *Y. pestis* strain 201 has an identical genome as strain 91,001, which is highly lethal to mice but avirulent to humans [14,15]. Different culture conditions were employed for *Y. pestis* and *E. coli* throughout the experiment. *Y. pestis* was cultivated in LB (Luria-Bertani) medium at a temperature of 26 °C to mimic the temperature of fleas (the vector of *Y. pestis* in the flea transmission process of plague) [16], whereas *E. coli* was cultured in LB medium at 37 °C. Chloramphenicol (34 μg/mL) was added during the cultivation of the complementation strain harboring the pACYC184 plasmid, while kanamycin (50 μg/mL) was necessary for the growth of the strain carrying the pET28a (+) plasmid. All bacterial experiments were performed in a biological safety cabinet.

Table 1. Bacterial strains and plasmids used in this study.

Strain or Plasmid	Genotype	Reference
	E. coli	
S17λ*pir*	Tp^r Sm^r *recA thi pro hsdR*$^-$*M*$^+$ (RP4-2-Tc::Mu: Kan^{r}Tn7) λ*pir*	[17]
S17-pDS132-*purR*-del	pDS132-*purR*-del was introduced into S17λpir	This study
DH5α	F- φ80lacZΔM15 Δ(*lacZYA*-arg F) U169 *endA*1 *recA*1 *hsdR*17(rk−, mk+) *supE*44 λ- thi-1 *gyrA*96 *relA*1 *phoA*	[17]
DH5α-pACYC184- *purR*	pACYC184-*purR* was introduced into DH5α	This study
DH5α-pET28a (+)-*purR*	pET28a (+)-*purR* was introduced into DH5α	This study
BL21(DE3)	F-ompT *hsdSB*($r_B^-m_B^-$) gal dcm (DE3)	[18]
BL21(DE3)-pET28a (+)-*purR*	pET28a (+)-*purR* was introduced into BL21 (DE3)	This study
	Y. pestis	
201-WT	*Y. pestis* biovar Microtus strain 201, WT	[14]
201-Δ*purR*	deleted *purR* based on strain 201	This study
201-Δ*purR*-Comp	201-Δ*purR* containing plasmid pACYC184-*purR*	This study
	plasmids	
pDS132	suicide vector, derived from pCVD442, without IS1 sequences. *bla* gene replaced by the *cat* gene	[17]
pACYC184	cloning vector, Cm^r Tet^r	[17,19]
pET28a (+)	overexpression vectors, carry an N-terminal His-Tag/T7-Tag configuration plus an optional C-terminal His-Tag sequence, Kan^r	[18]

2.2. DNA Extraction and Amplification

Genomic DNA and plasmids were extracted using the QIAamp DNA Mini Kit (Qiagen, Hilden, Germany) and QIAprep Spin Miniprep Kit (Qiagen, Hilden, Germany), respectively, following the manufacturer's instructions. The target segments were amplified using PCR with 1.1× GoldenStar mix (Green) (Tsingke Biotechnology Co., Ltd., Beijing, China).

2.3. Construction of the Mutant and Complementation Strain

The pDS132 was digested by incubating overnight at 37 °C with *Sph*I and *Sac*I enzymes (LMAI Bio, Shanghai, China) together. The upstream and downstream homology arms of *purR* were then ligated with the linearized pDS132 using 2× Seamless Cloning Mix (Biomed, Beijing, China) at 50 °C for 15 min with a molar ratio of vector to DNA of 1:3, and the recombinant vector was introduced into *E. coli* S17λpir to obtain S17-pDS132-*purR*-del for conjugation with *Y. pestis*.

S17-pDS132-*purR*-del and 201-WT strains were cultured in LB medium at either 37 °C or 26 °C until reaching an OD_{620} of 0.8. After centrifugation at 1900× *g* for 5 min, the pellet of S17-pDS132-*purR*-del (1.5 mL) and 201-WT (100 µL) cultures was resuspended and added dropwise onto a filter membrane (0.45 µm) placed on LB plates. The cells on the filter paper were then incubated overnight at 26 °C, and the resulting eluate was spread onto Yersinia Selective Agar Base plates (Oxoid, Basingstoke, UK) supplemented with chloramphenicol (6.8 µg/mL) and incubated at 26 °C. Conjugations were expected to occur under these conditions and were selected on LB plates containing 7% sucrose. The obtained colonies were further analyzed to confirm the expectant conjugant through PCR amplification and sequencing; the correct sequencing results indicate that *purR* has been successfully knocked out, and this strain has been named 201-Δ*purR*.

For complementation of *purR*, the pACYC184 plasmid was digested with *Hind* III and *BamH* I enzymes (LMAI Bio, Shanghai, China), and the *purR* fragment and linearized pACYC184 were then ligated as described previously. The resulting recombinant vector was introduced into competent *E. coli* DH5α cells. The expected recombinant plasmid was identified using PCR sequencing and extracted before being transferred into 201-Δ*purR*. The resulting transformant was confirmed using PCR sequencing and designated as 201-Δ*purR*-Comp.

2.4. Growth Rate Determination

The 201-WT and 201-Δ*purR* strains were cultured until they reached an optical density of 1.0 at OD_{620} (*ca.*2 × 10^8 CFU/mL). Subsequently, the bacterial cultures were inoculated at a ratio of 1:100 in Erlenmeyer flasks containing 60 mL of either fresh LB or a chemically defined TMH liquid medium [20]. The size of Erlenmeyer flasks was 150 mL. All Erlenmeyer flasks were then placed in a precision cell culture shaker (Zhicheng ZWYF-290, Shanghai, China) set at 26 °C with shaking at 200 rpm, and the OD_{600} values of the cultures were measured hourly. Once all strains entered the decline phase, data were collected and plotted for analysis. Each strain underwent three independent biological replications under identical conditions, and the results were expressed as the mean ± standard deviation of the three biological replicated experiments. The experimental procedure at 37 °C was similar to the above, with the exception that the temperature was adjusted accordingly.

2.5. Biofilm Formation Analysis

The 201-WT and 201-Δ*purR* strains were cultured until they reached an optical density of 1.0 at OD_{620} (*ca.*2 × 10^8 CFU/mL) and stored at 4 °C for 16 h. Subsequently, the cultures were diluted by a factor of 10 and transferred into a 24-well cell culture plate, with 1 mL per well. Six independent biological replicates were established for each strain under identical conditions. The plate was then shaken at either 26 °C or 37 °C for 24 h. After removing the bacterial cultures from the wells, OD_{620} was measured. The wells were gently washed twice with deionized water, and the biofilm was fixed at 80 °C for 15 min. Then, 3 mL of 0.1% crystal violet solution (Solarbio, Beijing, China) was added dropwise to each well.

The crystal violet solution was discarded after 15 min staining, and the wells were gently washed three times with deionized water. Subsequently, 2.8 mL of ethanol was added to each well and left at room temperature for 3 h. OD_{570} of the solution was determined after a five-fold dilution. The relative amount of biofilm formation was calculated using the formula: $100 \times OD_{570}/OD_{620}$.

2.6. *Survivability under Stressful Environments*

The strains 201-WT and 201-Δ*purR* were cultured until they reached an OD_{620} of 1.0. These strains were then exposed to various stressful conditions, including 0.5 M sorbitol for 0.5 h and 1.5 h, 7.5% NaCl for 1.5 h, cold shock for 24 h, or heat shock for 0.5 h. Bacterial numbers were counted before and after the stimulation to determine the survival rate of these two strains under different stressful conditions. Each strain was subjected to three independent biological replicates, and the results were reported as the mean standard deviation of the three experiments.

2.7. *Real-Time Cell Analysis (RTCA) Assay*

HeLa and 293 cells were cultured in Dulbecco's Modified Eagle's Medium (DMEM, Solarbio, Beijing, China) supplemented with 10% fetal bovine serum (FBS), at 37 °C and 5% CO_2. The baseline measurement was taken using 50 μL of DMEM with 10% FBS on the pre-incubated RTCA iCELLigence system (ACEA Biosciences, San Diego, CA, USA), maintained at same conditions. Subsequently, 5×10^3 cells were added to each well of an E-plate and incubated at room time for 30 min. The cells were then transferred to the RTCA iCELLigence system and incubated overnight [21]. The bottom of the cell culture plate compatible with the RTCA system has electrodes to record cell detachment as cell index (CI). Strains were cultivated in LB until they reached an optical density of 1.0 at OD_{620}, after removing the supernatant via centrifugation, the bacterial were resuspended in PBS to an optical density of 1.0 at OD620 (*ca.*2×10^8 CFU/mL). Subsequently, the appropriate volume of bacterial suspension was added at a ratio of MOI = 5 or 10 as calculated. Incubation was continued, and the cell index was measured every 15 min and normalized based on the time point at which bacteria were added. Each strain was subjected to three independent biological replicates under identical conditions.

2.8. *Survival Curves*

The 201-WT and 201-Δ*purR* strains were cultured until they reached an optical density of 1.0 at OD_{620} (*ca.*2×10^8 CFU/mL), and the concentration was adjusted to 3×10^4 CFU/mL with PBS. Female BALB/c mice, aged 8–10 weeks, were randomly divided into three groups (n = 10) and intraperitoneally challenged with a 100 μL diluted culture. The control group received an equal volume of PBS via the same injection route. Mouse mortality was monitored daily, and the survival curve was plotted.

2.9. *RNA-Seq and Quantitative Reverse Transcription PCR (qRT-PCR)*

201-WT and 201-Δ*purR* were cultured in LB medium at either 26 °C or 37 °C until reaching an OD_{620} of 1.0, and each strain had three biological replicates. Total RNA was extracted from the bacteria using the PureLink™ RNA Mini Kit (Tiangen, Beijing, China) following the manufacturer's instructions. After measuring the concentration of total RNA, they were sent to Beijing macro & micro- test Bio-Tech Co., Ltd. (Beijing, China) for sequencing. The company created a cDNA library and used Illumina NovaSeq 6000 for sequencing. The extracted RNA was used to create a cDNA library with at least 2 G raw data. The raw data were trimmed with Trimmomatric software fastp version 0.23.4 to filter adapters and low-quality reads (<Q20).

The genes for qRT-PCR were selected based on the results of the RNA-seq analysis under 26 °C culture conditions, and they were YP_RS00205, *purK*, *purE*, *purT*, *purF*, *cvpA*, *purL*, YP_RS13225, *purM*, *purN*, *purH*, YP_RS20395, *ybtE*, *ybtT*, *ybtU*, *irp1*, and YP_RS10830, respectively. The RNA samples used for RNA-seq were reverse transcribed into cDNA using SynScript III RT SuperMix (Tsingke Biotechnology Co., Ltd., Beijing, China). Linear regression analysis was employed to determine the correlations between the RNA-seq data and the outcomes of qRT-PCR

2.10. Expression and Purification of PurR

The pET28a (+) plasmid which contains T7 promoter and the target *purR* fragment were digested with *Hind* III and *BamH* I enzymes (LMAI Bio, Shanghai, China). After ligation of the *purR* fragment and linearized vector using T4 DNA ligase (Sangon, Shanghai, China), the resulting recombinant plasmid containing *purR* was introduced into *E. coli* DH5α. Subsequently, the recombinant plasmid was extracted and transferred into *E. coli* BL21(DE3), with the transformed strain designated as BL21(DE3)-pET28a (+)-*purR*.

For protein expression, BL21(DE3)-pET28a (+)-*purR* was cultured until reaching an OD_{600} of 0.6–1.0. Then, lactose induction (4 mM) was performed, and the culture was incubated at 16 °C under low-speed shaking for more than 12 h. The bacterial pellet was resuspended in a solution containing 300 mM NaCl, 50 mM NaH_2PO_4, and 10 mM imidazole, and pH adjusted to 7.0. Ultrasonication was used to disrupt the cells, and the supernatant was obtained via centrifugation at 10,000× *g* for 15 min. PurR protein in the supernatant was purified using a Ni-NTA resin column, and its presence was confirmed with SDS-PAGE and Western Blotting. After desalting using G25 rapid desalting column (Bersee, Beijing, China), a final concentration of 2.0 mg/mL PurR was obtained.

2.11. Motif Prediction of PurR

Six genes (*purH, purE, purT, purL, purF, purM*) were chosen from the *pur* operon based on findings from RNA-seq analysis and the existing literature. The promoter regions of these genes were used for PurR motif prediction using the online MEME website to identify potential DNA binding sites [22]. The predicted motif was then matched with the promoter regions of the *Y. pestis* genome using the FIMO module of the online MEME website to determine the genes that can be bound by PurR in the promoter region [23].

2.12. Electrophoretic Mobility Shift Assay (EMSA)

Genes with significant transcriptional changes and a high FIMO matching score were selected for EMSA analysis. Fragments containing the motif region or 500 bp upstream of the start codon were used as probes in EMSA. The probes were labeled using the EMSA Probe Biotin Labelling Kit (Beyotime Biotech. Inc., Shanghai, China). After obtaining the double-stranded probe, it was denatured to single-stranded probe at 95 °C for 5 min. The labeling system was prepared following the manufacturer's instructions, and biotin was added to the 3′ end by incubating at 37 °C for 30 min. The labeled probes were then mixed with chloroform–isopentanol (24:1) and gradually cooled to allow the single strands to reanneal into labeled double-stranded probes. Subsequently, 1/4 volume of 5 M ammonium acetate and 2 times the volume of anhydrous ethanol were added, and the mixture was precipitated at −20 °C overnight. After centrifugation and resuspension, purified labeled probes were obtained.

EMSA was performed using the Light Shift Chemiluminescent EMSA Kit (Beyotime Biotech. Inc., Shanghai, China). The EMSA binding reaction system was prepared according to the manufacturer's protocol and experimental demands, with gentle mixing at each step, and allowed to bind at room temperature for 30 min. Following this, loading buffer was added. Low-voltage electrophoresis was then performed in 0.5 × TBE buffer until the bromophenol blue dye migrated to approximately 2/3 to 3/4 of the gel length. Subsequently, the gel was transferred onto a nylon membrane with a positive charge, and cross-linked under UV light for 20 min. The membrane was then incubated at room temperature for 15 min in blocking buffer for blocking, followed by a 30 min reaction with conjugate/blocking buffer. The membrane was washed four times with wash buffer, gently agitated for 5 min in substrate equilibration buffer, and then incubated with substrate solution for visualization. Finally, the membrane was exposed and photographed. A negative control protein was the F1 antigen of *Y. pestis*, which is a non-transcriptional regulatory factor, and a labeled segment of 16 s rRNA gene served as a negative labeled probe in this experiment. The concentrations of PurR protein and targeted DNA fragments are listed in Supplementary Table S2.

2.13. Reverse Transcription PCR (RT-PCR)

Gene clusters exhibiting similar transcriptional changes were identified based on the RNA-seq results. Primers were designed to amplify the intergenic regions adjacent to two genes that were potentially part of the same operon. The DNA extraction procedure followed the protocol described in Section 2.2, while the RNA extraction procedures and reverse transcription of RNA were carried out following the protocol described in Section 2.9. The obtained DNA, RNA, and cDNA were utilized as templates for PCR amplification of the predicted operon intergenic regions, with deionized water serving as a negative control. Agarose gel electrophoresis was then employed to confirm the presence of the predicted operon intergenic sections in the amplified products.

2.14. Ethics Statement

All animal experiments adhered to the ethical guidelines for laboratory animals in China and were conducted in accordance with the regulations outlined in laboratory animal permit no. SCXK (Jing) 2021-0006, obtained from Beijing Vital River Laboratory Animal Technology Co., Ltd. (Beijing, China). The study was approved by the Institutional Review Board at the Beijing Institute of Microbiology and Epidemiology (IACUC-IME-2023-001).

2.15. Statistical Analysis

The mean and variation of each of the three experimental groups were computed from three independent experiments. A *t*-test was employed to evaluate differences in the data, assuming the prerequisites of normal distribution and homogenous variance were fulfilled. When data diverged from a normal distribution, a nonparametric analysis was conducted. A one-way analysis of variance was performed, and the Student–Newman–Keuls q test was applied for multiple comparisons. Evaluation of survival curves employed the Mantel–Cox test, emphasizing the utilization of log-rank analysis. Statistical significance was set as follows: * $p < 0.05$, ** $p < 0.01$, *** $p < 0.001$, **** $p < 0.0001$.

2.16. Data Availability Statement

The RNA-seq data generated and analyzed in this study have been deposited at the National Microbiology Data Center under the accession numbers of NMDC40041563-40041574 (https://nmdc.cn/resource/genomics/sra/detail/NMDC40041563, accessed on 7 November 2023).

3. Results

3.1. Deletion of the PurR Makes No Difference in Growth of Y. pestis

To assess the contribution of *purR* to the growth capacity of *Y. pestis*, a comparative analysis of growth curves was performed for both strains in LB and TMH media. The findings indicated no significant differences in growth rate between these strains at temperatures of 26 °C or 37 °C. Moreover, the deletion of *purR* had no discernible effect on the growth of *Y. pestis* under two nutritional conditions, including LB medium and the nutrient-limited TMH medium (Figure 1).

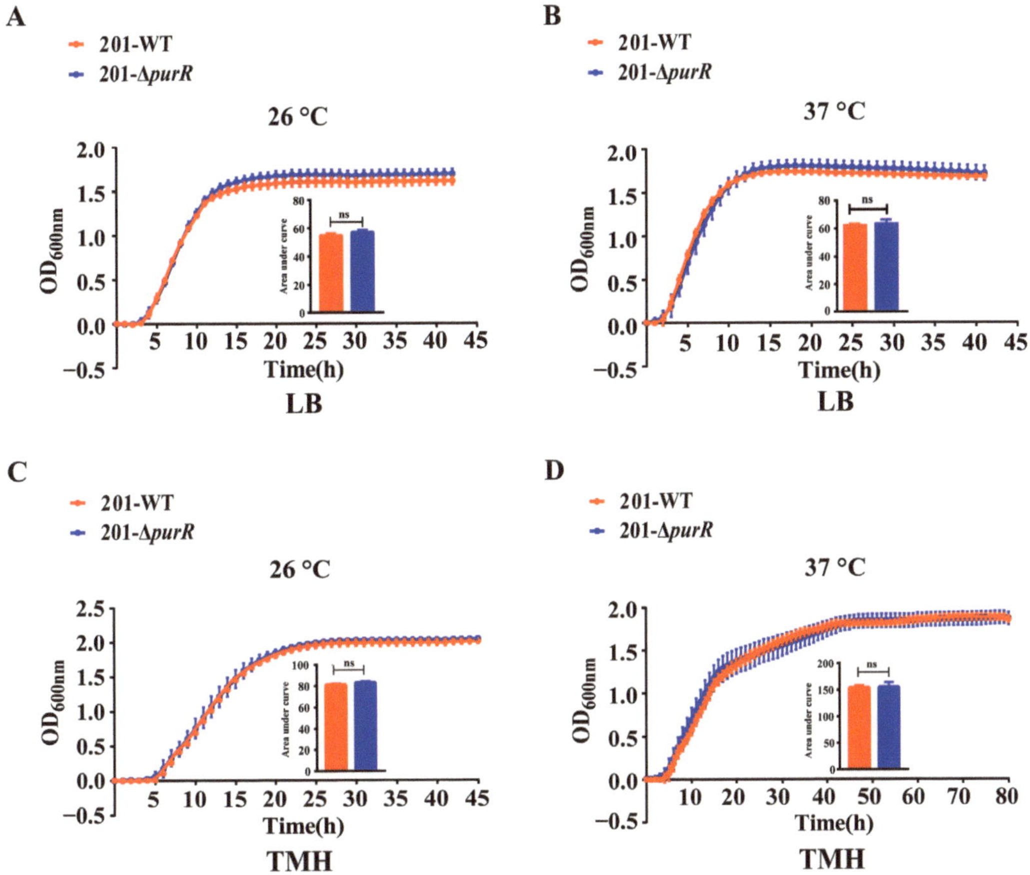

Figure 1. The growth curves of 201-WT and 201-Δ*purR*. The growth curves of 201-WT and 201-Δ*purR* were assessed under different culture conditions. The conditions included: (**A**) growth at 26 °C in LB medium, (**B**) growth at 37 °C in LB medium, (**C**) growth at 26 °C in TMH medium, and (**D**) growth at 37 °C in TMH medium. The bar graph presented below the growth curves illustrates the cumulative areas under the curves and is applied to statistical analysis. Each experiment included three independent biological replicates, and the results were expressed as mean ± standard deviation from three independent experiments. ns: not statistically significant.

3.2. No Differences Were Observed in the In Vitro Phenotypes of 201-WT and 201-ΔPurR

Biofilm formation plays a crucial role in the dissemination of *Y. pestis* by fleas. To investigate this phenomenon, we utilized crystal violet staining to evaluate biofilm formation in both 201-WT and 201-Δ*purR* strains. The results revealed no significant disparities in biofilm formation between the two strains at either 26 °C or 37 °C (Figure 2A). In order to simulate the environmental stresses that *Y. pestis* may encounter in natural environments, in this study, we exposed 201-WT and 201-Δ*purR* to various stressful in vitro conditions, including hypersaline, hypertonic, heat shock, and cold shock. The results demonstrated that there were no significant differences in survivability between 201-WT and 201-Δ*purR* under these environmental stresses (Figure 2B–E). These findings suggest that *purR* may not be critical for *Y. pestis* to withstand the simulated stressful environments.

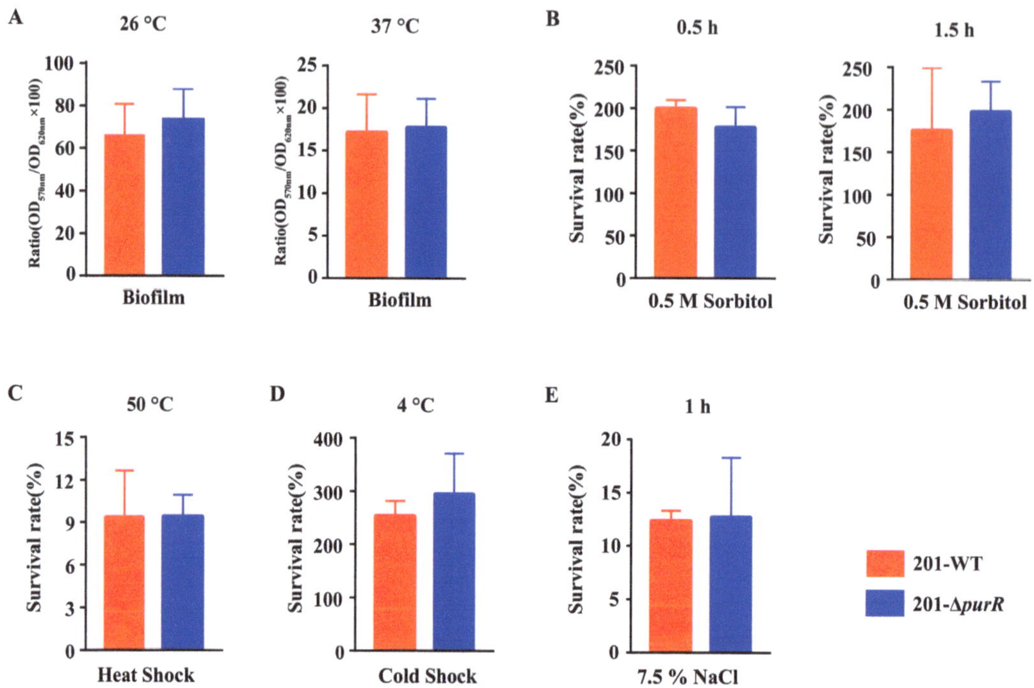

Figure 2. The in vitro phenotypes of 201-WT and 201-Δ*purR*. Biofilm formation of 201-WT and 201-Δ*purR* and a comparison of their survival rates in vitro under different simulated stress environments were assessed. (**A**) 0.1% crystal violet solution was used to quantify the relative amount of biofilm formation for both strains cultured at 26 °C or 37 °C (**B**). The survival rates of both strains after being stimulated by high osmotic pressure environment simulated by 0.5 M sorbitol after 30 min and after 1.5 h were compared. The survival rates of both strains after being stimulated by heat shock at 50 °C for 0.5 h (**C**) and cold shock at 4 °C for 24 h (**D**) were compared. (**E**) The survival rates of both strains after being stimulated by high salt environment simulated by 7.5% NaCl were compared after 1 h of stimulation. There were no significant differences in all results between 201-WT and 201-Δ*purR*. Each experiment was independently replicated three times for both strains, and statistical analysis was performed using a two-sample *t*-test for each comparison.

3.3. Deletion of the PurR Attenuated the Cytotoxicity to HeLa and 293 Cells of Y. pestis

RTCA was used to investigate any differences in the cytotoxicity of 201-WT and 201-Δ*purR* on HeLa or 293 cells. The cell index indicated that 201-Δ*purR* exhibited significantly lower cytotoxicity to cells compared to the 201-WT strain after 10 h of stimulation. These findings were further supported by the RTCA results for 201-Δ*purR*-Comp (Figure 3). The knockout of *purR* was found to weaken the cytotoxicity of *Y. pestis* strain 201 towards HeLa and 293 cells. However, the survival curve analysis suggested that both 201-WT and 201-Δ*purR* strains exhibited similar levels of virulence when tested on mice (Supplementary Figure S1).

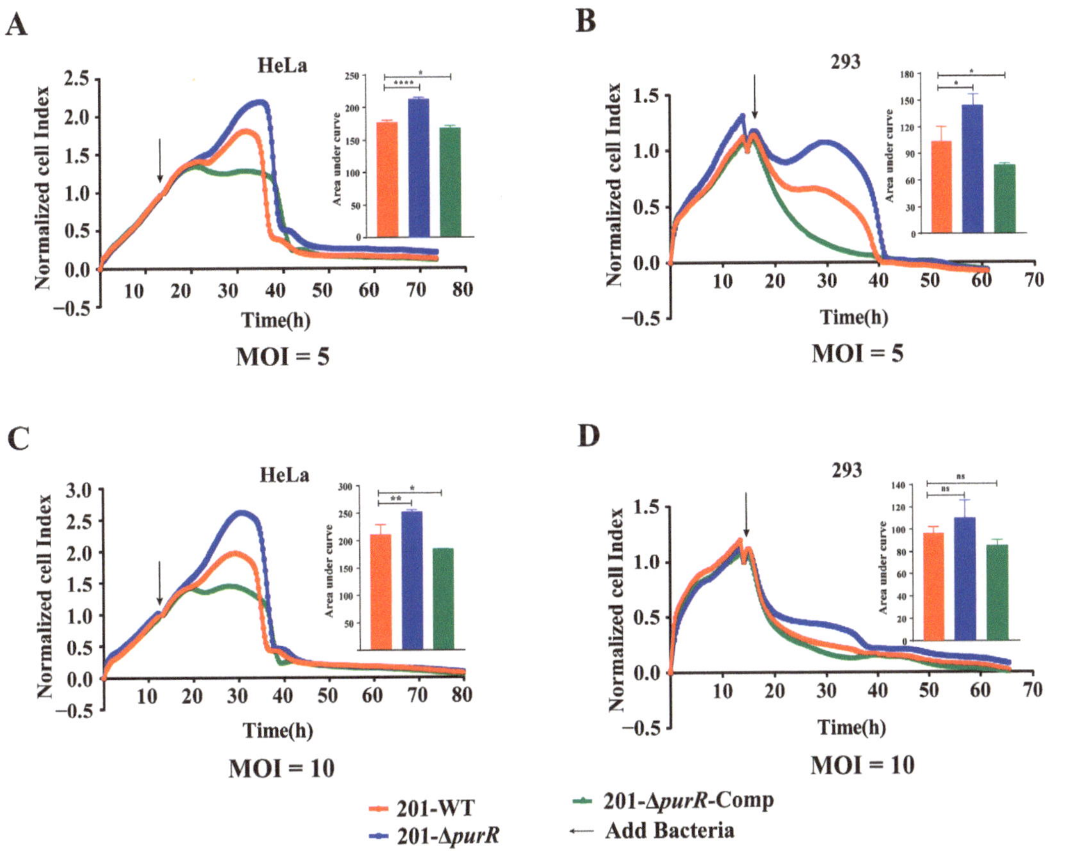

Figure 3. 201. *purR* showed cytotoxicity attenuation on HeLa cells and 293 cells. Cells were infected with 201-WT, 201-Δ*purR*, and 201-Δ*purR*-Comp at a specific multiplicity of infection (MOI). (**A**) Bacterial infection of HeLa cells at an MOI of 5. (**B**) Bacterial infection of 293 cells at an MOI of 5. (**C**) Bacterial infection of HeLa cells at an MOI of 10. (**D**) Bacterial infection of 293 cells at an MOI of 10. The cell index was measured every 15 min. The bar graph presented next to the curves illustrates the cumulative areas under the curves and is applied to statistical analysis. At the same time point, a larger "Normalized cell index" indicates more cells or a better cell status. The higher the curve, the larger the area under the curve, indicating that the bacterial strain has weaker cytotoxicity to the cells. Each experiment included three independent biological replicates, and the results were expressed as mean ± standard deviation from three independent experiments. * $p < 0.05$, ** $p < 0.01$, **** $p < 0.0001$, ns: not statistically significant.

3.4. Deletion of the PurR Significantly Alters Gene Expressions in *Y. pestis*

In many bacteria, the main identified role of PurR is a transcriptional repressor of purine biosynthesis [6]. In this study, we conducted RNA-seq analysis of 201-Δ*purR* to identify genes potentially associated with *purR* in *Y. pestis* strain 201. Differentially expressed genes were selected based on the criteria $|\log_2$ (fold change) $| > 0$ and p-adjust < 0.05. Comparing these genes with 201-WT, we found significant enrichment in pathways like ribosome, carbon metabolism, and others at 26 °C (Supplementary Figure S2A) and in pathways such as carbon metabolism, purine metabolism, sulfur metabolism, ABC transport, and other metabolic processes at 37 °C (Supplementary Figure S2B).

The RNA-seq results showed the up-regulation of genes involved in purine biosynthesis, including the *pur* operon, *guaB*, and *carA*, consistent with the role of *purR* as a purine repressor (Figure 4A, Tables 2 and 3). Additionally, genes indirectly involved in purine biosynthesis, such as glycine cleavage system genes (YP_RS13225, *gcvT*, *gcvH*), were also up-regulated. However, yersiniabactin (Ybt) siderophore-related genes, which are crucial for iron uptake and *Y. pestis* virulence, such as *ybtU* [24], were down-regulated (Supplementary Tables S3 and S4). At 26 °C, the type VI secretion system (T6SS) was up-regulated (Table 2), while genes encoding peroxidase and cytochrome (*katG*, *cybB*, and *cybC*) were down-regulated (Supplementary Table S3). At 37 °C, genes involved in sulfur metabolism (*ssuB*, *ssuC*, and *ssuD*) and taurine ABC transport permeases (*tauA*, *tauB*, and *tauC*) were additionally down-regulated (Table 3). Furthermore, among the up-regulated genes, *ppsA*, encoding phosphoenolpyruvate synthase, and *cytR*, involved in DNA transcription, were included (Supplementary Table S4).

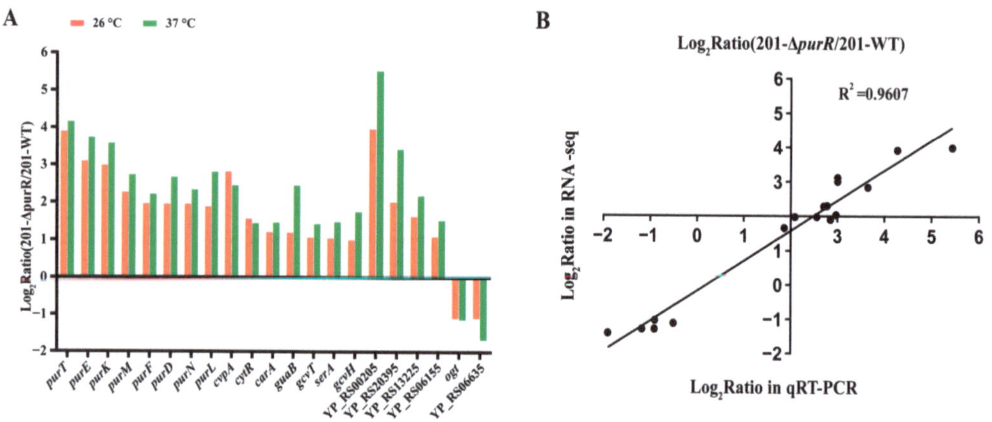

Figure 4. Common differentially expressed genes of 201-Δ*purR* cultured at 26 °C and 37 °C. The transcriptional level of 201-Δ*purR* was analyzed using RNA-seq and qRT-PCR under 26 °C and 37 °C culture conditions. (**A**) The 21 up-regulated or down-regulated genes that were shared under 26 °C and 37 °C culture conditions, screened with the criterion of $|\log_2(\text{FoldChange})| > 1.0$; all the selected genes had a *p*-adjust value of $< 10^{-5}$. (**B**) The correlation analysis between the 17 genes selected for qRT-PCR and these same 17 genes in the RNA-seq under 26 °C culture conditions, and the figure took point (2,2) as the origin. The selected genes were listed in the Section 2.9.

Table 2. Differential expressions of genes associated to purine biosynthesis and T6SS in 201-Δ*purR* in comparison with 201-WT at 26 °C.

Locus Tag *	Log^{2}FC	*p*-Adjust	Gene Name	Gene Description	Pathway Name
YP_RS08450	3.92	5.23×10^{-120}	*purT*	formate-dependent phosphoribosylglycinamide formyltransferase	Purine metabolism
YP_RS04375	3.12	4.25×10^{-46}	*purE*	5-(carboxyamino)imidazole ribonucleotide mutase	
YP_RS04370	3.01	2.65×10^{-52}	*purK*	5-(carboxyamino)imidazole ribonucleotide synthase	
YP_RS16125	2.30	1.50×10^{-46}	*purH*	bifunctional phosphoribosylaminoimidazolecarboxamide formyltransferase/IMP cyclohydrolase	
YP_RS14005	2.29	1.70×10^{-55}	*purM*	phosphoribosylformylglycinamidine cyclo-ligase	
YP_RS12405	1.98	2.79×10^{-52}	*purF*	amidophosphoribosyltransferase	
YP_RS16130	1.97	1.56×10^{-34}	*purD*	phosphoribosylamine--glycine ligase	
YP_RS14010	1.97	1.16×10^{-26}	*purN*	phosphoribosylglycinamide formyltransferase	
YP_RS13160	1.90	1.40×10^{-21}	*purL*	phosphoribosylformylglycinamidine synthase	
YP_RS14225	1.21	2.03×10^{-16}	*guaB*	IMP dehydrogenase	

Table 2. *Cont.*

Locus Tag *	Log^2FC	*p*-Adjust	Gene Name	Gene Description	Pathway Name
YP_RS19195	1.95	3.45×10^{-27}	-	Hcp family type VI secretion system effector	Bacterial secretion system
YP_RS19175	1.89	3.77×10^{-24}	*tssH*	type VI secretion system ATPase TssH	
YP_RS19210	1.83	3.03×10^{-21}	-	ImpA family type VI secretion system protein	
YP_RS19205	1.82	2.89×10^{-25}	*tssB*	type VI secretion system contractile sheath small subunit	
YP_RS19200	1.80	2.38×10^{-27}	*tssC*	type VI secretion system contractile sheath large subunit	
YP_RS19185	1.79	1.88×10^{-18}	*tssF*	type VI secretion system baseplate subunit TssF	
YP_RS19145	1.78	4.53×10^{-19}	*tssJ*	type VI secretion system lipoprotein TssJ	
YP_RS19180	1.72	8.93×10^{-18}	*tssG*	type VI secretion system baseplate subunit TssG	
YP_RS19140	1.71	1.51×10^{-25}	*tssK*	type VI secretion system baseplate subunit TssK	
YP_RS19135	1.68	5.57×10^{-37}	*icmH*	type IV secretion system protein IcmH/DotU	
YP_RS19170	1.62	1.09×10^{-15}	*vgrG*	type VI secretion system tip protein VgrG	
YP_RS19190	1.60	2.14×10^{-18}	-	type VI secretion system baseplate subunit TssE	
YP_RS19130	1.45	3.02×10^{-28}	*tssM*	type VI secretion system membrane subunit TssM	
YP_RS15960	1.44	2.50×10^{-5}	*tssE*	type VI secretion system baseplate subunit TssE	
YP_RS15965	1.11	1.88×10^{-15}	*tssC*	type VI secretion system contractile sheath large subunit	
YP_RS15970	1.03	2.09×10^{-6}	*tssB*	type VI secretion system contractile sheath small subunit	
YP_RS15955	0.96	8.61×10^{-4}	*tssF*	type VI secretion system baseplate subunit TssF	

*: NCBI reference sequence: ASM788v1.

Table 3. The genes of purine biosynthesis and sulfur metabolism system expression changes in 201-Δ*purR* in comparison with 201-WT at 37 °C.

Locus Tag *	Log^2FC	*p*-Adjust	Gene Name	Gene Description	KEGG
YP_RS08450	4.18	1.68×10^{-95}	*purT*	formate-dependent phosphoribosylglycinamide formyltransferase	Purine metabolism
YP_RS04375	3.76	4.12×10^{-46}	*purE*	5-(carboxyamino)imidazole ribonucleotide mutase	
YP_RS04370	3.60	5.14×10^{-64}	*purK*	5-(carboxyamino)imidazole ribonucleotide synthase	
YP_RS16125	2.97	1.89×10^{-64}	*purH*	bifunctional phosphoribosylaminoimidazolecarboxamide formyltransferase/IMP cyclohydrolase	
YP_RS13160	2.83	4.37×10^{-41}	*purL*	phosphoribosylformylglycinamidine synthase	
YP_RS14005	2.76	8.61×10^{-66}	*purM*	phosphoribosylformylglycinamidine cyclo-ligase	
YP_RS16130	2.70	2.68×10^{-33}	*purD*	phosphoribosylamine--glycine ligase	
YP_RS14225	2.47	3.44×10^{-23}	*guaB*	IMP dehydrogenase	
YP_RS14010	2.36	1.28×10^{-22}	*purN*	phosphoribosylglycinamide formyltransferase	
YP_RS12405	2.24	4.64×10^{-27}	*purF*	amidophosphoribosyltransferase	
YP_RS13925	1.95	8.44×10^{-11}	-	phosphoribosylaminoimidazolesuccinocarboxamide synthase	
YP_RS14215	1.75	1.04×10^{-14}	*guaA*	glutamine-hydrolyzing GMP synthase	
YP_RS09185	1.26	1.20×10^{-7}	*purB*	adenylosuccinate lyase	
YP_RS20415	−2.40	3.19×10^{-8}	*ssuD*	FMNH2-dependent alkanesulfonate monooxygenase	Sulfur metabolism
YP_RS20410	−1.80	2.48×10^{-4}	*ssuC*	aliphatic sulfonate ABC transporter permease SsuC	
YP_RS00940	−1.79	7.13×10^{-4}	*tauA*	taurine ABC transporter substrate-binding protein	
YP_RS00950	−1.79	1.23×10^{-3}	*tauC*	taurine ABC transporter permease TauC	
YP_RS00945	−1.59	1.27×10^{-3}	*tauB*	taurine ABC transporter ATP-binding subunit	
YP_RS20405	−1.35	5.37×10^{-2}	*ssuB*	aliphatic sulfonates ABC transporter ATP-binding protein	
YP_RS00425	−1.19	5.53×10^{-2}	-	sulfate ABC transporter substrate-binding protein	

*: NCBI reference sequence: ASM788v1.

Our investigation revealed that PurR regulates purine biosynthesis in *Y. pestis* as in other bacteria, but it may also have additional functions. To validate the RNA-seq data, we selected five down-regulated genes and twelve up-regulated genes for confirmation using qRT-PCR, and the results supported the findings of RNA-seq (Figure 4B).

3.5. Motif Prediction of PurR

Transcriptional regulators are DNA-binding proteins that can modulate gene transcription by interacting with specific promoter regions. To identify the DNA-binding domain of PurR, we conducted a search for the protein structure of *Y. pestis* PurR in Uniprot and identified an HTH (helix-turn-helix) domain at the N-terminal of PurR, known for its ability to bind to DNA (Figure 5A). HTH is a common DNA-binding motif found in prokaryotic transcription factors, consisting of a short chain connecting two helices in the structure [25].

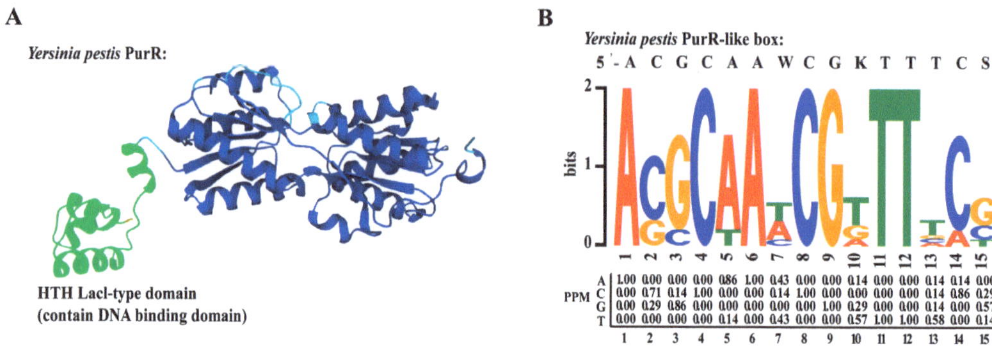

Figure 5. PurR structure and prediction of PurR motif. (**A**) The structure of *Y. pestis* PurR on Uniprot. Green color is the HTH lacI-type domain which contains DNA-binding domain in the front of PurR. (**B**) PurR motif predicted on the online MEME website, based on the promoter region of six genes (*purH*, *purE*, *purT*, *purL*, *purF*, *purM*) on the *pur* operon.

A motif is a specific DNA sequence that transcriptional regulators can recognize and bind to. In our study, we employed the online MEME tool to predict a 15-bp motif (5′-ACGCAAWCGKTTTCS-3′) for PurR (Figure 5B) [22], which exhibited high similarity to motifs previously predicted in other studies [26]. Notably, the PurR motif in *Y. pestis* was found to be similar to that of *E. coli* [7]. This observation might be attributed to the significant homology between PurR in *Y. pestis* strain 201 and PurR in *E. coli*, as the two share 82.4% identical amino acids. To identify potential promoter regions where PurR could bind, we conducted FIMO analysis on the complete promoter regions of the *Y. pestis* genome using this motif [23]. By employing a screening threshold of a p-value $\leq 1.0 \times 10^{-4}$, we predicted that the promoter regions of 486 genes in *Y. pestis* strain 201 contain a motif that closely resembles a specific sequence pattern (Supplementary Table S5).

3.6. PurR Regulates Potential Operons in Y. pestis Strain 201

As mentioned previously, this study confirms that PurR can regulate purine biosynthesis by binding to the promoter region of relevant genes in *Y. pestis*. Our findings demonstrate that several PurR-regulated genes involved in purine biosynthesis are co-transcribed. For example, the *purM-purN*, *purH-purD*, and *purE-purK* loci are part of the *pur* operon and contribute to IMP synthesis [27,28]. Additionally, *guaB-guaA* participates in the conversion of IMP to GMP and AMP (Figure 6). The *carA* gene, co-transcribed with *carB* (Supplementary Figure S3B), provides arginine and uracil, which are essential for bacterial growth [29,30]. Another gene, *gcvT*, encodes glycine lyase and is part of the GCV operon, which consists of *gcvT*, *gcvH*, and *gcvP* (Supplementary Figure S3C) [31].

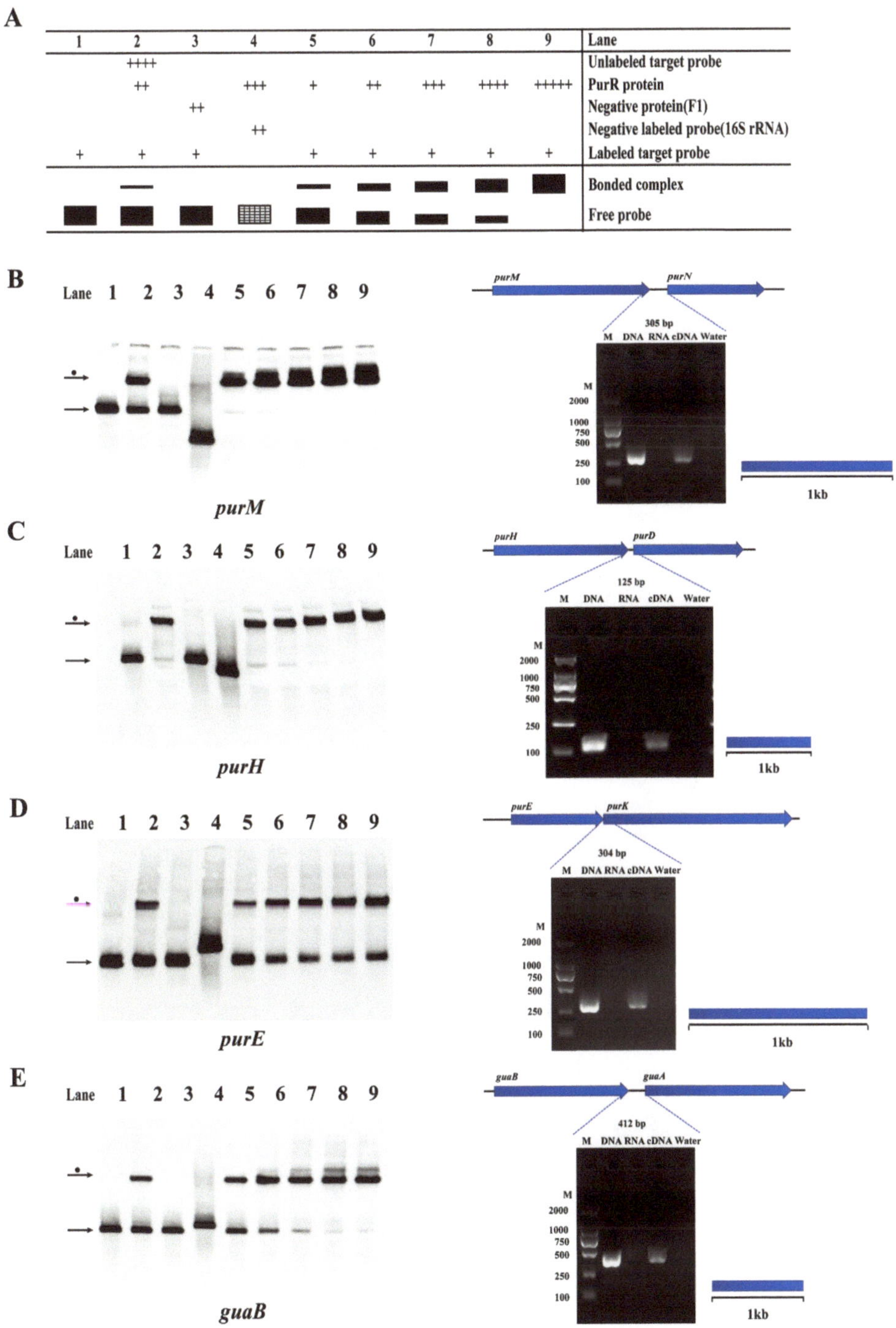

Figure 6. PurR may regulate some operons involved in purine biosynthesis of *Y. pestis* strain 201. The results of EMSA and RT-PCR experiments confirm the presence of operons and the regulation by

PurR in *Y. pestis* strain 201, involving genes related to purine biosynthesis. (**A**) The figure illustrates the experimental setup and schematic representation of each lane in the EMSA experiment. The filled black color represents the result of the experimental group, and the shaded grid indicates the result of the negative control group. The concentrations of the components added in each experimental channel are shown in Supplementary Table S2. The width of the bands reflects the quantity of binding between the tracer probe and PurR, while the number of '+' signs corresponds to the amount of the respective samples added. The identified operons include: (**B**) *purM-purN*; (**C**) *purH-purD*, (**D**) *purE-purK*; and (**E**) *guaB-guaA*. The left figure displays the EMSA results, while the right figure presents the RT-PCR results. In the RT-PCR results, the template for each gene intergenic region is indicated as DNA, RNA, cDNA, or water. The blue dotted line indicates the expected amplification fragment size for each gene intergenic region.

In addition to regulating the co-transcribed genes involved in purine biosynthesis, PurR was found to regulate two other operons, as confirmed in this study. This suggests that PurR may have additional roles in *Y. pestis* strain 201, consistent with previous findings of PurR having multiple functions in various bacterial species [12,32]. PurR was observed to interact with the promoter regions of *ssuE* and *katG*, which belong to the sulfur-starvation utilization (*ssu*) and *katG-cybC-cybB* operons [26,33], indicating that PurR could function as a regulator for these genes (Supplementary Figure S3D,E). In summary, these results imply that PurR may influence *Y. pestis* sulfur metabolism and suggest a connection between *purR* and *Y. pestis* energy metabolism.

3.7. PurR May Potentially Regulate Other Gene Expressions in *Y. pestis* Strain 201

We employed a combination of RNA-seq data and FIMO prediction to explore the potential regulatory roles of PurR in *Y. pestis*. Our findings revealed a correlation between significant transcriptional up- or down-regulation of certain genes and high matching scores in the FIMO prediction results. PurR was found to bind to the promoter regions of several other genes in the microorganism, including *pyrD* that can interact with PurR in other bacteria and contribute to the regulation of pyrimidine biosynthesis (Supplementary Figure S4C) [7]. Furthermore, PurR also exhibits self-regulation, adding an extra layer of security to its regulatory system (Supplementary Figure S4B).

Furthermore, our investigation unveiled previously unknown promoter regions of additional genes (*serA*, *ogt*, *fur*, *ybtA*, *djlA*) that PurR may bind to in *Y. pestis* strain 201 (Supplementary Figure S4D–H). Additionally, the results of EMSA and RT-PCR supported the presence of the *ybt* operon (*irp2-irp1-ybtU-ybtT-ybtE*) located on the high pathogenicity island (HPI) of *Y. pestis* (Supplementary Figure S5). These findings suggest that PurR might regulate various biological functions of *Y. pestis* beyond purine biosynthesis.

3.8. Potential Operons of *Y. pestis*

We performed an analysis of genes that potentially undergo co-transcription based on RNA-seq data. Our investigation confirmed the presence of two potential operons in *Y. pestis*, i.e., the T6SS gene cluster and YP_RS00935-*tauA-tauB-tauC-tauD* (Figure 7). Similar to the *ssu* operon, the *tauABCD* operon is involved in sulfur biosynthesis in *E. coli* [34]. The T6SS is a versatile secretion system observed in various Gram-negative bacteria, and it plays a role in multiple physiological functions in *Yersinia*, including host infection, bacterial competition, and stress responses [35,36]. These findings provide evidence for the first time that T6SS undergoes co-transcription in *Y. pestis*.

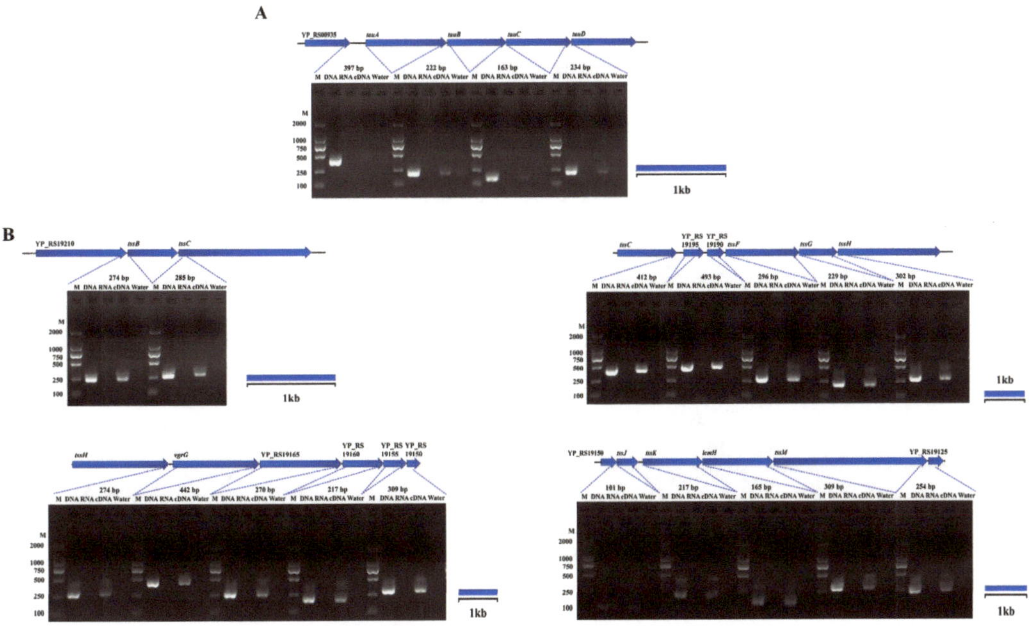

Figure 7. Other operon of *Y. pestis* strain 201. The RT-PCR results revealed the co-transcription of specific genes in the genome of *Y. pestis* strain 201, which was supported with RNA-seq analysis (excluding previously mentioned genes). The co-transcribed regions identified include: (**A**) YP_RS00935-*tauA-tauB-tauC-tauD* and (**B**) the type VI secretion system (T6SS).

4. Discussion

Y. pestis, as a multi-host pathogen, demonstrates remarkable adaptability to various environmental changes throughout its life cycle, allowing it to survive in nature, transmitted by fleas, and propagate within hosts. *Y. pestis* encounters stressful conditions, including decreased temperature, acidity, and hyperosmotic environments in nature and within fleas. Thus, we hypothesis that the loss of *purR* might plays a role in the fitness of *Y. pestis*.

As there is no significant variation in the growth rates of 201-WT and 201-Δ*purR* grown in different culture media (LB or TMH) at different temperatures (26 °C or 37 °C), it is reasonable to exclude the influence of strain growth rates in the other phenotypic tests (Figure 1). However, our study revealed that *purR* does not play a central role in *Y. pestis*' ability to adapt to tested stressful conditions in vitro (Figure 2), even though certain genes involved in the adaptation of *Y. pestis* to harsh environments showed significant changes in transcription levels for the 201-Δ*purR* mutant (Table 2). An effective transfer approach of *Y. pestis* necessitates the production of biofilms to facilitate its transmission from fleas to mammals [37], while the results showed that the loss of *purR* makes no difference in the formation of biofilms of *Y. pestis*. Data from the multi-omics online database for *Yersinia* suggest that the transcription of *purR* remains relatively stable in *Y. pestis* [38]. Interestingly, the knockout of *purR* did not impact the growth of *Y. pestis* 201 at different temperatures and nutritional levels, consistent with findings in *S. aureus* [39]. These suggest that the transcriptional regulation of *PurR* alone may not be sufficient to disrupt the adaptive mechanism of *Y. pestis*. As mature mechanisms are often complementary and mutually reinforcing, with each factor playing a distinct role, any gaps in the regulatory system may be supplemented by other aspects.

In our study, the knockout of *purR* resulted in reduced cytotoxicity of *Y. pestis* towards HeLa and 293 cells. However, the absence of *purR* did not affect the toxicity of *Y. pestis* in mice. Due to the complexity of the infection progress, the performance of bacteria in vivo and in vitro may be inconsistent. The result in cytotoxicity of 201-Δ*purR* stands in contrast to the findings in *S. aureus*, in which the virulence is increased after the deletion of *purR*, as the mutant expresses more virulence effectors [27], whereas the expression of virulence-related Ybt siderophore-related genes decreased in the *Y. pestis purR* knockout strains. We only performed preliminary animal challenge experiments in this study; it can be determined if the knockout of *purR* affects the toxicity of *Y. pestis* to animals based on the bacterial load and pathological changes of specific organs in a follow up study.

Previous research has identified the PurR protein as an HTH-type transcriptional repressor in bacteria, and has already confirmed its role in regulating purine biosynthesis by repressing the *pur* operon in other bacterial species [6,40,41]. In this study, we examined the purine regulatory function of PurR in *Y. pestis* 201. Our findings revealed that the absence of *purR* led to an upregulation of purine biosynthesis-related genes and that PurR exhibited binding affinity to the promoter region of nearly all these genes. Furthermore, we also identified eleven potential operons in *Y. pestis* 201 through RT-PCR analysis, eight of which could be bound by PurR, highlighting the role of PurR as a regulator of multiple operons and thereby establishing a potential regulon. Interestingly, PurR was shown to bind to the promoter region of *fur*, the gene encoding ferric uptake regulator Fur in *Y. pestis*, which was also found to regulate siderophore-associated operons including the *ybt* operon [42], indicating a cross-regulation of transcriptional regulators. This interconnection creates a vast and complex regulatory network for PurR in *Y. pestis*.

In addition to its role in regulating purine biosynthesis, we sought to explore other potential functions of *purR* in *Y. pestis*. Our analysis revealed that *purR* may also influence the virulence, sulfur metabolism, and energy synthesis of *Y. pestis* by directly or indirectly regulating specific operons, i.e., *irp2-irp1-ybtU-ybtT-ybtE*, *ssuE*-YP_RS20420-*ssuD-ssuC-ssuB*, and *katG-cybC-cybB*, respectively. Notably, we discovered that the T6SS (Type VI Secretion System) gene cluster was co-transcribed in *Y. pestis* 201, and the knockout of *purR* resulted in an increase in transcription of these genes. T6SS has been shown to play a crucial role in the interaction between *Y. pestis* and macrophages [43]. This finding provides valuable insights for studying the mechanism of interaction between *Y. pestis* and macrophages.

In summary, this study conducted a preliminary investigation into the function of *purR* in *Y. pestis* 201 and provided an initial analysis of the regulatory network of PurR. The findings laid the groundwork for future research, but a more comprehensive understanding of the underlying mechanism requires additional methods and robust evidence for validation.

Supplementary Materials: The following supporting information can be downloaded at: https://www.mdpi.com/article/10.3390/microorganisms11112801/s1, Figure S1: 201-WT and 201-ΔpurR showed virulence no differences in mice; Figure S2: KEGG maps of the differentially expressed genes of 201-ΔpurR culture at 26 °C and at 37 °C; Figure S3: PurR may regulate other operons of Y. pestis strain 201; Figure S4: PurR may regulate some genes of Y. pestis strain 201; Figure S5: The ybt operon in Y. pestis strain 201; Table S1: Primers used in the study; Table S2: EMSA settings for different target genes; Table S3: 201-ΔpurR differentially significantly expressed gene at 26 °C; Table S4: 201-ΔpurR differentially significantly expressed gene at 37 °C; Table S5: The results of FIMO analysis.

Author Contributions: Conceptualization: L.X., Y.S. and K.S.; formal analysis, X.Q., Y.C. (Yujun Cui) and Y.W.; methodology, L.X., J.J., Y.S., K.S., Z.S., Y.L., Y.Z., L.S., Y.C. (Yiming Cui) and W.Y.; software, X.Q., Y.C. (Yujun Cui), Y.W. and Z.X.; supervision, Y.S.; writing—original draft, L.X. and J.J.; writing—review and editing, Y.S., K.S. All authors have read and agreed to the published version of the manuscript.

Funding: This research was supported by National Natural Science Foundation of China (U22A20526).

Data Availability Statement: All data presented in this study are available in this published article. The RNA-seq data generated and analyzed are available in the National Microbiology Data Center under the accession numbers of NMDC40041563-40041574.

Conflicts of Interest: The authors declare no conflict of interest.

References

1. Yang, R.; Atkinson, S.; Chen, Z.; Cui, Y.; Du, Z.; Han, Y.; Sebbane, F.; Slavin, P.; Song, Y.; Yan, Y.; et al. *Yersinia pestis* and Plague: Some knowns and unknowns. *Zoonoses* **2023**, *3*, 5. [CrossRef]
2. Valles, X.; Stenseth, N.C.; Demeure, C.; Horby, P.; Mead, P.S.; Cabanillas, O.; Ratsitorahina, M.; Rajerison, M.; Andrianaivoarimanana, V.; Ramasindrazana, B.; et al. Human plague: An old scourge that needs new answers. *PLoS Neglected Trop. Dis.* **2020**, *14*, e0008251. [CrossRef] [PubMed]
3. Li, B.; Tan, Y.; Guo, J.; Cui, B.; Wang, Z.; Wang, H.; Zhou, L.; Guo, Z.; Zhu, Z.; Du, Z.; et al. Use of protein microarray to identify gene expression changes of *Yersinia pestis* at different temperatures. *Can. J. Microbiol.* **2011**, *57*, 287–294. [CrossRef] [PubMed]
4. Achtman, M.; Zurth, K.; Morelli, G.; Torrea, G.; Guiyoule, A.; Carniel, E. *Yersinia pestis*, the cause of plague, is a recently emerged clone of *Yersinia pseudotuberculosis*. *Proc. Natl. Acad. Sci. USA* **1999**, *96*, 14043–14048. [CrossRef]
5. Achtman, M. Evolution, population structure, and phylogeography of genetically monomorphic bacterial pathogens. *Annu. Rev. Microbiol.* **2008**, *62*, 53–70. [CrossRef] [PubMed]
6. Connolly, J.; Boldock, E.; Prince, L.R.; Renshaw, S.A.; Whyte, M.K.; Foster, S.J. Identification of *Staphylococcus aureus* Factors Required for Pathogenicity and Growth in Human Blood. *Infect. Immun.* **2017**, *85*, e00337-17. [CrossRef]
7. Cho, B.K.; Federowicz, S.A.; Embree, M.; Park, Y.S.; Kim, D.; Palsson, B.O. The PurR regulon in *Escherichia coli* K-12 MG1655. *Nucleic Acids Res.* **2011**, *39*, 6456–6464. [CrossRef] [PubMed]
8. Abdelwahed, E.K.; Hussein, N.A.; Moustafa, A.; Moneib, N.A.; Aziz, R.K. Gene Networks and Pathways Involved in *Escherichia coli* Response to Multiple Stressors. *Microorganisms* **2022**, *10*, 1793. [CrossRef]
9. Shi, T.; Wang, Y.; Wang, Z.; Wang, G.; Liu, D.; Fu, J.; Chen, T.; Zhao, X. Deregulation of purine pathway in *Bacillus subtilis* and its use in riboflavin biosynthesis. *Microb. Cell Factories* **2014**, *13*, 101. [CrossRef]
10. Liu, S.; Hu, W.; Wang, Z.; Chen, T. Rational Engineering of *Escherichia coli* for High-Level Production of Riboflavin. *J. Agric. Food Chem.* **2021**, *69*, 12241–12249. [CrossRef]
11. Jiménez, A.; Santos, M.A.; Pompejus, M.; Revuelta, J.L. Metabolic engineering of the purine pathway for riboflavin production in Ashbya gossypii. *Appl. Environ. Microbiol.* **2005**, *71*, 5743–5751. [CrossRef]
12. Shimizu, K.; Hayashi, S.; Doukyu, N.; Kobayashi, T.; Honda, H. Time-course data analysis of gene expression profiles reveals *purR* regulon concerns in organic solvent tolerance in *Escherichia coli*. *J. Biosci. Bioeng.* **2005**, *99*, 72–74. [CrossRef]
13. Alkam, D.; Jenjaroenpun, P.; Ramirez, A.M.; Beenken, K.E.; Spencer, H.J.; Smeltzer, M.S. The increased accumulation of *Staphylococcus aureus* virulence factors is maximized in a *purR* mutant by the increased production of SarA and decreased production of extracellular proteases. *Infect. Immun.* **2021**, *89*, e00718-20. [CrossRef]
14. Song, Y.; Tong, Z.; Wang, J.; Wang, L.; Guo, Z.; Han, Y.; Zhang, J.; Pei, D.; Zhou, D.; Qin, H.; et al. Complete genome sequence of *Yersinia pestis* strain 91001, an isolate avirulent to humans. *DNA Res.* **2004**, *11*, 179–197. [CrossRef] [PubMed]
15. Zhang, Q.; Wang, Q.; Tian, G.; Qi, Z.; Zhang, X.; Wu, X.; Qiu, Y.; Bi, Y.; Yang, X.; Xin, Y.; et al. *Yersinia pestis* biovar *Microtus* strain 201, an avirulent strain to humans, provides protection against bubonic plague in rhesus macaques. *Hum. Vaccines Immunother.* **2014**, *10*, 368–377. [CrossRef]
16. Barbieri, R.; Signoli, M.; Cheve, D.; Costedoat, C.; Tzortzis, S.; Aboudharam, G.; Raoult, D.; Drancourt, M. *Yersinia pestis*: The Natural History of Plague. *Clin. Microbiol. Rev.* **2021**, *34*, e00044-19. [CrossRef] [PubMed]
17. Chen, Y.; Song, K.; Chen, X.; Li, Y.; Lv, R.; Zhang, Q.; Cui, Y.; Bi, Y.; Han, Y.; Tan, Y.; et al. Attenuation of *Yersinia pestis fyuA* mutants caused by iron uptake inhibition and decreased survivability in macrophages. *Front. Cell. Infect. Microbiol.* **2022**, *12*, 874773. [CrossRef] [PubMed]
18. Zhang, Y.; Gao, H.; Wang, L.; Xiao, X.; Tan, Y.; Guo, Z.; Zhou, D.; Yang, R. Molecular characterization of transcriptional regulation of *rovA* by PhoP and RovA in *Yersinia pestis*. *PLoS ONE* **2011**, *6*, e25484. [CrossRef]
19. Chang, A.C.; Cohen, S.N. Construction and characterization of amplifiable multicopy DNA cloning vehicles derived from the P15A cryptic miniplasmid. *J. Bacteriol.* **1978**, *134*, 1141–1156. [CrossRef]
20. Straley, S.C.; Bowmer, W.S. Virulence genes regulated at the transcriptional level by Ca^{2+} in *Yersinia pestis* include structural genes for outer membrane proteins. *Infect. Immun.* **1986**, *51*, 445–454. [CrossRef] [PubMed]
21. Lisby, A.N.; Carlson, R.D.; Baybutt, T.R.; Weindorfer, M.; Snook, A.E. Evaluation of CAR-T cell cytotoxicity: Real-time impedance-based analysis. *Methods Cell Biol.* **2022**, *167*, 81–98. [CrossRef]
22. Bailey, T.L.; Elkan, C. Fitting a mixture model by expectation maximization to discover motifs in biopolymers. *Proc. Int. Conf. Intell. Syst. Mol. Biol.* **1994**, *2*, 28–36. [PubMed]
23. Grant, C.E.; Bailey, T.L.; Noble, W.S. FIMO: Scanning for occurrences of a given motif. *Bioinformatics* **2011**, *27*, 1017–1018. [CrossRef]
24. Miller, M.C.; Fetherston, J.D.; Pickett, C.L.; Bobrov, A.G.; Weaver, R.H.; DeMoll, E.; Perry, R.D. Reduced synthesis of the Ybt siderophore or production of aberrant Ybt-like molecules activates transcription of yersiniabactin genes in *Yersinia pestis*. *Microbiology* **2010**, *156*, 2226–2238. [CrossRef] [PubMed]
25. Perez-Rueda, E.; Collado-Vides, J.; Segovia, L. Phylogenetic distribution of DNA-binding transcription factors in bacteria and archaea. *Comput. Biol. Chem.* **2004**, *28*, 341–350. [CrossRef]
26. Han, Y.; Qiu, J.; Guo, Z.; Gao, H.; Song, Y.; Zhou, D.; Yang, R. Comparative transcriptomics in *Yersinia pestis*: A global view of environmental modulation of gene expression. *BMC Microbiol.* **2007**, *7*, 96. [CrossRef] [PubMed]

27. Sause, W.E.; Balasubramanian, D.; Irnov, I.; Copin, R.; Sullivan, M.J.; Sommerfield, A.; Chan, R.; Dhabaria, A.; Askenazi, M.; Ueberheide, B.; et al. The purine biosynthesis regulator PurR moonlights as a virulence regulator in *Staphylococcus aureus*. *Proc. Natl. Acad. Sci. USA* **2019**, *116*, 13563–13572. [CrossRef]
28. Anderson, B.W.; Schumacher, M.A.; Yang, J.; Turdiev, A.; Turdiev, H.; Schroeder, J.W.; He, Q.; Lee, V.T.; Brennan, R.G.; Wang, J.D. The nucleotide messenger (p)ppGpp is an anti-inducer of the purine synthesis transcription regulator PurR in *Bacillus*. *Nucleic Acids Res.* **2022**, *50*, 847–866. [CrossRef]
29. Jewett, M.W.; Lawrence, K.A.; Bestor, A.; Byram, R.; Gherardini, F.; Rosa, P.A. GuaA and GuaB are essential for *Borrelia burgdorferi* survival in the tick-mouse infection cycle. *J. Bacteriol.* **2009**, *191*, 6231–6241. [CrossRef]
30. Butcher, B.G.; Chakravarthy, S.; D'Amico, K.; Stoos, K.B.; Filiatrault, M.J. Disruption of the *carA* gene in *Pseudomonas syringae* results in reduced fitness and alters motility. *BMC Microbiol.* **2016**, *16*, 194. [CrossRef]
31. Tezuka, T.; Ohnishi, Y. Two glycine riboswitches activate the glycine cleavage system essential for glycine detoxification in *Streptomyces griseus*. *J. Bacteriol.* **2014**, *196*, 1369–1376. [CrossRef]
32. Goncheva, M.I.; Flannagan, R.S.; Heinrichs, D.E. De Novo Purine Biosynthesis Is Required for Intracellular Growth of *Staphylococcus aureus* and for the Hypervirulence Phenotype of a purR Mutant. *Infect. Immun.* **2020**, *88*, e00104-20. [CrossRef] [PubMed]
33. Musila, J.M.; Forbes, D.L.; Ellis, H.R. Functional Evaluation of the π-Helix in the NAD(P)H:FMN Reductase of the Alkanesulfonate Monooxygenase System. *Biochemistry* **2018**, *57*, 4469–4477. [CrossRef]
34. van der Ploeg, J.R.; Eichhorn, E.; Leisinger, T. Sulfonate-sulfur metabolism and its regulation in *Escherichia coli*. *Arch. Microbiol.* **2001**, *176*, 1–8. [CrossRef] [PubMed]
35. Jani, A.J.; Cotter, P.A. Type VI secretion: Not just for pathogenesis anymore. *Cell Host Microbe* **2010**, *8*, 2–6. [CrossRef]
36. Yang, X.; Pan, J.; Wang, Y.; Shen, X. Type VI Secretion Systems Present New Insights on Pathogenic *Yersinia*. *Front. Cell. Infect. Microbiol.* **2018**, *8*, 260. [CrossRef] [PubMed]
37. Hinnebusch, B.J.; Jarrett, C.O.; Bland, D.M. Molecular and Genetic Mechanisms That Mediate Transmission of *Yersinia pestis* by Fleas. *Biomolecules* **2021**, *11*, 210. [CrossRef]
38. Le-Bury, P.; Druart, K.; Savin, C.; Lechat, P.; Mas Fiol, G.; Matondo, M.; Becavin, C.; Dussurget, O.; Pizarro-Cerda, J. Yersiniomics, a Multi-Omics Interactive Database for *Yersinia* Species. *Microbiol. Spectr.* **2023**, *11*, e0382622. [CrossRef] [PubMed]
39. Goncheva, M.I.; Flannagan, R.S.; Sterling, B.E.; Laakso, H.A.; Friedrich, N.C.; Kaiser, J.C.; Watson, D.W.; Wilson, C.H.; Sheldon, J.R.; McGavin, M.J.; et al. Stress-induced inactivation of the *Staphylococcus aureus* purine biosynthesis repressor leads to hypervirulence. *Nat. Commun.* **2019**, *10*, 775. [CrossRef]
40. Weng, M.; Nagy, P.L.; Zalkin, H. Identification of the *Bacillus subtilis pur* operon repressor. *Proc. Natl. Acad. Sci. USA* **1995**, *92*, 7455–7459. [CrossRef]
41. Long, H.X.; Ma, W.J.; Qin, J.C.; Wang, A.Q. Regulation of purine biosynthetic genes expression in *Salmonella typhimurium*. *Yi Chuan Xue Bao* **2000**, *27*, 462–467. [PubMed]
42. Gao, H.; Zhou, D.; Li, Y.; Guo, Z.; Han, Y.; Song, Y.; Zhai, J.; Du, Z.; Wang, X.; Lu, J.; et al. The iron-responsive Fur regulon in *Yersinia pestis*. *J. Bacteriol.* **2008**, *190*, 3063–3075. [CrossRef] [PubMed]
43. Robinson, J.B.; Telepnev, M.V.; Zudina, I.V.; Bouyer, D.; Montenieri, J.A.; Bearden, S.W.; Gage, K.L.; Agar, S.L.; Foltz, S.M.; Chauhan, S.; et al. Evaluation of a *Yersinia pestis* mutant impaired in a thermoregulated type VI-like secretion system in flea, macrophage and murine models. *Microb. Pathog.* **2009**, *47*, 243–251. [CrossRef] [PubMed]

microorganisms

Article

Gene Expression of Ethanol and Acetate Metabolic Pathways in the *Acinetobacter baumannii* EmaSR Regulon

Yu-Weng Huang [1], Hung-Yu Shu [2] and Guang-Huey Lin [3,4,*]

1 Department of Biomedical Sciences and Engineering, School of Medicine, Tzu Chi University, Hualien 970374, Taiwan; 110726102@gms.tcu.edu.tw
2 Department of Bioscience Technology, Chang Jung Christian University, Tainan 711301, Taiwan; hyshu@mail.cjcu.edu.tw
3 Master Program in Biomedical Sciences, School of Medicine, Tzu Chi University, Hualien 970374, Taiwan
4 International College, Tzu Chi University, Hualien 970374, Taiwan
* Correspondence: veronica@gms.tcu.edu.tw; Tel.: +886-3-8565301 (ext. 2104)

Abstract: Background: Previous studies have confirmed the involvement of EmaSR (ethanol metabolism a sensor/regulator) in the regulation of *Acinetobacter baumannii* ATCC 19606 ethanol and acetate metabolism. RNA-seq analysis further revealed that *DJ41_568-571*, *DJ41_2796*, *DJ41_3218*, and *DJ41_3568* regulatory gene clusters potentially participate in ethanol and acetate metabolism under the control of EmaSR. Methods: This study fused the EmaSR regulon promoter segments with reporter genes and used fluorescence expression levels to determine whether EmaSR influences regulon expression in ethanol or acetate salt environments. The enzymatic function and kinetics of significantly regulated regulons were also studied. Results: The EmaSR regulons P_{2796} and P_{3218} exhibited > 2-fold increase in fluorescence expression in wild type compared to mutant strains in both ethanol and acetate environments, and P_{emaR} demonstrated a comparable trend. Moreover, increases in DJ41_2796 concentration enhanced the conversion of acetate and succinyl-CoA into acetyl-CoA and succinate, suggesting that DJ41_2796 possesses acetate: succinyl-CoA transferase (ASCT) activity. The k_{cat}/K_M values for DJ41_2796 with potassium acetate, sodium acetate, and succinyl-CoA were 0.2131, 0.4547, and 20.4623 $mM^{-1}s^{-1}$, respectively. Conclusions: In *A. baumannii*, EmaSR controls genes involved in ethanol and acetate metabolism, and the EmaSR regulon DJ41_2796 was found to possess ASCT activity.

Keywords: two-component system; ethanol metabolism; acetate metabolism; *Acinetobacter baumannii*; acetate: succinyl-CoA transferase (ASCT)

check for updates

Citation: Huang, Y.-W.; Shu, H.-Y.; Lin, G.-H. Gene Expression of Ethanol and Acetate Metabolic Pathways in the *Acinetobacter baumannii* EmaSR Regulon. *Microorganisms* **2024**, *12*, 331. https://doi.org/10.3390/microorganisms12020331

Academic Editor: Tomohiro Shimada

Received: 4 January 2024
Revised: 31 January 2024
Accepted: 2 February 2024
Published: 4 February 2024

1. Introduction

Ethanol is a widely-used disinfectant, but previous research has shown that sub-inhibitory concentrations of ethanol are not only unable to eradicate bacteria but can conversely promote biofilm formation to increase bacterial resistance [1–4]. Ethanol concentrations of 2.3–3.5% enhance the ability of *Staphylococcus aureus* to form biofilms, while 1–2% concentrations are sufficient to induce biofilm formation in *Pseudomonas aeruginosa* [1–4]. The presence of alcohol dehydrogenases (Adhs) can further help bacteria to degrade alcohols, reducing ethanol-induced damage and transforming ethanol into a carbon source for growth instead [5,6].

Acetate is a type of short-chain fatty acid (SCFA) produced during ethanol metabolism. Previous research has shown that SCFAs inhibit toxin-producing bacteria such as *Bacillus cereus* and *Clostridium difficile*, as well as foodborne pathogens such as *Campylobacter* and *Salmonella* [7]. Additionally, it has been found that acetate increases the activity of macrophages to enhance their lethality against *Streptococcus pneumoniae* [8]. However, many types of bacteria can also metabolize acetate; for example, *Escherichia coli* can convert acetate into acetyl-CoA via acetate kinase (AckA) and phosphotransacetylase (Pta), or through acetyl-CoA synthetase [9]. Both of these pathways require the consumption of ATP, but in *Pseudomonas* sp., *Klebsiella pneumoniae*, and *Acetobacter*, it has been observed that acetate:succinyl-CoA transferase (ASCT) can transfer the CoA from succinyl-CoA to acetate to produce acetyl-CoA without consuming ATP [10].

Bacteria often use a two-component system (TCS) to detect and respond to changes in the environment. A typical two-component system consists of a histidine sensor kinase and a response regulator. Previous research has identified an ErdSR TCS that regulates ethanol metabolism [11,12]; in addition, MxtR/ErdR in *Pseudomonas putida* KT2440 was found to regulate the metabolism of acetate and pyruvate, partly through enzymes such as ScpC (PP_0154), which functions as an ASCT [13]. Recent research on *Acinetobacter baumannii*, *K. pneumoniae*, and *S. aureus* has also confirmed that many of their TCSs have regulatory functions in metabolism [14–16].

A. baumannii is a common pathogen in nosocomial infections. Previous research identified seven Adh genes in the genome of *A. baumannii* ATCC 19606, of which Adh exhibits a higher affinity for ethanol over other alcohols; the absence of Adh4 specifically hindered bacterial growth in low concentrations of ethanol, propanol, and butanol [17]. Subsequently, it was confirmed that the absence of the *emaSR* TCS in *A. baumannii* leads to an inability to grow in low concentrations of ethanol, as EmaSR was shown to regulate ethanol metabolism. Additionally, it was discovered that EmaSR shares approximately 40% or more amino acid sequence similarity with ErdSR and MxtR/ErdR, which regulate ethanol metabolism in *P. aeruginosa* PAO1. Furthermore, the loss of EmaSR was found to lead to growth defects in *A. baumannii* when exposed to a 20 mM acetate environment [18]. However, in low-concentration ethanol environments, transcriptome analysis revealed that EmaSR significantly upregulated several genes annotated for involvement in acetate and pyruvate metabolic pathways (Figure 1), including *DJ41_568-571*, *DJ41_2796*, *DJ41_3218*, and *DJ41_3568* [18]. Considering that the presence of ethanol, which is widely used in hospital environments as a disinfectant, has been reported to induce stress responses in *A. baumannii* that can lead to increased virulence [17], understanding ethanol and acetate metabolic pathways can be crucial to formulating new strategies aimed at preventing or mitigating nosocomial and opportunistic infections.

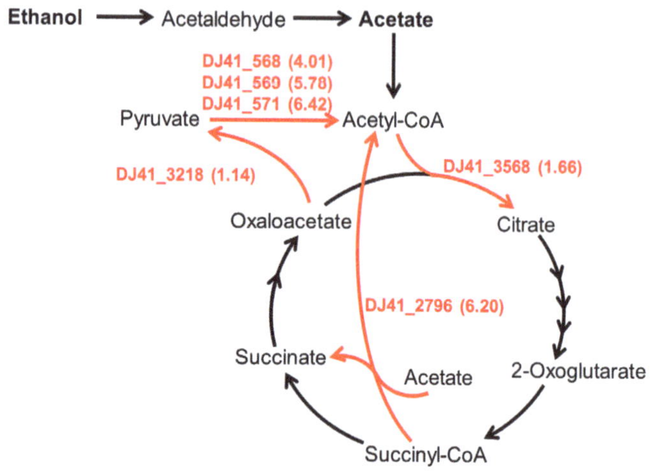

Figure 1. The carbon metabolism pathway regulated by EmaSR regulons. Based on gene function annotations from the National Center for Biotechnology Information (NCBI) database for *A. baumannii* ATCC 19606 and the metabolic pathways of the relevant Kyoto Encyclopedia of Genes and Genomes (KEGG) pathway. The numbers listed in parentheses beside the EmaSR regulons represent the log2 fold-change in gene expression between strains with and without *emaSR* in 0.5% ethanol, as derived from EmaSR transcriptome analysis.

In this study, we investigated the expression of a reporter gene fused with the promoter of EmaSR regulons, in order to confirm the metabolic pathways through which EmaSR regulates bacterial metabolism of ethanol and acetate. We also conducted an enzymatic analysis to identify EmaSR regulons with crucial enzymatic activity in the *A. baumannii* metabolism of ethanol and acetate. Our results provide further insight into the role of the EmaSR TCS in *A. baumannii* resilience and may have implications for other pathogenic bacterial species as well.

2. Materials and Methods

2.1. Bacterial Strains, Plasmids, Primers, and Growth of Strains

All strains of *E. coli* and *A. baumannii* were grown in LB medium at 37 °C with shaking to increase the bacterial count; however, the fluorescence expression experiments were carried out with cultures using an M9 medium (composition described in Section 2.2). Based on the antibiotic resistance of each strain, the culture medium was supplemented with a final concentration of 50 μg/mL ampicillin, 50 μg/mL kanamycin, or 12.5 μg/mL tetracycline. Bacterial strains and plasmids are described in Table 1 and Table 2, respectively. Primers used for the construction and verification of the recombinant strains are presented in Table 3.

Table 1. Bacterial strains used in this study.

No.		Description	References or Source
		***Escherichia coli* (*E. coli*)**	
1	DH5α	F^-, *supE44, hsdR17, reA1, gyrA96, endA1, thi-1, relA1, delR*, λ^-	ATCC 53868
2	DH5α/pQE80LK-DJ41_2796	DH5α contain pQE80LK-DJ41_2796, Km^r	This study
3	DH5α/TCSG	DH5α contain TCSG, Ap^r	This study
4	DH5α/TCS-P_{emaS}G	DH5α contain TCS-P_{emaS}G, Ap^r	This study
5	DH5α/TCS-P_{emaR}G	DH5α contain TCS-P_{emaR}G, Ap^r	This study
6	DH5α/TCS-P_{571}G	DH5α contain TCS-P_{571}G, Ap^r	This study
7	DH5α/TCS-P_{2796}G	DH5α contain TCS-P_{2796}G, Ap^r	This study
8	DH5α/TCS-P_{3218}G	DH5α contain TCS-P_{3218}G, Ap^r	This study
9	DH5α/TCS-P_{3568}G	DH5α contain TCS-P_{3568}G, Ap^r	This study
10	DH5α/pWH1266	DH5α contain pWH1266, Ap^r; Tet^r	[19]
11	DH5α/pWH1266G	DH5α contain pWH1266G, Tet^r	This study
12	DH5α/pWH1266-P_{emaS}G	DH5α contain pWH1266-P_{emaS}G, Tet^r	This study
13	DH5α/pWH1266-P_{emaR}G	DH5α contain pWH1266-P_{emaR}G, Tet^r	This study
14	DH5α/pWH1266-P_{571}G	DH5α contain pWH1266-P_{571}G, Tet^r	This study
15	DH5α/pWH1266-P_{2796}G	DH5α contain pWH1266-P_{2796}G, Tet^r	This study
16	DH5α/pWH1266-P_{3218}G	DH5α contain pWH1266-P_{3218}G, Tet^r	This study
17	DH5α/pWH1266-P_{3568}G	DH5α contain pWH1266-P_{3568}G, Tet^r	This study
		***Acinetobacter baumannii* (*A. baumannii*)**	
18	ATCC 19606	Ap^r, clinical isolate, wild type	[20]
19	ATCC 19606/pWH1266G	ATCC 19606 contain pWH1266G, Ap^r, Tet^r	This study
20	ATCC 19606/pWH1266-P_{emaS}G	ATCC 19606 contain pWH1266-P_{emaS}G, Ap^r, Tet^r	This study
21	ATCC 19606/pWH1266-P_{emaR}G	ATCC 19606 contain pWH1266-P_{emaR}G, Ap^r, Tet^r	This study
22	ATCC 19606/pWH1266-P_{2796}G	ATCC 19606 contain pWH1266-P2796G, Apr, Tet^r	This study

Table 1. *Cont.*

No.		Description	References or Source
23	ATCC 19606/pWH1266-P_{3218}G	ATCC 19606 contain pWH1266-P_{3218}G, Ap^r, Tet^r	This study
24	ATCC 19606/pWH1266-P_{3568}G	ATCC 19606 contain pWH1266-P_{3568}G, Ap^r, Tet^r	This study
25	Δ*emaS*	ATCC 19606, Δ*emaS*, Ap^r	[18]
26	Δ*emaS*/pWH1266G	Δ*emaS* contain pWH1266G, Ap^r, Tet^r	This study
27	Δ*emaS*/pWH1266-P_{emaS}G	Δ*emaS* contain pWH1266-P_{emaS}G, Ap^r, Tet^r	This study
28	Δ*emaS*/pWH1266-P_{emaR}G	Δ*emaS* contain pWH1266-P_{emaR}G, Ap^r, Tet^r	This study
29	Δ*emaS*/pWH1266-P_{571}G	Δ*emaS* contain pWH1266-P_{571}G, Ap^r, Tet^r	This study
30	Δ*emaS*/pWH1266-P_{2796}G	Δ*emaS* contain pWH1266-P_{2796}G, Ap^r, Tet^r	This study
31	Δ*emaS*/pWH1266-P_{3218}G	Δ*emaS* contain pWH1266-P_{3218}G, Ap^r, Tet^r	This study
32	Δ*emaS*/pWH1266-P_{3568}G	Δ*emaS* contain pWH1266-P_{3568}G, Ap^r, Tet^r	This study
33	Δ*emaR*	ATCC 19606, Δ*emaR*, Ap^r	[18]
34	Δ*emaR*/pWH1266G	Δ*emaR* contain pWH1266G, Ap^r, Tet^r	This study
35	Δ*emaR*/pWH1266-P_{emaS}G	Δ*emaR* contain pWH1266-P_{emaS}G, Ap^r, Tet^r	This study
36	Δ*emaR*/pWH1266-P_{emaR}G	Δ*emaR* contain pWH1266-P_{emaR}G, Ap^r, Tet^r	This study
37	Δ*emaR*/pWH1266-P_{571}G	Δ*emaR* contain pWH1266-P_{571}G, Ap^r, Tet^r	This study
38	Δ*emaR*/pWH1266-P_{2796}G	Δ*emaR* contain pWH1266-P_{2796}G, Ap^r, Tet^r	This study
39	Δ*emaR*/pWH1266-P_{3218}G	Δ*emaR* contain pWH1266-P_{3218}G, Ap^r, Tet^r	This study
40	Δ*emaR*/pWH1266-P_{3568}G	Δ*emaS* contain pWH1266-P_{3568}G, Ap^r, Tet^r	This study
41	Δ*emaSR*	ATCC 19606, Δ*emaSR*, Ap^r	[18]
42	Δ*emaSR*/pWH1266G	Δ*emaSR* contain pWH1266G, Ap^r, Tet^r	This study
43	Δ*emaSR*/pWH1266-P_{emaS}G	Δ*emaSR* contain pWH1266-P_{emaS}G, Ap^r, Tet^r	This study
44	Δ*emaSR*/pWH1266-P_{emaR}G	Δ*emaSR* contain pWH1266-P_{emaR}G, Ap^r, Tet^r	This study
45	Δ*emaSR*/pWH1266-P_{571}G	Δ*emaSR* contain pWH1266-P_{571}G, Ap^r, Tet^r	This study
46	Δ*emaSR*/pWH1266-P_{2796}G	Δ*emaSR* contain pWH1266-P_{2796}G, Ap^r, Tet^r	This study
47	Δ*emaSR*/pWH1266-P_{3218}G	Δ*emaSR* contain pWH1266-P_{3218}G, Ap^r, Tet^r	This study
48	emaSR/pWH1266-P_{3568}G	Δ*emaSR* contain pWH1266-P_{3568}G, Ap^r, Tet^r	This study

Table 2. Plasmids used in this study.

No		Description	References or Source
1	pQE80LK	expression vector, pUC ori; Km^r	This study
2	pQE80LK-DJ41_2796	pQE80LK/*DJ41_2796*; Km^r	This study
3	TCSG	pGFPuv/*emaS*; *emaR*; *gfpuv*; Ap^r	This study
4	TCS-P_{emaS}G	pGFPuv/*emaS*; *emaR*; P_{emaS}-*gfpuv*; Ap^r	This study
5	TCS-P_{emaR}G	pGFPuv/*emaS*; *emaR*; P_{emaR}-*gfpuv*; Ap^r	This study
6	TCS-P_{571}G	pGFPuv/*emaS*; *emaR*; P_{571}-*gfpuv*; Ap^r	This study
7	TCS-P_{2796}G	pGFPuv/*emaS*; *emaR*; P_{2796}-*gfpuv*; Ap^r	This study
8	TCS-P_{3218}G	pGFPuv/*emaS*; *emaR*; P_{3218}-*gfpuv*; Ap^r	This study
9	TCS-P_{3568}G	pGFPuv/*emaS*; *emaR*; P_{3568}-*gfpuv*; Ap^r	This study
10	pWH1266	*E. coli–A. baumannii* shuttle vector; Ap^r; Tet^r	[19]

Table 2. *Cont.*

No		Description	References or Source
11	pWH1266G	pWH1266G/*gfpuv*; Tet^r	This study
12	pWH1266-P_{emaS}G	pWH1266G/P_{emaS}-*gfpuv*; Tet^r	This study
13	pWH1266-P_{emaR}G	pWH1266G/P_{emaR}-*gfpuv*; Tet^r	This study
14	pWH1266-P_{571}G	pWH1266G/P_{571}-*gfpuv*; Tet^r	This study
15	pWH1266-P_{2796}G	pWH1266G/P_{2796}-*gfpuv*; Tet^r	This study
16	pWH1266-P_{3218}G	pWH1266G/P_{3218}-*gfpuv*; Tet^r	This study
17	pWH1266-P_{3568}G	pWH1266G/P_{3568}-*gfpuv*; Tet^r	This study

Table 3. Primers used in this study.

No.		Sequences (5′-3′)	References or Source
1	pQE80LK-DJ41_2796-eF	AAGGATCCATGTCTTTAAGTCGTATT	This study
2	pQE80LK-DJ41_2796-eR	AACTGCAGTTAAGCTGATTTTGCAAC	This study
3	TCSG-P_{emaS}-gF	GGCTGCAGTAACCAAACTCCTTACATAG	This study
4	TCSG-P_{emaR}-gF	GGCTGCAGCCCGAATAGTTGATTTTTAT	This study
5	TCSG-P_{571}-gF	GGCTGCAGTTTGTTTTACAAGTATATGA	This study
6	TCSG-P_{2796}-gF	GGCTGCAGGCGCTATTTTAAACCTCAAA	This study
7	TCSG-P_{3218}-gF	GGCTGCAGGTCAGATATAGAGTTGAGAA	This study
8	TCSG-P_{3568}-gF	GGCTGCAGTCCCGATGCAGTGATTTTAC	This study
9	pWH-P_{gfp}-F	ACGTTGTTGCCATTGCTGCAAAAAATCTAATGCATGCCTGC	This study
10	TCSG-P_{emaS}-gR	CCTCTAGAATGACCTACATAAGTGAAAC	This study
11	TCSG-P_{emaR}-gR	CCTCTAGAAAGCGATTAAAGTAATCTTG	This study
12	TCSG-P_{571}-gR	CCTCTAGAATCCTTTTCCTTTATTATCT	This study
13	TCSG-P_{2796}-gR	CCTCTAGATGGACATCCTCAATATTGTC	This study
14	TCSG-P_{3218}-gR	CCTCTAGATGCCTATCTCATTTTCCAGC	This study
15	TCSG-P_{3568}-gR	CCTCTAGATCAATAGATCTCCTGTCCTG	This study
16	pWH-P_{gfp}-reR	GATAAGCTGTCAAACATGAGCATTATTTGTAGAGCTCATCCA	This study
17	GFP-34R	TTCACCCTCTCCACTGACAGA	This study
18	TcR	GATGCGTCCGGCGTAGAG	This study
19	P-Ab-ITSB	AGAGCACTGTGCACTTAAG	[21]
20	P-Ab-ITSF	CATTATCACGGTAATTAGTG	[21]

2.2. Green Florescence Assay

TCSG represents the abbreviation of the two-component system with *gfp* gene. This reporter system utilized the *Pst*I and *Xba*I restriction enzyme sites to combine the promoter fragment with *gfpuv*, which contains three amino acid modifications that significantly increase fluorescence emissions without affecting emission wavelength [22]. Each promoter-*gfpuv* fragment were then recombined into pWH1266 using Gibson assembly (E2611, NEB, Ipswich, MA, USA). The recombinant plasmids were introduced into *A. baumannii* strains by electroporation at 1.8 kV. The construction of each strain was confirmed by a colony PCR, and the analysis results are presented in Figures S1–S5.

Strains with the pWH1266G reporter plasmid were cultured in an M9 medium (33.7 mM Na_2PO_4, 22 mM KH_2PO_4, 8.55 mM NaCl, 9.35 mM NH_4Cl, 1 mM $MgSO_4$,

0.3 mM $CaCl_2$), With 5 mM citrate, supplemented with 0.5% ethanol or 20 mM potassium acetate. Each strain was grown at 37 °C with an initial OD_{600} of 0.1, with continuous shaking. The samples were collected periodically to analyze fluorescence and bacterial growth. Samples were diluted 10-fold in 0.9% NaCl, and fluorescence expression was measured using a Varioskan LUX Multimode Microplate Reader (3020-80145, ThermoFisher, Waltham, MA, USA) with excitation at 395 nm and emission at 509 nm. Samples for growth curve determination were collected simultaneously, diluted in 0.9% NaCl, and then assessed for changes in OD_{600} using the same instrument.

2.3. *Purification of DJ41_2796 and Enzymatic Assay*

Following the methodology described in previous research [18], the *DJ41_2796* gene was cloned into pQE80LK, which was derived by substituting the ampicillin resistance gene of pQE80L (Qiagen, Hilden, Germany) with kanamycin. A gene fragment of *DJ412796*, 1515 bp in length, was ligated with a pQE80LK cutting by *Bam*HI and *Pst*I. The resulting construct, pQE80LK-DJ41_2796, was then transformed into *E. coli* DH5α for protein expression and purification (Figure S6). A total of 100 mL of bacteria were cultured, with OD_{600} of 0.1 at initiation. Protein expression was induced by 0.5 mM isopropyl ß-D-1-thiogalactopyranoside (IPTG) after OD_{600} reached 0.6 for three hours. The cultures were then harvested and lysed using a high-pressure homogenizer CF1 (Constant Systems Ltd., Daventry, UK), and purification of DJ41_2796 from the supernatant was conducted at 4 °C using NGC Chromatography Systems (Bio-rad, Hercules, CA, USA). In this system, the supernatant was passed through a nickel-affinity column (Nuvia IMAC Ni-Charge, Bio-rad, Hercules, CA, USA). The histidine-tag fusion protein was purified by increasing the concentration of imidazole from 5 mM to 1 M in the buffer [17].

The purified protein was utilized for enzyme activity analysis in accordance with the experimental design of previous studies [23,24]. Each 1 mL reaction sample for testing the dose response of DJ41_2796 contained a final concentration of 100 mM phosphate buffer solution, 10 mM potassium acetate, 0.2 mM succinyl-CoA, 0~10 μM DJ41_2796, and 0.1 mM Ellman's Reagent (5,5′-dithio-bis-(2-nitrobenzoic acid), DTNB). After the reaction between DTNB and CoA, TNB (extinction coefficient: 14.15 $M^{-1}cm^{-1}$) was produced [25]. The changes in OD_{412} were monitored every 30 s within a 1.5 min timeframe, using the Multiskan SkyHigh Microplate Spectrophotometer (ThermoFisher, Waltham, MA, USA). Enzyme activity was then determined according to the conversion rate of acetate into acetyl-CoA, in nanomoles per minute.

3. Results

3.1. *Genes of EmaSR Regulons Were Upregulated in a Low-Concentration Ethanol Environment*

Binding-box analysis was conducted on sequences of approximately 250 base pairs upstream of each gene, and a sequence of AAxCTTAxxxxTAxxxTTxxxx upstream of the EmaSR regulon was found to be highly conserved (Table 4). This same sequence was identified upstream of *emaS* and *emaR*, thereby suggesting that these genes might be regulated by EmaR. This study constructed plasmid TCSG to confirm that EmaSR regulons were indeed upregulated in low-ethanol conditions. TCSG enables the expression of *emaS* and *emaR*, along with *gfpuv* (Figure S1). Approximately 250 bp upstream of each EmaSR regulon were combined with the *gfpuv* of the TCSG reporter plasmid to construct derivative plasmids, and after confirming its successful construction and transformation using colony PCR (Figure S2), each clone was observed for fluorescence expression in LB medium containing 1% ethanol (Figure S7). However, differences in fluorescence expression among all strains, including the control group, did not exceed 6% of the average fluorescence value. There was no significant difference in the expression levels of each EmaSR regulon compared to the control group (Figure S7). Since gene expression in *E. coli* cultured in LB medium differs from the growth of *A. baumannii* in low-concentration ethanol conditions, we inferred that the TCSG reporter system in *E. coli* is unsuitable for analyzing EmaSR regulon expression; therefore, DNA fragments combining the regulon promoters with *gfpuv* were transferred from TCSG to the *E. coli* and *A. baumannii* shuttle vector pWH1266 (Figure S3) for further analysis. The successful establishment of each pWH1266 reporter strain was subsequently confirmed by the colony PCR (Figures S4 and S5).

Table 4. EmaR binding boxes on EmaSR regulons.

Name	Ratio	Strand	Position *	Sequence (5′-3′) †
DJ41_566-571	2.85~6.42	+	−109 to −130	AAACTTATTTAAAACTTTTTAG
DJ41_3218	1.14	+	−64 to −85	AGTCTTAAGCTTACGCATACAA
DJ41_3568	1.66	+	−74 to −95	AATCTTATAGCAAATTTTGACA
DJ41_2796	6.21	-	−105 to −126	TAAAAATCATAAAAATAAGTTA
emaR	2.94	-	−14 to −35	GGCCATACTTATGCTCAAGATT
emaS	2.30	-	−18 to −39	GGGTTATGATAGGCATAAGGTT

* Start codon (Met) set as 0. † Underlined base pairs perfectly match AAxCTTAxxxxTAxxxTTxxxx, while **Bold** base pairs are identical to the highly conserved sequence that was found in every sequence analyzed here; analysis was conducted to 250 bp upstream of the start codon.

During the analysis, it was observed that *A. baumannii* wild-type strains carrying reporter plasmids, such as *DJ41_566-571* (P_{571}), *DJ41_2796* (P_{2796}), *DJ41_3218* (P_{3218}), and *DJ41_3568* (P_{3568}), exhibited slightly higher fluorescence levels after 24 h compared to wild-type strains carrying an empty reporter plasmid (P_{gfp}) under citrate conditions (Figure 2A). In addition, under 0.5% ethanol culture conditions, fluorescence expression levels of EmaSR regulons increased by four to six times over the control group (Figure 2C). Although growth curves leveled out after 24 h for all strains, statistically significant differences in fluorescence expression continued to be observed between different strains, indicating that the differences in fluorescence were unlikely to be influenced by growth fluctuation (Figure 2B,D). The fluorescence expression of P_{2796} and P_{3218} significantly increased under 0.5% ethanol culture conditions, respectively, reaching approximately 3-fold and 2-fold higher than the control (Figure 2C). By contrast, fluorescence expression levels for P_{571} and P_{3568} were not significantly different from P_{gfp} at 48 h. These results suggest that *DJ41_2796* and *DJ41_3218* are regulated by EmaSR in ethanol-containing culture conditions. P_{emaR} also exhibited significant differences in fluorescence expression compared to the control from 24 h onwards in 0.5% ethanol-containing cultures, but unexpectedly, there were no significant differences in the fluorescence expression for P_{emaS} in comparison to the control. These results confirm that P_{emaR}, P_{2796}, and P_{3218} are induced in the presence of ethanol (Figure 2C).

To further confirm the upregulation of EmaSR regulons, the expression levels of P_{2796} and P_{3218} were analyzed in wild-type and mutant strains (Δ*emaS*, Δ*emaR*, and Δ*emaSR*), cultured under low-concentration ethanol conditions. Similar expression levels at 48 h were observed in the wild-type strain as previously noted (Figure 3 black and gray slash, Figures S8A and S9A). However, in mutants Δ*emaS*, Δ*emaR*, and Δ*emaSR*, there was no significant elevation in fluorescence expression for $P_{2796}$and P_{3218} (Figure 3 black and gray slash, Figures S8A and S9A). Although the wild-type strain grew approximately 4- to 6-fold higher than mutant strains initially, all strains reached similar growth levels after 48 h (Figures S8B and S9B). This confirmed that at least by 48 h, the difference in expression levels of P_{2796} and P_{3218} between wild-type and mutant strains was not due to growth defects and demonstrated that the increase in fluorescence expression levels in low-concentration ethanol culture conditions was indeed regulated by EmaSR.

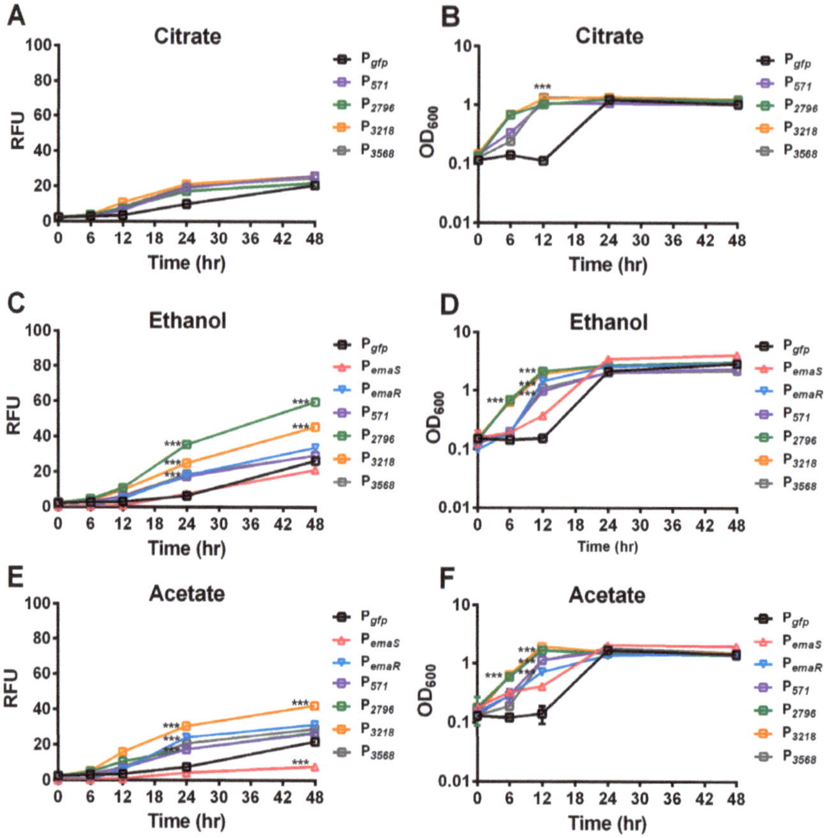

Figure 2. Mean fluorescence expression levels and growth of EmaSR regulon-transformed *A. baumannii* strains under 0.5% ethanol and 20 mM potassium acetate culture conditions. Wild-type strains of *A. baumannii* containing pWH1266G (black), pWH-P_{571}G (purple), pWH-P_{2796}G (green), pWH-P_{3218}G (orange), pWH-P_{3568}G (gray), pWH-P_{emaS}G (red triangle), and pWH-P_{emaR}G (blue inverted triangle) were grown in final concentrations of (**A**,**B**) 5 mM citrate; (**C**,**D**) 5 mM citrate and 0.5% ethanol; and (**E**,**F**) 5 mM citrate and 20 mM potassium acetate as carbon sources. Fluorescence expression (**A**,**C**,**E**) and growth curves (**B**,**D**,**F**) are shown. *** $p < 0.0001$, multiple *t*-test, derived from the comparison of strains with promoter and empty vector pWH1266G. RFU, relative fluorescence units. OD_{600}, optical density. Data were collected from at least three replicates.

3.2. EmaSR Regulates DJ41_2796 and DJ41_3218 in Acetate Culture Conditions

In previous research, it was discovered that EmaSR may be involved not only in the regulation of ethanol metabolism, but also in acetate metabolism [18]. To confirm whether EmaSR regulates acetate metabolism via regulons, the fluorescence expression of EmaSR regulons was observed under low-concentration acetate culture conditions. After 24 h, fluorescence expression levels of EmaSR regulons were observed to be 3-fold higher than the P_{gfp} control after 24 h, and the fluorescence expression for P_{3218} reached 5-fold higher than P_{gfp} (Figure 2E). A growth curve analysis confirmed no significant differences in the growth of each strain (Figure 2F). Comparable fluorescence expression levels of P_{emaR} in acetate culture conditions to ethanol culture conditions were observed, with the fluorescence expression of P_{emaR} significantly increased; however, there was a significant decrease in the fluorescence expression of P_{emaS} compared to P_{gfp} (Figure 2E), suggesting that *emaR* can be induced in the presence of acetate, but *emaS* cannot.

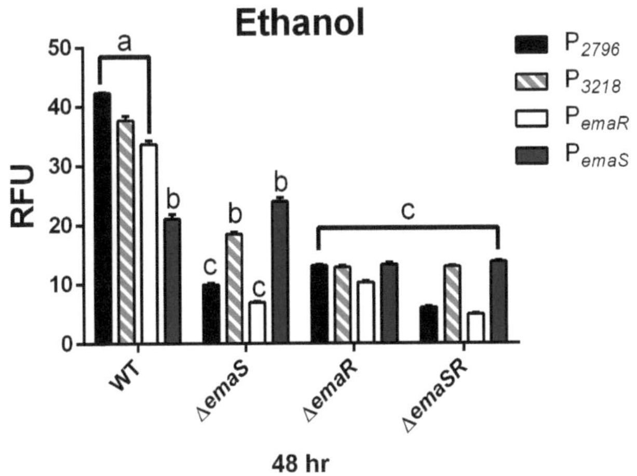

Figure 3. EmaSR regulon mean fluorescence expression levels in different *A. baumannii* strains cultured in low-concentration ethanol after 48 h. Changes in fluorescence levels at 48 h were observed in wild-type, Δ*emaS*, Δ*emaR*, and Δ*emaSR* strains of *A. baumannii*, each carrying the pWH-P_{2796}G (black), pWH-P_{3218}G (gray slash), pWH-P_{emaR}G (white) or pWH-P_{emaS}G (dark gray) reporter plasmid. Strains were cultured with 0.5% ethanol. The a, b, and c labels indicate the comparisons for which $p < 0.0001$, multiple *t*-test. RFU, relative fluorescence units. Data were derived from at least three independent experiments.

Fluorescence expression levels of P_{2796} and P_{3568} were observed in both wild-type and mutant *A. baumannii* strains, and as in ethanol-containing cultures, there was no significantly elevated fluorescence expression observed for P_{2796} and P_{3568} in all mutants. During the 24 h culture period, the fluorescence expression levels of P_{2796} and P_{3218} in the wild-type strain were both 4-fold higher than that of the mutant strains (Figure 4 black and gray slash, Figures S8C and S9C). The growth curve analysis also confirmed that this difference was not due to growth differences (Figure 4 black and gray slash, Figures S8D and S9D). This indicates that in low-concentration acetate culture conditions, EmaSR indeed regulates the expression of EmaSR regulons.

3.3. *EmaR Enhances the Expression of P_{emaS} and P_{emaR} in Acetate*

TCS are known to not only activate the regulation of downstream genes but also to control the expression of their own promoters. Previous results indicated significantly elevated expression of P_{emaR} under low-concentration ethanol culture conditions [18], but here it was found that in a similar environment, P_{emaR} was barely expressed in all three mutant strains (Figure 3 white, Figure S10A). This difference was shown to be unaffected by growth (Figure S10B) and confirmed that the expression of P_{emaR} in low-concentration ethanol environments is also regulated by EmaSR. Similar results were observed under low-concentration acetate culture conditions (Figure 4 white, Figure S10C), and the findings were not influenced by differences in growth (Figure S10D).

By contrast, P_{emaS} showed no significant increase in expressions in both ethanol and acetate, and even had a significantly lower expression in the acetate environment compared to the P_{gfp} control. However, it was observed that in the Δ*emaR* and Δ*emaSR* mutants, fluorescence expression of P_{emaS} was almost half that of wild type and Δ*emaS* in low concentrations of ethanol (Figure 3 dark gray, Figure S11A). There was no significant variation in tye fluorescence expression levels for all strains in the acetate-containing medium (Figure 4 dark gray, Figure S11C). The growth curve analysis confirmed that this result was not due to differences in growth rates (Figure S11B,D). These results show that although the expression of P_{emaS} was not induced by ethanol or acetate, its expression was still regulated by EmaR, suggesting that in strains with intact *emaR*, the expression of P_{emaS} can still be enhanced. Additionally, although not significantly different, slightly higher expression levels of P_{emaS} were observed in Δ*emaS* under both low ethanol and acetate conditions, as compared to the wild type. There may be other regulatory factors that suppress P_{emaS} expression, of which are present in the wild-type strain but not Δ*emaS*.

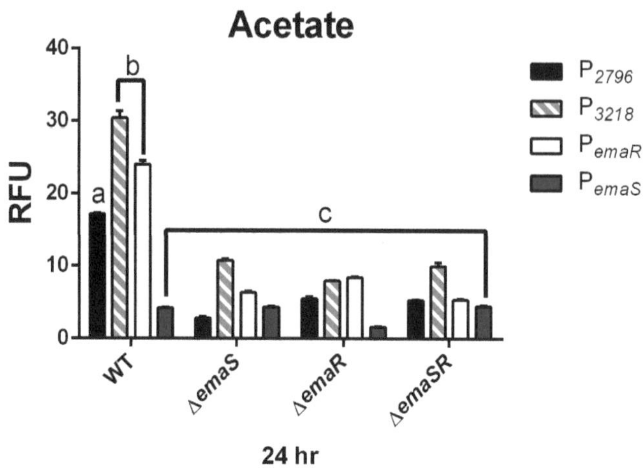

Figure 4. EmaSR regulon mean fluorescence expression levels in different *A. baumannii* strains cultured in low-concentration actetae after 24 h. Changes in fluorescence levels at 24 h were observed in wild-type, Δ*emaS*, Δ*emaR*, and Δ*emaSR* strains of *A. baumannii*, each carrying the pWH-P_{2796}G (black), pWH-P_{3218}G (gray slash), pWH-P_{emaR}G (white) or pWH-P_{emaS}G (dark gray) reporter plasmid. Strains were cultured with 20 mM acetate. The a, b, and c labels indicate the comparisons for which $p < 0.0001$, multiple *t*-test. RFU, relative fluorescence units. Data were derived from at least three independent experiments.

3.4. DJ41_2796 Possesses Acetate: Succinyl-CoA Transferase Enzymatic Activity

Both transcriptome analysis and reporter assays confirmed the potentially significant role of DJ41_2796 in EmaSR-regulated ethanol and acetate metabolism, and a functional prediction using the amino acid sequence suggests that DJ41_2796 may have ASCT activity. The homologous genes of this enzyme have been proven to play a crucial role in the metabolic regulation of MxtR/ErdR [13]. The enzymatic activity of DJ41_2796 was subsequently assessed through interaction with potassium acetate and succinyl-CoA, to ascertain if acetyl-CoA and succinate would be produced. DJ41_2796 was expressed in *E. coli*, using the vector pQE80LK-DJ41_2796, and was purified through nickel affinity chromatography. Different concentrations of DJ41_2796 were tested, and as concentrations increased from 2 μM to 10 μM, the reaction rate also increased from 0.9611 U/g to 4.9823 U/g (Figure 5). This confirmed that DJ41_2796 possesses ASCT enzymatic activity to convert potassium acetate and succinyl-CoA into acetyl-CoA and succinate.

3.5. Kinetic Characterization of DJ41_2796

To better understand the enzymatic performance of DJ41_2796, reactions were tested under different pH and temperature conditions. It was observed that DJ41_2796 had the highest reaction rate (34.9 U/g) in a pH 8.0 phosphate buffer; however, in pH 8.0 Tris-Cl buffer, reaction rates were lower despite being at the same pH. Additionally, a decrease in reaction rates was noted in conditions with the pH exceeding 8.0 (Figure 6A). Subsequently, all enzymatic reactions with DJ41_2796 were carried out in the pH 8.0 phosphate buffer. Interestingly, it was noted that as the temperature increased from 4 °C to 55 °C, reaction rates also increased from 1.4 U/g to 21.7 U/g (Figure 6B), suggesting that DJ41_2796 may exhibit better reaction activity at higher temperatures.

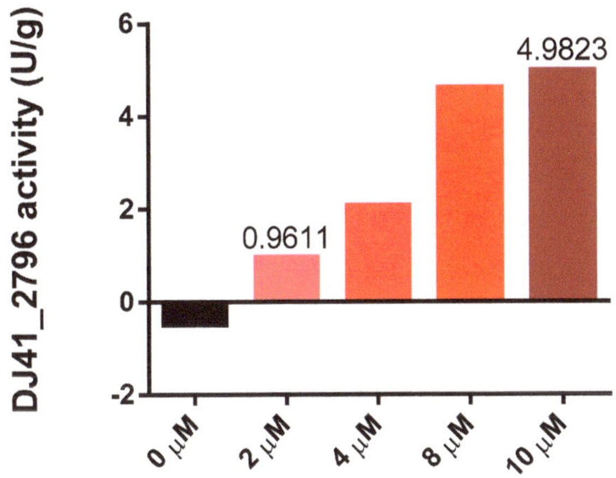

Figure 5. Enzymatic activity of DJ41_2796 at different protein concentrations. The y–axis represents the enzymatic activity of DJ41_2796 within one minute after subtracting the background value (U/g), and the x–axis represents the protein concentration of DJ41_2796. The reaction mixture contains a final concentration of 10 mM potassium acetate, 0.2 mM succinyl-CoA, 0.1 mM DTNB, and 0 to 10 μM DJ41_2796. The results were obtained from one experiment.

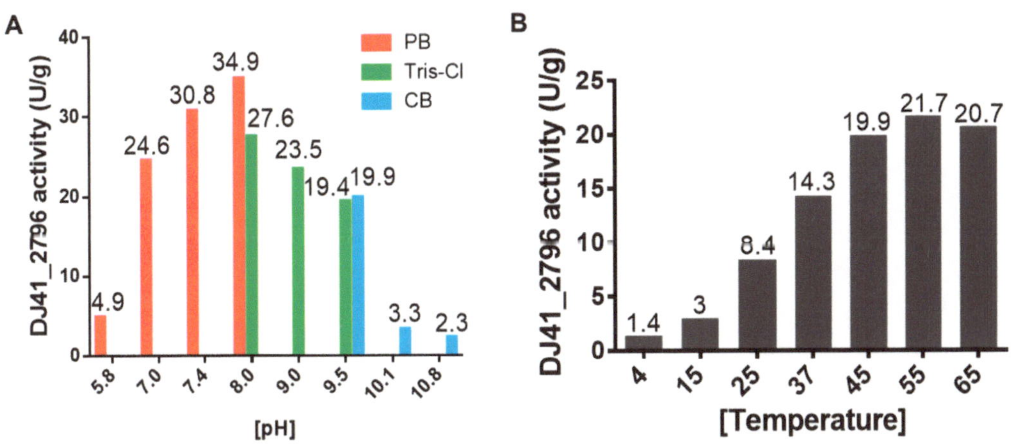

Figure 6. Assessing the optimal enzymatic activity of DJ41_2796. (**A**) Enzymatic activity of DJ41_2796 in buffer solutions with different pH values. Red bars represent phosphate buffer (PB), green bars represent Tris-HCl buffer (Tris-Cl), and blue bars represent carbonate–bicarbonate buffer (CB). The y–axis indicates the DJ41_2796 specific activity (U/g), and the numbers on top of each bar represent the activity of DJ41_2796 in that environment. (**B**) Enzymatic activity of DJ41_2796 under different temperature conditions, with the x–axis representing tested reaction temperatures (°C), and the y–axis indicating DJ41_2796 specific activity (U/g). Each number above the bars represents the enzymatic activity of DJ41_2796 in that environment. The results were obtained from one experiment.

The enzyme kinetics of DJ41_2796 were assessed, using either potassium acetate or sodium acetate as substrates, in order to ascertain whether there was a difference in affinity for these two types of acetate. The K_M for potassium acetate was found to be approximately twice as high as that for sodium acetate, indicating that a higher concentration of potassium acetate is needed to reach half of V_{max}. However, a further assessment of k_{cat}/K_M revealed no significant difference in the reaction efficiency between these two substrates. This suggests that DJ41_2796 does not exhibit a significant difference in affinity for these two substrates (Table 5). The enzyme kinetics of DJ41_2796 with succinyl-CoA were also assessed, and it was observed that DJ41_2796 exhibited greater ASCT activity compared to a homologous protein in *Acetobacter aceti* (Table 5).

Table 5. Enzyme kinetics of DJ41_2796 for different substrates in pH 8.0 phosphate buffer.

	Potassium Phosphate Buffer, pH = 8.0		
	Potassium Acetate *	**Sodium Acetate ***	**Succinyl-CoA ***
V_{max} (nmole·min^{-1})	41.55	33.51	34.97
K_M (mM)	39	14.74	0.3418
k_{cat} (s^{-1})	8.31	6.702	6.994
k_{cat}/K_M (mM^{-1}s^{-1})	0.2131	0.4547	20.4623

* The results were obtained from one experiment.

4. Discussion

4.1. Expression of Emas in Ethanol and Acetate May Be Inhibited by Other Regulatory Proteins

Previous research showed that EmaSR plays a role in *A. baumannii* ethanol metabolism [18]. In this study, the wild-type strain exhibited an increase in the expression levels of various promoters, including P_{emaR} and the promoter of EmaSR regulons, in ethanol-containing environments. In acetate-containing environments, there was also an upward trend in expression levels, except for P_{emaS}. Furthermore, it was observed that expression levels of P_{emaS} in Δ*emaS* were higher than in the wild-type strain, regardless of whether the environment contained ethanol or acetate. In the absence of EmaS to transfer phosphate, EmaR may regulate downstream genes by receiving phosphate from other factors. There may also be other factors that act to inhibit EmaS in the wild-type strain, which could result in higher expression of P_{emaS} in Δ*emaS* compared to the wild type. A similar scenario has been observed in *P. aeruginosa*, where it was found that the regulatory protein HapZ interacts with the REC domain of the sensor protein SagS when bound to cyclic di-GMP, thereby inhibiting phosphate transfer between SagS and downstream proteins [26]. Through SMART (Simple Modular Architecture Research Tool) analysis of its secondary structure, we found that the sensor protein EmaS also possesses a REC domain within its C-terminal structure, which can receive signals from sensor proteins and bind with DNA. Therefore, based on the functions predicted in the NCBI database, EmaS is also annotated as a hybrid sensor histidine kinase/response regulator, similar to SagS. From this, it can be inferred that EmaS may also have the ability to interact with other regulatory proteins, and in ethanol- or acetate-containing environments, we hypothesize that other factors may be involved in the suppression of signal transduction from EmaS to EmaR, and these may act via binding to the REC structural domain of EmaS.

4.2. DJ41_2796 Plays a Crucial Role in the Ethanol and Acetate Metabolism of A. baumannii ATCC19606

In a study on the toxicity of *A. pasteurianus* against acetate accumulation, it was confirmed that the expression levels of four proteins increased when exposed to high concentrations of acetate [27]. Overall, three of these proteins were annotated as TCS sensor proteins (two from the NtrB family and one from the SsrA family), and one was annotated as a BaeS and OmpR family regulatory protein [27]. It was also found that the expression of *ackA* (acetate kinase), *acs* (acetyl-CoA synthetase), *pta* (phosphate acetyltransferase), and *aarC* (ASCT) genes increased in high acetate environments [27]. Although it has not been confirmed that TCSs can directly regulate these genes, the results are indicative of the important role that TCSs and genes involved in acetate metabolism have with regard to *A. pasteurianus* resistance against acetic acid stress [27,28]. Interestingly, succinyl-CoA synthetase, which can convert succinyl-CoA into succinate and acts in a similar way to ASCT, is retained in many *Proteobacteria* species, as revealed by genome analysis; however, unlike *Proteobacteria*, most bacteria in the *Actinobacteria* and *Bacteroidetes* phyla predominantly possess only the gene for ASCT [29]. This suggests that ASCT-only bacteria may be using the ASCT enzyme to replace succinyl-CoA synthetase [29,30]. Genomic analysis of *A. aceti* has shown that the species lacks succinyl-CoA synthetase, and previous research showed that AarC6 in *A. aceti* may have replaced the function of succinyl-CoA synthetase [31]. DJ41_2796 and AarC in *A. aceti* exhibit the same enzymatic activity, and although the genome of *A. baumannii* ATCC 19606 retains a gene annotated as succinyl-CoA synthetase (*DJ41_3576*), transcriptional analysis of EmaSR confirmed that this gene is not regulated by EmaSR in low concentrations of ethanol [18]. While it is not clear whether DJ41_3576 functions as a succinyl-CoA synthetase, the results of this study affirm that DJ41_2796 may take on an important role in ethanol and acetate metabolism, replacing succinyl-CoA synthetase.

4.3. EmaSR May Influence the Pyruvate Metabolism by Regulating the Metabolism of Ethanol and Acetate

Previous research has demonstrated that in *Bacillus subtilis*, the enzyme MaeA is highly expressed only in the presence of malate, which is homologous to DJ41_3218. The study confirmed that *maeA* is regulated by YufL/YufM, which belongs to the CitA/CitB family of TCS and shares homologous proteins with CitA/CitB in *E. coli* and YdbF/YdbG in *K. pneumoniae* [32]. These TCSs have been previously shown to regulate the expression of genes involved in the citric acid (TCA) cycle. It was also confirmed that MaeA possesses enzymatic activity as a malate dehydrogenase, thereby substantiating the involvement of YufL/YufM in the regulation of the *B. subtilis* TCA cycle [32]. Subsequent studies further confirmed that LytS/LytT can increase the expression of the pyruvate transporter protein PftAB. However, when MaeA causes an accumulation of pyruvate within the bacterial cells, this inhibits LytS/LytT, thereby decreasing the expression of *pftAB* and preventing the loss of pyruvate from secondary carbon sources when bacteria deplete their preferred carbon sources, such as glucose and malate [33]. In this study, although the enzymatic function of *DJ41_3218* has not been confirmed, it is annotated to participate in pyruvate metabolism. The functional annotation of *DJ41_568-571* also indicates its involvement in pyruvate metabolism. These genes exhibit increased expression in *A. baumannii* when ethanol is used as a carbon source, and the results resemble the regulation observed in *B. subtilis* [33]. *A. baumannii* metabolizes ethanol and acetate as carbon sources to produce pyruvate, and when ethanol and acetate in the environment are depleted, this may upregulate pyruvate metabolism. A previous study showed that *B. subtilis* regulates pyruvate metabolism through two pairs of TCSs [32], and transcriptional analysis in our previous research [18] also found that more than ten genes annotated with regulatory protein functions are under the control of EmaSR [18]. Therefore, EmaSR may play an important regulatory role not just in ethanol and acetate metabolism but can also influence other metabolic processes.

4.4. EmaSR Regulation of Pyruvate Metabolism and the Relationship with Virulence

In this study, we observed that in ethanol-containing environments, the expression of P_{2796} is higher than that of P_{3218}. Conversely, in acetate-containing environments, the expression of P_{3218} is significantly higher than that of P_{571}, P_{2796} and P_{3568}. *DJ41_3218* is annotated as an NAD-dependent malic enzyme, primarily capable of converting malate into pyruvate. Previous research has also confirmed that this enzyme catalyzes the conversion of oxaloacetate to pyruvate in *E. coli* and *Lactobacillus arabinosus* [34]. On the other hand, within the EmaSR regulated gene cluster *DJ41_566-571*, the three genes *DJ41_568*, *DJ41_569*, and *DJ41_571* have been respectively annotated to encode dihydrolipoyl dehydrogenase, 2-oxo acid dehydrogenase subunit E2, and thiamine pyrophosphate (TPP)-dependent dehydrogenase E1, respectively. These three enzymes collectively work in the metabolic pathway to convert pyruvate into acetyl-CoA. In previous studies, it was found that *P. putida* KT2440 lost the ability to grow by using pyruvate as a single carbon source when *mxtR* and *erdR* were deleted [13]. Furthermore, YufL/YufM also participates in the regulation of pyruvate metabolism in *B. subtilis* [33]. Previous studies have already demonstrated that MxtR/ErdR and YufL/YufM can regulate pyruvate metabolism. Therefore, we speculate that EmaSR may regulate pyruvate metabolism via DJ41_3218.

Moreover, previous research has shown that the deletion of enzymes in the pyruvate metabolism pathway does not lead to a reduction in adhesins, but decreased bacterial virulence in *Streptococcus pneumoniae* [35]. In this study, it is suggested that this decrease may be indirectly caused by a lack of energy. Our previous study has also found that the loss of *emaS* and *emaR* led to a decrease in the virulence of *A. baumannii* when infecting *Galleria mellonella* larvae [18]. Taken together, it is possible that EmaSR may indirectly affect virulence by regulating pyruvate metabolism in *A. baumannii*.

5. Conclusions

In this study, we found that EmaSR increased the expression levels of *DJ41_566-571*, *DJ41_2796*, *DJ41_3218*, and *DJ41_3568*, as well as *emaR*, in ethanol- and acetate-containing environments. Additionally, biochemical functional analysis confirmed the role of DJ41_2796 in the carbon metabolic pathway. Results from the reporter assays showed that EmaSR increased the expression of DJ41_2796, leading to the conversion of more acetate to acetyl-CoA in the ethanol. These results furthered our understanding of EmaSR regulation in *A. baumannii*. When strains were cultured in a low-concentration ethanol or acetate environment, EmaSR was able to increase the expression of DJ41_2796, aiding in the metabolism of ethanol and acetate. This enables *A. baumannii* to increase the survival in ethanol and acetate while utilizing them as carbon sources for growth.

Supplementary Materials: The following supporting information can be downloaded at: https://www.mdpi.com/article/10.3390/microorganisms12020331/s1, Figure S1: Map of the TCSG plasmid indicating genes and restriction enzyme sites; Figure S2: Confirmation of *E. coli* transformation strains carrying promoter-TCSG; Figure S3: Map of the pWH1266G plasmid; Figure S4: Confirmation of *E. coli* transformation strains carrying pWH1666-derived reporter plasmids; Figure S5: PCR product analysis by 2% agarose gel electrophoresis; Figure S6: Analysis of expressed and purified DJ41_2796 protein by 10% SDS-PAGE; Figure S7: Mean expression levels of EmaSR regulons in *E. coli* cultured in LB containing 1% ethanol; Figure S8: Mean P_{2796} fluorescence expression levels in different *A. baumannii* strains cultured in low-concentration ethanol and acetate; Figure S9: Mean P_{3218} fluorescence expression levels in different *A. baumannii* strains cultured in low-concentration ethanol and acetate; Figure S10: Mean P_{emaR} fluorescence expression levels in different *A. baumannii* strains cultured in low-concentration ethanol and acetate; Figure S11: Mean P_{emaS} fluorescence expression levels in different *A. baumannii* strains cultured in low-concentration ethanol and acetate.

Author Contributions: Conceptualization, H.-Y.S. and G.-H.L.; methodology, H.-Y.S. and G.-H.L.; software, H.-Y.S.; validation, Y.-W.H., H.-Y.S. and G.-H.L.; formal analysis, H.-Y.S. and G.-H.L.; investigation, Y.-W.H. and G.-H.L.; resources, G.-H.L.; data curation, Y.-W.H.; writing—original draft preparation, Y.-W.H. and G.-H.L.; writing—review and editing, H.-Y.S., Y.-W.H. and G.-H.L.; visualization, Y.-W.H. and G.-H.L.; supervision, G.-H.L.; project administration, G.-H.L.; funding acquisition, G.-H.L. All authors have read and agreed to the published version of the manuscript.

Funding: This study was funded by the Taiwan National Science and Technology Council under Grant No. MOST111-2635B320-001-MY2 to Guang-Huey Lin, and by the Buddhist Tzu Chi Medical Foundation under Grant No. TCMF11139A to Guang-Huey Lin.

Data Availability Statement: The data presented in this study are available upon reasonable request from the corresponding author.

Acknowledgments: The authors are grateful for the support from the Core Facility Center, Tzu Chi University.

Conflicts of Interest: The authors declare no conflicts of interest. The funders had no role in the design of the study; in the collection, analyses, or interpretation of data; in the writing of the manuscript; or in the decision to publish the results.

References

1. Lo, E.A.-G.; Law, L.S.-C.; Tan, K.; Ashokka, B. A review of the science and clinical use of alcohol-based hand rubs. *Int. J. Infect. Control* **2022**, *18*, 20611. [CrossRef]
2. Yeung, Y.W.S.; Ma, Y.; Liu, S.Y.; Pun, W.H.; Chua, S.L. Prevalence of alcohol-tolerant and antibiotic-resistant bacterial pathogens on public hand sanitizer dispensers. *J. Hosp. Infect.* **2022**, *127*, 26–33. [CrossRef]
3. Kampf, G. Ethanol. In *Antiseptic Stewardship: Biocide Resistance and Clinical Implications*; Kampf, G., Ed.; Springer: Cham, Switzerland, 2018; pp. 9–35.
4. Tashiro, Y.; Inagaki, A.; Ono, K.; Inaba, T.; Yawata, Y.; Uchiyama, H.; Nomura, N. Low concentrations of ethanol stimulate biofilm and pellicle formation in *Pseudomonas aeruginosa*. *Biosci. Biotechnol. Biochem.* **2014**, *78*, 178–181. [CrossRef]
5. Reid, M.F.; Fewson, C.A. Molecular characterization of microbial alcohol dehydrogenases. *Crit. Rev. Microbiol.* **1994**, *20*, 13–56. [CrossRef]
6. Shortall, K.; Djeghader, A.; Magner, E.; Soulimane, T. Insights into Aldehyde Dehydrogenase Enzymes: A Structural Perspective. *Front. Mol. Biosci.* **2021**, *8*, 659550. [CrossRef]
7. Xiong, R.G.; Zhou, D.D.; Wu, S.X.; Huang, S.Y.; Saimaiti, A.; Yang, Z.J.; Shang, A.; Zhao, C.N.; Gan, R.Y.; Li, H.B. Health Benefits and Side Effects of Short-Chain Fatty Acids. *Foods* **2022**, *11*, 2863. [CrossRef]
8. Machado, M.G.; Patente, T.A.; Rouillé, Y.; Heumel, S.; Melo, E.M.; Deruyter, L.; Pourcet, B.; Sencio, V.; Teixeira, M.M.; Trottein, F. Acetate Improves the Killing of *Streptococcus pneumoniae* by Alveolar Macrophages via NLRP3 Inflammasome and Glycolysis-HIF-1α Axis. *Front. Immunol.* **2022**, *13*, 773261. [CrossRef]
9. El-Mansi, M. Control of central metabolism's architecture in *Escherichia coli*: An overview. *Microbiol. Res.* **2023**, *266*, 127224. [CrossRef]
10. Mutyala, S.; Kim, J.R. Recent advances and challenges in the bioconversion of acetate to value-added chemicals. *Bioresour. Technol.* **2022**, *364*, 128064. [CrossRef]
11. Hempel, N.; Görisch, H.; Mern, D.S. Gene *ercA*, encoding a putative iron-containing alcohol dehydrogenase, is involved in regulation of ethanol utilization in *Pseudomonas aeruginosa*. *J. Bacteriol.* **2013**, *195*, 3925–3932. [CrossRef] [PubMed]
12. Badal, D.; Jayarani, A.V.; Kollaran, M.A.; Prakash, D.; Monisha, P.; Singh, V. Foraging Signals Promote Swarming in Starving *Pseudomonas aeruginosa*. *mBio* **2021**, *12*, e0203321. [CrossRef] [PubMed]
13. Henríquez, T.; Hsu, J.S.; Hernandez, J.S.; Kuppermann, S.; Eder, M.; Jung, H. Contribution of Uncharacterized Target Genes of MxtR/ErdR to Carbon Source Utilization by *Pseudomonas putida* KT2440. *Microbiol. Spectr.* **2023**, *11*, e0292322. [CrossRef] [PubMed]
14. Casella, L.G.; Torres, N.J.; Tomlinson, B.R.; Shepherd, M.; Shaw, L.N. The novel two-component system AmsSR governs alternative metabolic pathway usage in *Acinetobacter baumannii*. *Front. Microbiol.* **2023**, *14*, 1139253. [CrossRef] [PubMed]
15. Wang, L.; Huang, X.; Jin, Q.; Tang, J.; Zhang, H.; Zhang, J.R.; Wu, H. Two-Component Response Regulator OmpR Regulates Mucoviscosity through Energy Metabolism in *Klebsiella pneumoniae*. *Microbiol. Spectr.* **2023**, *11*, e0054423. [CrossRef] [PubMed]
16. Wittekind, M.A. The Small Protein ScrA Is a Novel Regulator of *Staphylococcus aureus* Virulence Acting as an Intermediary between the ArlRS and SaeRS Two-Component Systems. Ph.D. Dissertation, Ohio University, Athens, OH, USA. Available online: https://www.proquest.com/dissertations-theses/small-protein-scra-is-novel-regulator-em/docview/2856749705/se-2 (accessed on 5 May 2023).
17. Lin, G.H.; Hsieh, M.C.; Shu, H.Y. Role of Iron-Containing Alcohol Dehydrogenases in *Acinetobacter baumannii* ATCC 19606 Stress Resistance and Virulence. *Int. J. Mol. Sci.* **2021**, *22*, 9921. [CrossRef] [PubMed]
18. Shu, H.Y.; Huang, Y.W.; Tsai, P.Y.; Hsieh, K.S.; Lin, G.H. Role of EmaSR in Ethanol Metabolism by *Acinetobacter baumannii*. *Int. J. Mol. Sci.* **2022**, *23*, 12606. [CrossRef]
19. Hunger, M.; Schmucker, R.; Kishan, V.; Hillen, W. Analysis and nucleotide sequence of an origin of DNA replication in *Acinetobacter calcoaceticus* and its use for *Escherichia coli* shuttle plasmids. *Gene* **1990**, *87*, 45–51. [CrossRef]
20. Bouvet, P.J.M.; Grimont, P.A.D. Taxonomy of the genus *Acinetobacter* with the recognition of *Acinetobacter baumannii* sp. nov., *Acinetobacter haemolyticus* sp. nov., *Acinetobacter johnsonii* sp. nov., and *Acinetobacter junii* sp. nov. and emended descriptions of *Acinetobacter calcoaceticus* and *Acinetobacter lwoffii*. *Int. J. Syst. Evol. Microbiol.* **1986**, *36*, 228–240.
21. Chen, T.L.; Siu, L.K.; Wu, R.C.; Shaio, M.F.; Huang, L.Y.; Fung, C.P.; Lee, C.M.; Cho, W.L. Comparison of one-tube multiplex PCR, automated ribotyping and intergenic spacer (ITS) sequencing for rapid identification of *Acinetobacter baumannii*. *Clin. Microbiol. Infect.* **2007**, *13*, 801–806. [CrossRef]
22. Crameri, A.; Whitehorn, E.A.; Tate, E.; Stemmer, W.P. Improved green fluorescent protein by molecular evolution using DNA shuffling. *Nat. Biotechnol.* **1996**, *14*, 315–319. [CrossRef]
23. Mullins, E.A.; Kappock, T.J. Crystal structures of *Acetobacter aceti* succinyl-coenzyme A (CoA): Acetate CoA-transferase reveal specificity determinants and illustrate the mechanism used by class I CoA-transferases. *Biochemistry* **2012**, *51*, 8422–8434. [CrossRef] [PubMed]

24. Mochizuki, K.; Inaoka, D.K.; Mazet, M.; Shiba, T.; Fukuda, K.; Kurasawa, H.; Millerioux, Y.; Boshart, M.; Balogun, E.O.; Harada, S.; et al. The ASCT/SCS cycle fuels mitochondrial ATP and acetate production in *Trypanosoma brucei*. *Biochim. Biophys. Acta Bioenerg.* **2020**, *1861*, 148283. [CrossRef] [PubMed]
25. Eyer, P.; Worek, F.; Kiderlen, D.; Sinko, G.; Stuglin, A.; Simeon-Rudolf, V.; Reiner, E. Molar absorption coefficients for the reduced Ellman reagent: Reassessment. *Anal. Biochem.* **2003**, *312*, 224–227. [CrossRef] [PubMed]
26. Römling, U.; Cao, L.Y.; Bai, F.W. Evolution of cyclic di-GMP signalling on a short and long term time scale. *Microbiology* **2023**, *169*, 001354. [CrossRef] [PubMed]
27. Xia, K.; Han, C.; Xu, J.; Liang, X. Transcriptome response of *Acetobacter pasteurianus* Ab3 to high acetic acid stress during vinegar production. *Appl. Microbiol. Biotechnol.* **2020**, *104*, 10585–10599. [CrossRef]
28. Yang, H.; Chen, T.; Wang, M.; Zhou, J.; Liebl, W.; Barja, F.; Chen, F. Molecular biology: Fantastic toolkits to improve knowledge and application of acetic acid bacteria. *Biotechnol. Adv.* **2022**, *58*, 107911. [CrossRef]
29. Kwong, W.K.; Zheng, H.; Moran, N.A. Convergent evolution of a modified, acetate-driven TCA cycle in bacteria. *Nat. Microbiol.* **2017**, *2*, 17067. [CrossRef]
30. Zhang, B.; Lingga, C.; Bowman, C.; Hackmann, T.J. A New Pathway for Forming Acetate and Synthesizing ATP during Fermentation in Bacteria. *Appl. Environ. Microbiol.* **2021**, *87*, e0295920. [CrossRef]
31. Arai, H.; Sakurai, K.; Ishii, M. Metabolic Features of *Acetobacter aceti*. In *Acetic Acid Bacteria: Ecology and Physiology*; Matsushita, K., Toyama, H., Tonouchi, N., Okamoto-Kainuma, A., Eds.; Springer: Tokyo, Japan, 2016; pp. 255–271.
32. Doan, T.; Servant, P.; Tojo, S.; Yamaguchi, H.; Lerondel, G.; Yoshida, K.I.; Fujita, Y.; Aymerich, S. The *Bacillus subtilis ywkA* gene encodes a malic enzyme and its transcription is activated by the YufL/YufM two-component system in response to malate. *Microbiology* **2003**, *149*, 2331–2343. [CrossRef]
33. Charbonnier, T.; Le Coq, D.; McGovern, S.; Calabre, M.; Delumeau, O.; Aymerich, S.; Jules, M. Molecular and Physiological Logics of the Pyruvate-Induced Response of a Novel Transporter in *Bacillus subtilis*. *mBio* **2017**, *8*, e00976-17. [CrossRef] [PubMed]
34. Yamaguchi, M. Studies on regulatory functions of malic enzymes. IV. Effects of sulfhydryl group modification on the catalytic function of NAD-linked malic enzyme from *Escherichia coli*. *J. Biochem.* **1979**, *86*, 325–333. [CrossRef] [PubMed]
35. Echlin, H.; Frank, M.; Rock, C.; Rosch, J.W. Role of the pyruvate metabolic network on carbohydrate metabolism and virulence in *Streptococcus pneumoniae*. *Mol. Microbiol.* **2020**, *114*, 536–552. [CrossRef] [PubMed]

microorganisms

MDPI

Article

The LiaSR Two-Component System Regulates Resistance to Chlorhexidine in *Streptococcus mutans*

Shan Huang [1,2], Jing Huang [1], Jingyun Du [1], Yijun Li [1], Minjing Wu [1], Shuai Chen [1], Ling Zhan [3,*] and Xiaojing Huang [1,*]

1 Fujian Key Laboratory of Oral Diseases, Fujian Provincial Engineering Research Center of Oral Biomaterial, Stomatological Key Lab of Fujian College and University, School and Hospital of Stomatology, Fujian Medical University, 246 Yangqiao Zhong Road, Fuzhou 350002, China; shanhuangfjmu@163.com (S.H.); huangjing56@163.com (J.H.); dujingyun2003@163.com (J.D.); liyijun1923@163.com (Y.L.); minjingwu@foxmail.com (M.W.); drchenshuai@163.com (S.C.)

2 Department of Stomatology, Zhongshan Hospital Affiliated to Xiamen University, Xiamen 361004, China

3 Division of Pediatric Dentistry, Department of Orofacial Sciences, University of California, San Francisco, CA 94143, USA

* Correspondence: ling.zhan@ucsf.edu (L.Z.); xiaojinghuang@fjmu.edu.cn (X.H.)

Abstract: Chlorhexidine (CHX) is widely considered to be the gold standard for preventing dental caries. However, it is possible to induce resistance to CHX. The LiaSR two-component system has been identified that contributed to CHX resistance in *Streptococcus mutans*, which is one of the major pathogens in dental caries. However, the underlying mechanisms remain unclear. In this study, an MIC assay and a viability assessment demonstrated that after deleting the *liaS* and *liaR* genes, the sensitivity of mutants could increase. The Nile Red efflux assay exhibited that the efflux rates of mutants were significantly decreased. The RT-qPCR results indicated that the LiaSR two-component system-mediating influence on the expression of *lmrB* in *S. mutans* contributed to the efflux rate. The hydrophobicity assay and membrane potential assay showed that the mutants had higher levels of hydrophobicity and depolarization, suggesting that their membranes were more easily disturbed. The TEM graphs revealed that the border of the cell membrane was unclear in mutants compared with the wild-type strain, indicating that the cell envelope's stress response may have been inhibited. While the surface charge of mutants showed no significant difference in the wild-type strain according to the result of cytochrome c-based charged determination. This study provides valuable novel insights into the mechanisms of the LiaSR two-component system in the CHX resistance of *S. mutans*.

Keywords: *Streptococcus mutans*; LiaSR two-component system; chlorhexidine resistance; efflux pump; cells envelope stress responses

Citation: Huang, S.; Huang, J.; Du, J.; Li, Y.; Wu, M.; Chen, S.; Zhan, L.; Huang, X. The LiaSR Two-Component System Regulates Resistance to Chlorhexidine in *Streptococcus mutans*. *Microorganisms* **2024**, *12*, 468. https://doi.org/10.3390/microorganisms12030468

Academic Editor: Tomohiro Shimada

Received: 30 January 2024
Revised: 14 February 2024
Accepted: 16 February 2024
Published: 26 February 2024

1. Introduction

Controlling microbial biofilm infections is a global challenge [1,2]. This is partly because the bacteria are embedded in the extracellular matrix, which can provide bacteria with a protective environment and promote the bacteria tightly adhering to the interface [3,4]. Dental caries is one of the most prevalent oral infectious diseases. The accumulation of biofilms on teeth can lead to the destruction of dental tissue. *Streptococcus mutans* (*S. mutans*) is one of the etiological agents of dental caries due to its strong capacity for the synthesis of exopolysaccharides, acid production and acid tolerance [5–7].

In clinical dental practice, mechanical methods are not sufficient for the removal of biofilms. Therefore, chlorhexidine (CHX) is widely applied because of its broad-spectrum antibacterial effect. CHX is a bis-biguanide agent that carries a positive charge and can be attracted to the negatively charged cell surface, resulting in membrane disruption and leakage of the cell's contents [8]. It is considered to be the gold standard for dental caries prevention [8].

Although CHX is considered relatively safe, its side effects cannot be ignored, particularly with regard to resistance. Studies have confirmed that continuous exposure to CHX can induce CHX resistance in *S. mutans* [9,10]. In recent years, researchers have focused on developing CHX-incorporated materials to prevent dental caries [11,12]. The release of CHX molecules may increase the exposure time of *S. mutans*, leading to CHX resistance. The emergence of CHX-resistant *S. mutans* has presented new challenges for dental caries treatment and prevention because the cariogenicity of CHX-resistant *S. mutans* appears to be similar to that of the wild-type strains [9], indicating that even under CHX pressure on *S. mutans*, resistant strains can still play an important role in dental caries. However, the related mechanisms of CHX resistance remain unclear, which is crucial for the development of CHX-incorporated materials.

The LiaSR two-component system in *S. mutans* comprises a membrane-bound histidine kinase receptor (LiaS) and a cognate response regulator (LiaR) [13–15]. When the LiaS senses the environmental changes, it undergoes autophosphorylation and then transfers the phosphoryl group to the LiaR, altering the LiaR's affinity to bind to the promoter regions of the target genes and regulate their expression [16]. When the stress disappears, the LiaF can "switch off" this system [16]. The LiaSR two-component system in this study was found to be homologous to the LiaSR two-component system in *Bacillus subtilis* (*B. subtilis*) [17] and the VraSR two-component system in *Staphylococcus aureus* (*S. aureus*) [18], which has been linked to antiseptic resistance. In *S. aureus*, the VraSR two-component system can be activated by CHX to modulate cell envelope stress responses (CESRs), contributing to vancomycin resistance [19]. The LiaSR two-component system plays an important role in CESRs in *B. subtilis* [17]. In *S. mutans*, the deletion of *liaS* and *liaR* may reduce the sensitivity to various antiseptics, including bacitracin and CHX [16,20]. The phenomenon implied that the LiaSR two-component system was associated with CHX-sensitivity in *S. mutans*. However, the underlying mechanisms remain unclear. Therefore, this study aimed to explore whether the LiaSR two-component system is involved in CHX sensitivity in *S. mutans* and to elaborate on the underlying mechanisms. Based on the antimicrobial mechanisms of CHX, bacteria could activate CESRs [19] and change the surface charge to resist CHX [9]. These mechanisms, combined with the fact that CHX can be detoxified by efflux pumps, have led to the hypothesis that the LiaSR two-component system confers CHX resistance on *S. mutans* by mediating its efflux rates, CESRs and surface charge.

2. Methods

2.1. Bacterial Strains and Culture Conditions

All bacterial strains used in this study are listed in Table 1. *S. mutans* 593 (serotype C), which was isolated from a caries-active adult (number of decayed and filled teeth (DFT) = 10, no missing teeth, and three non-restored cavities) and has been proven to possess high cariogenicity [21]. *S. mutans* 593 liaS$^-$ and *S. mutans* 593 liaR$^-$ were constructed in a previous study and were defined as *liaS* and *liaR* deletion mutants [22]. *S. mutans* strains were grown in brain–heart infusion (BHI) broth or BHI agar plates under anerobic conditions (37 °C, 80% N_2, 10% CO_2, 10% H_2) in an anerobic incubator (DG250, Don Whitley Scientific, Shipley, UK).

Table 1. Bacterial strains used in this study.

Strain or Plasmid	Relevant Characteristics	Source
S. mutans 593	Wild-type, serotype c	[21]
S. mutans 593 liaS$^-$	Defined *liaS* deletion mutant	[22]
S. mutans 593 liaR$^-$	Defined *liaR* deletion mutant	

2.2. Minimum Inhibitory Concentration (MIC) Assay

A minimum inhibitory concentration assay was performed as previously described [9]. Briefly, CHX was dissolved in BHI broth before serial dilution. Overnight cultures of

S. mutans were adjusted to 1×10^6 CFU/mL, then 100 μL of a bacterial suspension was used to inoculate the same volume of BHI broth containing the CHX. BHI broth without bacteria was used as the control. The 96-well microplates were incubated at 37 °C for 24 h, and the lowest concentration that inhibited visible bacterial growth was defined as the MIC.

2.3. Biofilm Formation and Viability Assessment

Construction of the biofilm was performed in accordance with the method previously described [22]. Mid-exponential phase cells were adjusted to 1×10^8 CFU/mL with BHI broth. Then the bacterial suspension was diluted (1:20) using the BHI broth supplemented with 1% sucrose before being plated into 96-well microplates. The plates were cultured anaerobically at 37 °C.

After 24 h of incubation, the biofilms were rinsed with phosphate-buffered saline (PBS) to remove the planktonic cells. The biofilms were then treated with 0.2% CHX for 0, 15 and 30 min. After treatment, the biofilms were collected using PBS and serially diluted in a PBS solution. The samples were inoculated onto BHI agar plates and incubated for 48 h. Finally, the number of colonies were calculated to analyze the disinfection effect.

2.4. Nile Red Efflux Assay

Nile Red is a fluorescent marker whose polarity is similar to that of many antiseptics [23]. Based on the fact that Nile Red is weakly fluorescent in aqueous solutions but strongly fluoresces in a non-polar environment, the variation in fluorescence can be monitored when it enters and exits bacterial cells [24]. The Nile Red efflux assay was conducted according to a protocol published earlier [25], with slight modifications. The construction of *S. mutans* biofilms was performed as previously described. A 20 mM potassium phosphate buffer (pH = 7.0) containing 1 mM $MgCl_2$ (PPB) was used to remove the planktonic cells of the samples before being collected with PPB. After centrifugation (4000× *g* for 5 min; Eppendorf, Hamburger, Germany), the cells were pelleted and then resuspended in PPB to an absorbance at 600 nm of 1.0. Carbonyl cyanide m-chlorophenylhydrazone was added to a final concentration of 5 mM. After 15 min, Nile Red was added to a final concentration of 5 μM and incubated on a shaker (140 rpm; 37 °C; NRT-100B, Shanghai, China) in the dark. After incubation, the cells were pelleted by centrifugating (4000× *g* for 5 min; Eppendorf, Hamburger, Germany) before resuspension. Next, 200 μL of the solution was transferred to a 96-well plate. Fluorescence intensity of each well was measured at an excitation wavelength of 525 nm and an emission wavelength of 636 nm using a SpectraMax iD3 Multi-Mode Microplate Reader (Molecular devices iD3, San Jose, CA, USA) for 200 s. Nile Red was triggered by rapid energization with 1 M glucose and monitored for another 820 s.

2.5. Transmission Electron Microscopy (TEM) Observation

TEM observation was performed as previously described [9]. Exponential-phase cells were fixed with 2.5% glutaraldehyde overnight. The specimens were prepared for examination. The bacterial ultrastructure was photographed using TEM at magnifications ranging from ×2.5 k to ×50 k at 80 kV.

2.6. Membrane Potential Assay

The biofilm was constructed as described above. After incubation, deionized water was used to rinse the biofilm. The biofilm was then incubated in 200 μL deionized water containing 10 μM Bis-(1,3-dibutylbarbituric acid) trimethine oxonol (DiBAC4(3)) in the dark for 30 min. After that, the dye was discarded, and the biofilm was rinsed twice with deionized water. The biofilm was then treated with 0.2% CHX for 15 min. Thereafter, the CHX was discarded. PBS was used to wash the biofilm to remove residual CHX and resuspend the cells. Aliquots of the suspension were serially diluted and were spread on the BHI agar. BHI agar was incubated at 37 °C before counting. The remaining suspension was transferred into a new dark 96-well plate for measurement of the fluorescence. The fluorescence intensity was measured with the SpectraMax iD3 Multi-Mode Microplate Reader (Molecular Devices iD3, San Jose, CA, USA) (λ_{ex}: 493 nm and λ_{em}: 516 nm).

2.7. Assay for Hydrophobicity

To measure the hydrophobicity, a modified protocol was used [9]. Briefly, the overnight culture was pelleted by centrifugation. A phosphate–urea–magnesium sulfate (PUM) buffer was used to resuspend the cells, and the initial optical density at 600 nm was recorded as $OD_{600\ Before}$. Next, 5 mL of the bacterial suspension was mixed vigorously with 500 μL n-hexadecane for 30 s before incubation at room temperature for 10 min. After incubation, the OD_{600} of the lower aqueous phase was measured and recorded as $OD_{600\ After}$. Hydrophobicity was calculated as follows: $(OD_{600\ Before} - OD_{600\ After})/OD_{600\ Before} \times 100\%$.

2.8. Cytochrome C-Based Charge Determination of Biofilm-Grown Cells

Construction of the biofilm was conducted as previously described. To measure the charge of the biofilm-grown cells, a modified protocol described by Martin et al. was performed [26]. For this, 3-(*N*-morpholino)propanesulfonic acid (20 mM MOPS, pH = 7.0) was used to harvest the biofilm-grown cells, and the cell suspension was adjusted to $OD_{600} = 1.0$. Cytochrome c was added to a final concentration of 0.5 mg/mL. Thereafter, the samples were incubated for 15 min at room temperature and were then centrifuged. The amount of cytochrome c remaining in the supernatant was measured by the absorbance at 530 nm.

2.9. RNA Isolation and Reverse-Transcription Quantitative PCR (RT-qPCR)

Total RNA was extracted from the samples according to the manufacturer's instructions (Qiagen, Valencia, CA, USA). The purity and quantity of the RNA were evaluated by NanoDrop 2000 (Thermo Scientific, Waltham, Massachusetts, USA) according to the A_{260}/A_{280} ratio. Agarose gel electrophoresis was used to validate the integrity of the RNA. To remove DNA contamination and reverse transcription, the kit from Takara Bio (Otsu, Japan) was used according to the manufacturer's instructions. RT-qPCR was conducted using SYBR® Premix Ex Taq™ (Tli RNaseH Plus) (Takara, Beijing, China) in a LightCycler 480 quantitative PCR instrument. The primers specific to the target genes are listed in Table 2. The genes' expression levels were calculated by the normalizing target genes to the respective reference genes on the basis of the $2^{-\Delta\Delta Ct}$ method.

Table 2. Primers used in this study.

Primers	Sequence (5′→3′)
dltC-F	ACCAAGGCTGTACTTAATCAGG
dltC-R	AGGATGCAGATAAAGGAAGCC
dagK-F	GGGATCCTTCTCATGTTCTCTC
dagK-R	CATCTGCATTGACTTCCATCATT
murB-F	GCAGTGCTAAGACTCCCGAATC
murB-R	TTGCGGAAGTGTGGAGATTGGC
rgpB-F	GGAATCGTTAGAAGCAAAAGGTG
rgpB-R	TGACCAAAGCAGTAACCCTC
lmrB-F	CGTCATCTCTTTATCGCAAGCATC
lmrB-R	CAGTATGTTCCAGCACTTCAGCC
SMU_113-F	AAATACGGTGACACTTGCTGGTAAG
SMU_113-R	TCAATAGATGGCACAACGGTAATC
SMU_1611-F	ACCGCTATGCTAAGGCTCTC
SMU_1611-R	AACCTATCTGAAGACGCTTGTC
SMU_1286-F	TGCCACAGGATGCAGAAGG
SMU_1286-R	TGTTGCACATCCTTACCCGT
16S-F	AGCGTTGTCCGGATTTATTG
16S-R	CTACGCATTTCACCGCTACA

2.10. Statistical Analysis

Each experiment was carried out in triplicate. Data were statistically analyzed using the Statistical Package for Social Sciences (SPSS, Inc., Chicago, IL, USA), version 21.0. One-way ANOVA was used to detect the significant effects of the variables. Data were considered significantly different when the two-tailed p-value was <0.05.

3. Results

3.1. Sensitivity of *S. mutans* to CHX

As shown in Figure 1A, the MIC of the wild-type strain was significantly higher than that of the *liaS* and *liaR* deletion mutants. For the biofilm, significant decreases were observed in bacterial number of the *liaS* and *liaR* deletion mutants after 15 min of treatment with 0.2% CHX when compared with the wild-type strain. After 30 min of treatment, no viable cells were observed in the *S. mutans* biofilm (Figure 1B).

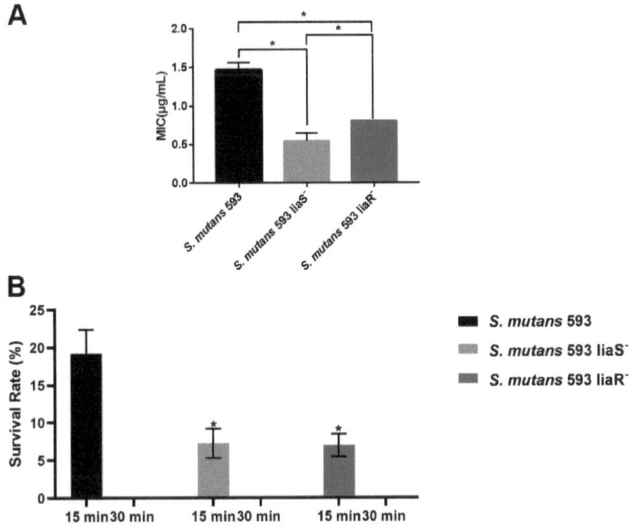

Figure 1. The sensitivity of *S. mutans* to CHX. (**A**) The MICs of *S. mutans*. Values are presented as the mean ± SD. *, $p < 0.05$. (**B**) The survival rates of *S. mutans* after 15 min and 30 min of CHX treatment. *, $p < 0.05$ compared with *S. mutans* 593.

3.2. Efflux Activity of *S. mutans*

Figure 2A presents the efflux activity of *S. mutans* strains. Nile Red is an intracellular dye that is a common substrate for efflux pumps. The results showed that the rates of extrusion of the dye from the cells of the *liaS* and *liaR* deletion mutants were reduced, suggesting that the efflux pumps were suppressed. Figure 2B exhibited the expression of efflux-pump-encoding genes in *S. mutans* [27]. The results showed that the *lmrB* gene was downregulated in the two mutants when compared with the wild-type strain. SMU_1611c were upregulated in the two mutants when compared with the wild-type strain. The expression level of SMU_113c showed no significant difference in the mutants. The expression level of SMU_1286c in the *liaR* deletion mutant was higher than in the wild-type strain. While in the *liaS* deletion mutant, no significant difference was observed.

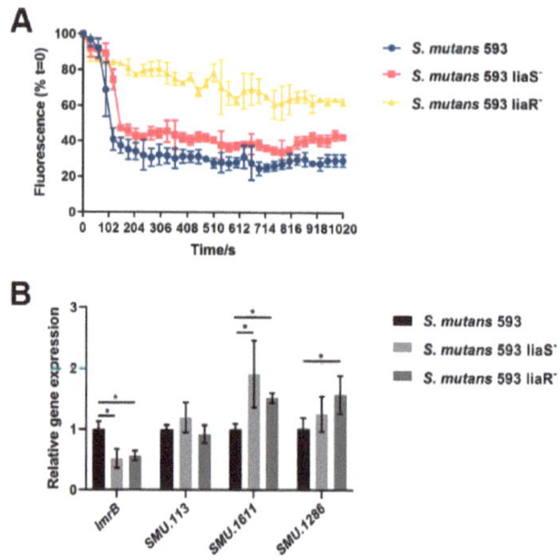

Figure 2. The efflux pump of *S. mutans*. (**A**) The efflux rates of *S. mutans*. (**B**) Expression levels of *lmrB*, *SMU. 113*, *SMU. 1611* and *SMU. 1286*. Values are presented as the mean ± SD. *, $p < 0.05$.

3.3. The Surface Charge of S. mutans

Positively charged cytochrome c was used to assess the charge of the biofilm-grown cells based on the fact that it could bind to negatively charged cells [28]. The OD_{530} value represented the proportion of unbounded cytochrome c. The higher the value, the more positively charged the cell surface. As shown in Figure 3A, no significant difference was observed between the wild-type strain and the two mutants. The expression level of *dltC*, which might correlate with CHX resistance in *S. mutans*, was higher in the *liaR* deletion mutant. However, there was no significant difference in the expression of the *dltC* gene between the wild-type strain and *liaS* deletion mutant (Figure 3B).

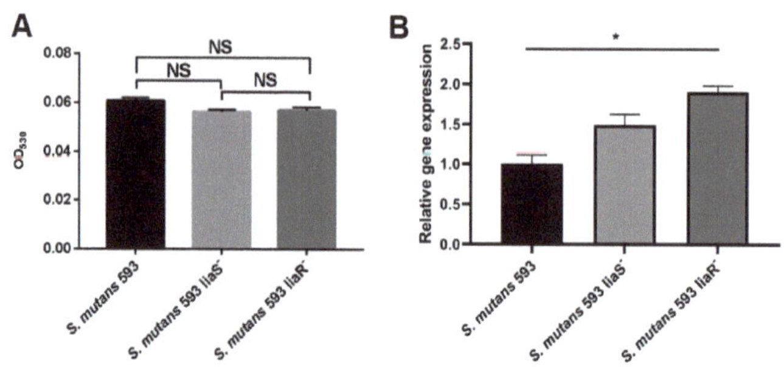

Figure 3. The surface charge of *S. mutans*. (**A**) The surface charge of *S. mutans*. (**B**) Expression level of *dltC*. Values are presented as the mean ± SD. *, $p < 0.05$. NS, $p > 0.05$.

3.4. The Ability of CESRs in S. mutans

To detect the ability of CESRs in *S. mutans*, we first performed TEM to observe the changes in the membrane, as the cell membrane is essential for resistance to CHX. As shown in Figure 4A–F, the border of the cell membrane was unclear in the mutants compared with the wild-type strain.

The ability to depolarize the cell membrane and the hydrophobicity are the indicators of cell disruption [29]. To monitor changes in the membrane's polarization, we first used DiBAC4(3), which is an anionic potential-response membrane probe. It can be excluded from healthy cells. In damaged cells whose membrane polarization is altered, the dye can enter the cells and bind to intracellular proteins and membranes, accompanied by increased fluorescence. Figure 4G shows the evidently increased intensity of fluorescence in the *liaS* and *liaR* deletion mutants compared with the wild-type strain with or without CHX treatment. All strains with CHX treatment exhibited stronger fluorescence intensity than the cells without CHX treatment.

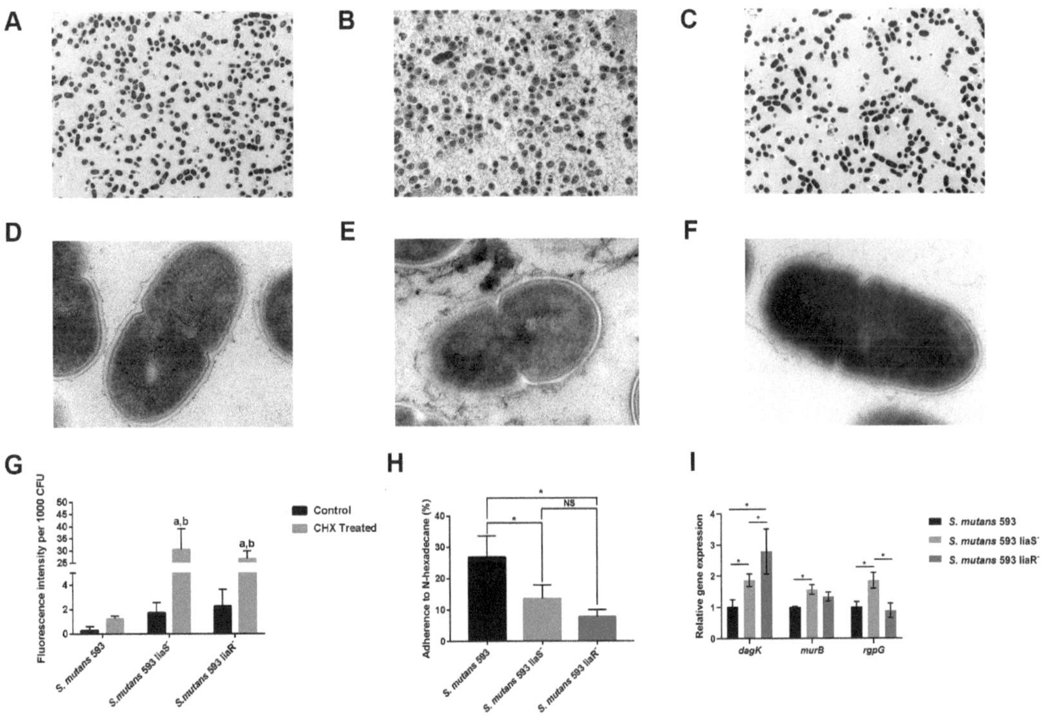

Figure 4. The CESRs of *S. mutans*. (**A–F**) TEM images showing the morphology of cells from *S. mutans* 593 ((**A**), ×2.5 k and (**D**), ×50 k), *S. mutans* 593 liaS$^-$ ((**B**), ×2.5 k and (**E**), ×50 k) and *S. mutans* 593 liaR$^-$ ((**C**), ×2.5 k and (**F**), ×50 k). (**G**) Cell surface hydrophobicity of *S. mutans*. (**H**) The percentages of positive cells stained with the DiBAC4(3) probe. (**I**) Expression levels of *dagK*, *murB* and *rgpG*. Values are presented as the mean ± SD. [a], significant differences at $p < 0.05$ compared to the control group; [b], significant differences at $p < 0.05$ compared to *S. mutans* 593; *, $p < 0.05$; NS, $p > 0.05$.

In terms of hydrophobicity, the mutants showed significantly lower hydrophobicity when compared with the wild-type strain (Figure 4H). In summary, the higher depolarization and reduced hydrophobicity of the mutants indicated that the membrane of the *liaS* and *liaR* gene deletion mutants in *S. mutans* may be more easily disrupted.

The high level of depolarization and hydrophobicity implied that the cell membranes of the mutants were more likely to be disturbed. Therefore, we performed the RT-qPCR to analyze the expression levels of *dagK*, *murB* and *rgpD*, which are involved in membrane biosynthesis, remodeling and modification. Figure 4I shows that *dagK* was upregulated in the mutants. *murB* and *rgpD* were upregulated in the *liaS* deletion mutant. But in the *liaR* deletion mutant, the expression of *murB* and *rgpD* showed no significant difference compared with the wild-type strain.

4. Discussion

CHX is considered to be the gold standard for the prevention of dental caries. Previous studies have shown that the deletion of the LiaSR two-component system may lead to increased sensitivity to CHX [16]. Our study verified the fact again and investigated the underlying mechanisms. To our best knowledge, this is the first in vitro study to demonstrate the relationship between the LiaSR two-component system and CHX resistance.

To analyze the role of the LiaSR two-component system in *S. mutans* under CHX treatment, we performed an MIC assay and a viability assessment. The results exhibited that the *liaS* and *liaR* deleted mutants were more sensitive to CHX, which was in accordance with the study by Suntharalingam et al. [16]. The increased sensitivity of the mutants prompted us to explore the underlying mechanisms.

Multidrug efflux pumps are membrane proteins that contain multiple transmembrane domains that form channels to remove the CHX from the cytoplasm and reduce the intracellular CHX concentrations [30]. In *S. aureus* [31], *Pseudomonas aeruginosa* (*P. aeruginosa*) [32] and *Klebsiella pneumoniae* (*K. pneumoniae*) [33], the major facilitator superfamily (MFS) was one of the multidrug efflux pumps that contributed to CHX resistance. It had two domains that underwent a conformational switch between the inward-open and outward-open states while facilitating efflux activities [34]. On the basis of this, we speculated that the LiaSR two-component system could modulate the efflux of *S. mutans*, leading to a decrease in the accumulation of CHX, thereby facilitating the cells' survival under CHX treatment. Our study showed that after deletion of the *liaS* and *liaR* genes, the efflux rates of *S. mutans* were significantly reduced. To explore the putative regulated efflux-pump-encoding genes, we analyzed the expression levels of *lmrB*, SMU_113c, SMU_1611c and SMU_1286c, which might be related to CHX resistance in *S. mutans* [27]. Consistently, the expression of *lmrB* was downregulated in the mutants, indicating that the LiaSR two-component system may regulate the *lmrB* gene to induce *S. mutans* to pump out CHX, which may reduce sensitivity to CHX. SMU_113c, SMU_1611c and SMU_1286c in the mutants showed upregulation or no significant difference compared with the wild-type strain. The results tied in well with a previous study [27]. In that study, SMU_113c, SMU_1611c and SMU_1286c were upregulated in the *lmrB* deletion mutant, accompanied by increases in exopolysaccharides and the extracellular matrix [27]. The increase in exopolysaccharides and the extracellular matrix could also be observed in the *liaS* and *liaR* deletion mutants [35], suggesting that this two-component system may regulate the *lmrB* gene indirectly.

Martin et al. [26] found that *S. mutans* could upregulate the expression of the *dlt* operon to increase the positive charge, which might contribute to the resistance to gentamicin. The bactericidal mechanism of CHX is similar to that of gentamicin. It can attach to the negatively charged cell's surface due to its positive charge, leading to leakage of the cell's components. A previous study has shown that the expression of *dltC* was upregulated in CHX-resistant *S. mutans* strains [9]. Given these findings, we speculated that the LiaSR two-component system may modulate the *dltC* gene to increase the positively charged cells' surface and the ability to resist CHX in *S. mutans*. However, our results showed no significant difference in the cell surface charge of mutants compared with the wild-type strain. The expression of *dltC* was upregulated in the mutants. These findings did not correlate with an increased sensitivity to CHX in this study. On the basis of these findings, we speculated that the LiaSR two-component system did not enable the *dltC* gene to resist CHX.

The cell membrane is the outer barrier that is critical for viability in bacteria. CHX is a membrane-targeting antiseptic that can attach to and disrupt the cell membrane, resulting in the formation of ion channels, electrochemical degree destruction and membrane depolarization, which ultimately lead to bacterial death [36]. Therefore, we first carried out TEM observations. The images showed that although the integrity of the cell membrane in the mutants was maintained, the boundary was unclear in the mutants compared with the wild-type strain. The phenomenon may partly explain why the sensitivity of mutants to CHX increased.

The degree of bacterial membrane depolarization represents the state of membrane perturbation [29]. Our results showed that the mutants exhibited higher membrane depolarization after the CHX treatment. Hydrophobicity is another indicator of membrane disturbance [29]. A previous study confirmed that CHX-resistant *S. mutans* strains' hydrophobicity was significantly elevated [9]. In this study, after deleting the *liaS* and *liaR* genes in *S. mutans*, the mutant strains' hydrophobicity was reduced. This result was in accordance with the previous study [9]. Taken together, the membrane of the mutants was more likely to be disrupted, which was consistent with a previous study showing that the LiaSR two-component system participated in membrane homeostasis [22,37].

In *B. subtilis*, the LiaSR two-component system, which is homologous to *S. mutans*' LiaSR two-component system, contributed to the cell envelope stress responses (CESRs) [38–40]. When bacterial cells sense a disturbance, the ability of the envelope and modification of the membrane will be activated to maintain homeostasis and repair the damage. Suntharalingam et al. found that after 10 min of bacitracin treatment, *dagK*, *rgpG* and *murB*, which are related to membrane biosynthesis, remodeling and modification in the *liaFSR* deletion mutant, were reduced [16]. To determine whether *dagK*, *rgpG* and *murB* were regulated by the LiaSR two-component system to affect CESRs in *S. mutans*, we performed RT-qPCR. However, our study was inconsistent with the study of Suntharalingam et al. [16]. The expression levels of *dagK*, *rgpG* and *murB* were elevated or not significantly different in mutants, which indicated that these genes were not the target genes for the LiaSR two-component system in CESRs. Previous studies have validated that the LiaSR two-component system exerts positive effects on the expression of *spxA2* to maintain the membrane's homeostasis, which is also essential for CHX resistance [16,22,37]. Accordingly, we speculated that the LiaSR two-component system mediation of CHX resistance relied on the expression of the *spxA2* gene. However, further studies are needed to explore the other target genes regulated by this two-component system.

In conclusion, our study demonstrated that the LiaSR two-component system is involved in CHX resistance in *S. mutans*. This system could increase the *lmrB* gene, which plays an important role in detoxifying intracellular CHX. It is also required for CESRs, which contributes to the decreased sensitivity to CHX. This study provides valuable insights into the mechanisms of the LiaSR two-component system in CHX-resistant *S. mutans*. However, more detailed mechanistic studies must be performed to prove the relationship between the LiaSR two-component system and the *lmrB* gene, which may be beneficial for the development of the LiaSR two-component system or efflux inhibitors as potential adjunct methods to therapeutically intervene against biofilms. CESRs would be activated when bacteria encounter many types of stress, such as mechanical forces and antiseptics. The CESR-related genes regulated by the LiaSR two-component system must be further investigated, which may contribute to the development of CESR-targeting antiseptics.

Author Contributions: X.H., L.Z. and S.H. designed the experiments; X.H. and L.Z. provided the funding; S.H., Y.L., J.H. and J.D. carried out the experiments; M.W., J.D. and S.C. analyzed the experimental results; X.H., S.H., J.H. and J.D. wrote and revised the manuscript. All authors have read and agreed to the published version of the manuscript.

Funding: This work was supported by National Natural Science Foundation of China (grant number 30500564) and the Scientific Research Foundation for Minjiang Scholars (grant number 2018-KQMJ-02).

Data Availability Statement: Data are contained within the article.

Conflicts of Interest: The authors declare no conflicts of interest.

References

1. Sharma, S.; Mohler, J.; Mahajan, S.D.; Schwartz, S.A.; Bruggemann, L.; Aalinkeel, R. Microbial Biofilm: A Review on Formation, Infection, Antibiotic Resistance, Control Measures, and Innovative Treatment. *Microorganisms* **2023**, *11*, 1614. [CrossRef] [PubMed]
2. Hanson, K.E.; Banerjee, R.; Doernberg, S.B.; Evans, S.R.; Komarow, L.; Satlin, M.J.; Schwager, N.; Simner, P.J.; Tillekeratne, L.G.; Patel, R.; et al. Priorities and Progress in Diagnostic Research by the Antibacterial Resistance Leadership Group. *Clin. Infect. Dis.* **2023**, *77* (Suppl. 4), S314–S320. [CrossRef] [PubMed]
3. Zheng, T.; Jing, M.; Gong, T.; Yan, J.; Wang, X.; Xu, M.; Zhou, X.; Zeng, J.; Li, Y. Regulatory mechanisms of exopolysaccharide synthesis and biofilm formation in *Streptococcus mutans*. *J. Oral Microbiol.* **2023**, *15*, 2225257. [CrossRef] [PubMed]
4. Cugini, C.; Shanmugam, M.; Landge, N.; Ramasubbu, N. The Role of Exopolysaccharides in Oral Biofilms. *J. Dent. Res.* **2019**, *98*, 739–745. [CrossRef] [PubMed]

5. Du, J.; Huang, S.; Wu, M.; Chen, S.; Zhou, W.; Zhan, L.; Huang, X. Dlt operon regulates physiological function and cariogenic virulence in *Streptococcus mutans*. *Future Microbiol.* **2023**, *18*, 225–233. [CrossRef] [PubMed]
6. Lemos, J.A.; Palmer, S.R.; Zeng, L.; Wen, Z.T.; Kajfasz, J.K.; Freires, I.A.; Abranches, J.; Brady, L.J. The Biology of *Streptococcus mutans*. *Microbiol. Spectr.* **2019**, *7*, 7. [CrossRef] [PubMed]
7. Lamont, R.J.; Koo, H.; Hajishengallis, G. The oral microbiota: Dynamic communities and host interactions. *Nat. Rev. Microbiol.* **2018**, *16*, 745–759. [CrossRef] [PubMed]
8. Brookes, Z.; McGrath, C.; McCullough, M. Antimicrobial Mouthwashes: An Overview of Mechanisms—What Do We Still Need to Know? *Int. Dent. J.* **2023**, *73* (Suppl. 2), S64–S68. [CrossRef]
9. Huang, S.; Wu, M.; Li, Y.; Du, J.; Chen, S.; Jiang, S.; Huang, X.; Zhan, L. The dlt operon contributes to the resistance to chlorhexidine in *Streptococcus mutans*. *Int. J. Antimicrob. Agents* **2022**, *59*, 106540. [CrossRef]
10. Kaspar, J.R.; Godwin, M.J.; Velsko, I.M.; Richards, V.P.; Burne, R.A. Spontaneously Arising *Streptococcus mutans* Variants with Reduced Susceptibility to Chlorhexidine Display Genetic Defects and Diminished Fitness. *Antimicrob. Agents Chemother.* **2019**, *63*, 7. [CrossRef]
11. Wang, S.; Fang, L.; Zhou, H.; Wang, M.; Zheng, H.; Wang, Y.; Weir, M.D.; Masri, R.; Oates, T.W.; Cheng, L.; et al. Silica nanoparticles containing nano-silver and chlorhexidine respond to pH to suppress biofilm acids and modulate biofilms toward a non-cariogenic composition. *Dent. Mater.* **2024**, *40*, 179–189. [CrossRef]
12. Boaro, L.C.C.; Campos, L.M.; Varca, G.H.C.; dos Santos, T.M.R.; Marques, P.A.; Sugii, M.M.; Saldanha, N.R.; Cogo-Müller, K.; Brandt, W.C.; Braga, R.R.; et al. Antibacterial resin-based composite containing chlorhexidine for dental applications. *Dent. Mater.* **2019**, *35*, 909–918. [CrossRef]
13. Li, Y.-H.; Lau, P.C.Y.; Tang, N.; Svensäter, G.; Ellen, R.P.; Cvitkovitch, D.G. Novel two-component regulatory system involved in biofilm formation and acid resistance in *Streptococcus mutans*. *J. Bacteriol.* **2002**, *184*, 6333–6342. [CrossRef]
14. Chong, P.; Drake, L.; Biswas, I. *LiaS* regulates virulence factor expression in *Streptococcus mutans*. *Infect. Immun.* **2008**, *76*, 3093–3099. [CrossRef]
15. Perry, J.A.; Levesque, C.M.; Suntharaligam, P.; Mair, R.W.; Bu, M.; Cline, R.T.; Peterson, S.N.; Cvitkovitch, D.G. Involvement of *Streptococcus mutans* regulator *RR11* in oxidative stress response during biofilm growth and in the development of genetic competence. *Lett. Appl. Microbiol.* **2008**, *47*, 439–444. [CrossRef]
16. Suntharalingam, P.; Senadheera, M.D.; Mair, R.W.; Lévesque, C.M.; Cvitkovitch, D.G. The *LiaFSR* system regulates the cell envelope stress response in *Streptococcus mutans*. *J. Bacteriol.* **2009**, *191*, 2973–2984. [CrossRef]
17. Kesel, S.; Mader, A.; Höfler, C.; Mascher, T.; Leisner, M. Immediate and heterogeneous response of the LiaFSR two-component system of *Bacillus subtilis* to the peptide antibiotic bacitracin. *PLoS ONE* **2013**, *8*, e53457. [CrossRef]
18. Fernandes, P.B.; Reed, P.; Monteiro, J.M.; Pinho, M.G. Revisiting the Role of VraTSR in *Staphylococcus aureus* Response to Cell Wall-Targeting Antibiotics. *J. Bacteriol.* **2022**, *204*, e0016222. [CrossRef]
19. Baseri, N.; Najar-Peerayeh, S.; Bakhshi, B. The effect of subinhibitory concentration of chlorhexidine on the evolution of vancomycin-intermediate *Staphylococcus aureus* and the induction of mutations in walKR and vraTSR systems. *Infect. Genet. Evol.* **2021**, *87*, 104628. [CrossRef]
20. Zhang, J.; Biswas, I. A phenotypic microarray analysis of a *Streptococcus mutans liaS* mutant. *Microbiology* **2009**, *155 Pt 1*, 61–68. [CrossRef]
21. Huang, X.; Liu, T.; Chen, G. Typing of *Streptococcus mutans* (serotype C) by arbitrarily primed polymerase chain reaction. *Zhonghua Kou Qiang Yi Xue Za Zhi* **2001**, *36*, 281–284.
22. Huang, S.; Du, J.; Li, Y.; Wu, M.; Chen, S.; Jiang, S.; Zhan, L.; Huang, X. LiaSR two-component system modulates the oxidative stress response in *Streptococcus mutans*. *Microb. Pathog.* **2023**, *185*, 106404. [CrossRef]
23. Jenic, D.; Waller, H.; Collins, H.; Erridge, C. Reversal of Tetracycline Resistance by Cepharanthine, Cinchonidine, Ellagic Acid and Propyl Gallate in a Multi-drug Resistant *Escherichia coli*. *Nat. Prod. Bioprospect.* **2021**, *11*, 345–355. [CrossRef]
24. Schuster, S.; Vavra, M.; Greim, L.; Kern, W.V. Exploring the Contribution of the AcrB Homolog MdtF to Drug Resistance and Dye Efflux in a Multidrug Resistant *E. coli* Isolate. *Antibiotics* **2021**, *10*, 503. [CrossRef]
25. Bohnert, J.A.; Karamian, B.; Nikaido, H. Optimized Nile Red efflux assay of AcrAB-TolC multidrug efflux system shows competition between substrates. *Antimicrob. Agents Chemother.* **2010**, *54*, 3770–3775. [CrossRef]
26. Nilsson, M.; Rybtke, M.; Givskov, M.; Høiby, N.; Twetman, S.; Tolker-Nielsen, T. The dlt genes play a role in antimicrobial tolerance of *Streptococcus mutans* biofilms. *Int. J. Antimicrob. Agents* **2016**, *48*, 298–304. [CrossRef]
27. Liu, J.; Zhang, J.; Guo, L.; Zhao, W.; Hu, X.; Wei, X. Inactivation of a putative efflux pump (LmrB) in *Streptococcus mutans* results in altered biofilm structure and increased exopolysaccharide synthesis: Implications for biofilm resistance. *Biofouling* **2017**, *33*, 481–493. [CrossRef]
28. Aleksandrowicz, A.; Kolenda, R.; Baraniewicz, K.; Thurston, T.L.M.; Suchański, J.; Grzymajlo, K. Membrane properties modulation by SanA: Implications for xenobiotic resistance in Salmonella Typhimurium. *Front. Microbiol.* **2023**, *14*, 1340143. [CrossRef]
29. Feng, Z.; Luo, J.; Lyu, X.; Chen, Y.; Zhang, L. Selective antibacterial activity of a novel lactotransferrin-derived antimicrobial peptide LF-1 against *Streptococcus mutans*. *Arch. Oral Biol.* **2022**, *139*, 105446. [CrossRef]
30. Cieplik, F.; Jakubovics, N.S.; Buchalla, W.; Maisch, T.; Hellwig, E.; Al-Ahmad, A. Resistance Toward Chlorhexidine in Oral Bacteria—Is There Cause for Concern? *Front. Microbiol.* **2019**, *10*, 587. [CrossRef]

31. Horner, C.; Mawer, D.; Wilcox, M. Reduced susceptibility to chlorhexidine in *Staphylococci*: Is it increasing and does it matter? *J. Antimicrob. Chemother.* **2012**, *67*, 2547–2559. [CrossRef]
32. Morita, Y.; Tomida, J.; Kawamura, Y. Responses of *Pseudomonas aeruginosa* to antimicrobials. *Front. Microbiol.* **2014**, *4*, 422. [CrossRef]
33. Wand, M.E.; Bock, L.J.; Bonney, L.C.; Sutton, J.M. Mechanisms of Increased Resistance to Chlorhexidine and Cross-Resistance to Colistin following Exposure of *Klebsiella pneumoniae* Clinical Isolates to Chlorhexidine. *Antimicrob. Agents Chemother.* **2017**, *61*, 1. [CrossRef]
34. Du, D.; Wang-Kan, X.; Neuberger, A.; Van Veen, H.W.; Pos, K.M.; Piddock, L.J.V.; Luisi, B.F. Multidrug efflux pumps: Structure, function and regulation. *Nat. Rev. Microbiol.* **2018**, *16*, 523–539. [CrossRef]
35. Huang, S.; Du, J.Y.; Li, Y.J.; Wu, M.J.; Chen, S.; Jiang, S.; Huang, X.J. Role and related mechanisms of LiaSR two-component system in acid tolerance and biofilm formation of *Streptococcus mutans*. *Zhonghua Kou Qiang Yi Xue Za Zhi* **2024**, *59*, 54–63.
36. Schrank, C.L.; Wilt, I.K.; Ortiz, C.M.; Haney, B.A.; Wuest, W.M. Using membrane perturbing small molecules to target chronic persistent infections. *RSC Med. Chem.* **2021**, *12*, 1312–1324. [CrossRef]
37. Baker, J.L.; Saputo, S.; Faustoferri, R.C.; Quivey, R.G., Jr. *Streptococcus mutans SpxA2* relays the signal of cell envelope stress from *LiaR* to effectors that maintain cell wall and membrane homeostasis. *Mol. Oral Microbiol.* **2020**, *35*, 118–128. [CrossRef] [PubMed]
38. Cho, T.H.; Pick, K.; Raivio, T.L. Bacterial envelope stress responses: Essential adaptors and attractive targets. *Biochim. Biophys. Acta Mol. Cell Res.* **2023**, *1870*, 119387. [CrossRef] [PubMed]
39. Saha, S.; Lach, S.R.; Konovalova, A. Homeostasis of the Gram-negative cell envelope. *Curr. Opin. Microbiol.* **2021**, *61*, 99–106. [CrossRef]
40. Hews, C.L.; Cho, T.; Rowley, G.; Raivio, T.L. Maintaining Integrity Under Stress: Envelope Stress Response Regulation of Pathogenesis in Gram-Negative Bacteria. *Front. Cell. Infect. Microbiol.* **2019**, *9*, 313. [CrossRef]

microorganisms

Article

HigA2 (Rv2021c) Is a Transcriptional Regulator with Multiple Regulatory Targets in *Mycobacterium tuberculosis*

Mingyan Xu [1], Meikun Liu [1], Tong Liu [1], Xuemei Pan [1], Qi Ren [1], Tiesheng Han [1,*] and Lixia Gou [2,*]

[1] Hebei Province Key Laboratory of Occupational Health and Safety for Coal Industry, School of Public Health, North China University of Science and Technology, Tangshan 063210, China; xumingyan0518@163.com (M.X.); liumeikun2019@126.com (M.L.); liutt_1206@163.com (T.L.); 18861999241@163.com (X.P.); renqi@ncst.edu.cn (Q.R.)

[2] School of Life Science, North China University of Science and Technology, Tangshan 063210, China

* Correspondence: ts_han@ncst.edu.cn or ts_han@163.com (T.H.); goulixia@ncst.edu.cn or lixia_gou@163.com (L.G.)

Abstract: Toxin-antitoxin (TA) systems are the major mechanism for persister formation in *Mycobacterium tuberculosis* (*Mtb*). Previous studies found that HigBA2 (Rv2022c-Rv2021c), a predicted type II TA system of *Mtb*, could be activated for transcription in response to multiple stresses such as anti-tuberculosis drugs, nutrient starvation, endure hypoxia, acidic pH, etc. In this study, we determined the binding site of HigA2 (Rv2021c), which is located in the coding region of the upstream gene *higB2* (*Rv2022c*), and the conserved recognition motif of HigA2 was characterized via oligonucleotide mutation. Eight binding sites of HigA2 were further found in the *Mtb* genome according to the conserved motif. RT-PCR showed that HigA2 can regulate the transcription level of all eight of these genes and three adjacent downstream genes. DNA pull-down experiments showed that twelve functional regulators sense external regulatory signals and may regulate the transcription of the HigBA2 system. Of these, Rv0903c, Rv0744c, Rv0474, Rv3124, Rv2603c, and Rv3583c may be involved in the regulation of external stress signals. In general, we identified the downstream target genes and possible upstream regulatory genes of HigA2, which paved the way for the illustration of the persistence establishment mechanism in *Mtb*.

Keywords: toxin–antitoxin; *Mycobacterium tuberculosis*; transcriptional regulation; persistence

check for updates

Citation: Xu, M.; Liu, M.; Liu, T.; Pan, X.; Ren, Q.; Han, T.; Gou, L. HigA2 (Rv2021c) Is a Transcriptional Regulator with Multiple Regulatory Targets in *Mycobacterium tuberculosis*. *Microorganisms* **2024**, *12*, 1244. https://doi.org/10.3390/microorganisms12061244

Academic Editor: Tomohiro Shimada

Received: 25 May 2024
Revised: 17 June 2024
Accepted: 18 June 2024
Published: 20 June 2024

1. Introduction

At present, tuberculosis (TB) remains a global infectious disease that is difficult to fully cure and is the second leading cause of death from a single pathogen [1]. TB is caused by *Mtb* infection. *Mtb* possesses a survival mechanism known as persistence, which means it can enter a persist state under external stress conditions such as antibiotics, with basic metabolic activities ceasing, rendering antimicrobial drugs ineffective [2]. After persistence is generated, most bacteria are sensitive to antibiotics, and only a very small part survive. After re-inoculation culture, most of the persistent bacteria were sensitive to antibiotics, and only a very small part survived [3]. Presently, almost all of the antibiotics used are aimed at growing bacteria and are basically ineffective for persisters. Since the formation of persistent bacteria does not require genetic mutation, but rather only the need to enter a "dormant" state, theoretically, all bacteria can be persistent, resulting in prolonged illness and easier relapse. At the same time, the persistent bacteria have long-term resistance to the environment of antibacterial drugs and are prone to gene mutation and drug resistance [3,4].

The bacterial TA system is a key factor in inducing the formation of persisters [5]. There are at least 80 TA systems in *Mtb*, and type II is the main type [6]. Type II TA systems are selectively degraded by host proteases such as Lon and ClpCP under stress conditions, allowing the toxin proteins to act. Antitoxin proteins of type II systems such as MqsRA, RelBE, HigBA, and other families are transcriptional regulators [7,8].

It has been found that certain antitoxins can regulate the transcription of several genes outside their own TA systems, including other TA systems [9]. The TA system is therefore likely to have transcriptional regulators. Thus, there may be transcriptional regulatory networks of TA systems that can generate complex stress responses to multiple external stresses. Among the many TA systems, HigBA2 has long been predicted as a pair of TA systems [10], but studies have not identified a toxic phenotype for the toxin HigB2 [11] and protein interactions between HigB2 and HigA2 have not been reported. One of the antitoxin genes, *higA2*, was predicted to be a transcriptional regulator, and ChIP-seq analysis showed that it may regulate the transcription of up to 25 genes [12]. After stressing *Mtb* with anti-TB drugs such as isoniazid, rifampicin, streptomycin, and ciprofloxacin, the transcript levels of at least 10 TA systems were found to be up-regulated up to 3-fold [13]. Of these, the HigBA2 system was identified as the key site that was activated under all conditions tested, and it could be starved by starvation conditions in addition to responding to anti-TB drugs [14], sustained hypoxia [15], and acidic environments [16]. Other TB foci and macrophage-like environments activate transcription, and all of these conditions are known to induce the formation of persisters. Therefore, the HigBA2 system may play a key role in the formation of *Mtb* persisters. In addition, the protein crystal structure of HigA2 has been resolved and was found to have N-terminal autocleavage activity that spontaneously removes the N-terminal's 30 amino acids [17]. Although the three-dimensional protein structure of HigA2 has been characterized, its DNA recognition site has not been identified. In this study, we report the TA activity of HigBA2 in *Mtb* and the binding site of HigA2 to find its upstream and downstream regulatory genes.

2. Materials and Methods

2.1. Bacterial Strains, Plasmids, and Growth Conditions

E. coli Top10, the BL21(DE3) pLysS strain, and the *Mycobacterium smegmatis* (*M. smegmatis*) mc^2155 strain were kept in the laboratory. The pJV53 plasmid and the pRSFDuet-1-*higA2* plasmid were purchased from GenScript (Nanjing, China). *E. coli* BL21(DE3) pLysS was used for protein expression and *E. coli* Top10 was used for gene cloning. Growth and virulence experiments were performed with *M. smegmatis* mc^2155 as described previously [18]. *M. smegmatis* mc^2155 was grown at 37 °C in 7H9 supplemented with 0.2% glycerol, 0.05% Tween-80, and 10% Middlebrook OADC. Expression was induced using 0.2% acetamide when grown in a liquid medium or on a 7H10 solid medium containing 0.5% glycerol and 10% OADC. *E. coli* strains were grown on liquid LB or solid LB medium at 37 °C and expression was induced using 0.8 M Isopropyl β-D-1-Thiogalactopyranoside (IPTG). Middlebrook 7H9, 7H10, and OADC were from BD (New Jersey, CA, USA).

2.2. Construction of mc^2155 Heterologous Expression Vector

higA2, *higB2*, and *higBA2* were cloned between the *Nde*I and *EcoR*I (New England Biolabs, Hitchin, UK) sites of the pJV vector to construct the *Mycobacterium* shuttle plasmid using the Seamless Cloning Kit (Beyotime, Shanghai, China). The Seamless Cloning PCR primers used to amplify the target genes are listed in Table S1. The plasmid was added to the pJV vector along with 200 μL of mc^2155 receptor cells together in a 2 mm electroshock cup (parameters: 2.5 kV, 1000 Ω, 25 vf). The 7H9 medium was added immediately after electric shock by Harvard Apparatus BTX (Holliston, MA, USA). The medium was resuscitated at 37 °C and 160 rpm for 3–4 h, and then coated on a 7H10 plate containing 50 ng/μL kanamycin (Kan, Vernon, CA, USA).

2.3. HigB2-HigA2 Neutralization Assay

M. smegmatis containing either individual or combined inducible expression plasmids was cultured at 37 °C for 2 days, and the OD_{600} was adjusted to 0.2 by a 10-fold gradient dilution as described in previous studies [18,19]. Each drop of 2.5 μL bacterial suspension was placed on a 100 mL 7H10 solid plate with or without the 0.2% acetamide inducer and the appropriate amount of antibiotics. The plates were incubated for 2 days at 37 °C, and the growth phenotypes were observed.

2.4. Protein Purification

The sequence of the coding region of the HigA2 protein was cloned into the *E. coli* expression vector pRSF-Duet-1 between the *Nde*I and *Xho*I sites and transformed into *E. coli* BL21(DE3) pLysS for protein overexpression using the calcium chloride method. Cell growth, induction, and harvesting were performed as described previously, followed by sonication and elution [20]. Purified proteins were desalted by a His Trap desalting column containing buffer (20 mM Tris-HCl, 150 mM NaCl, pH 8.0) and measured using SDS-PAGE. Proteins were stored in a stock buffer containing 20% glycerol at −30 °C.

2.5. EMSA

The DNA-binding capacity of proteins was assessed using modified EMSA as previously described [21]. DNA substrates for EMSA were obtained by PCR from *Mtb* H37Rv genomic DNA, and oligo fragments and primer sequences are shown in Table S1. The purified HigA2 protein (1 μM) and DNA substrate were added to the EMSA reaction buffer (20 mM Tris-HCl, 150 mM NaCl, pH 8.0) (10 μL), and the reaction was performed at 37 °C for 5 min. Then, 2 μL of the 6 × DNA loading buffer was added, mixed well, and directly subjected to agarose gel electrophoresis or native-PAGE.

2.6. RT-PCR Assays

The cDNAs of H37Rv wild-type and Δ*higA2* knockout strains (Shanghai Gene-optimal Science & Technology, Shanghai, China) were used as templates, and primers are shown in Table S1. The *sigA* gene was selected as the internal reference, and a fragment of approximately 200 bp of possible downstream regulatory genes of HigA2 was subjected to RT-PCR.

2.7. DNA Pull-Down

Mtb H37Rv genomic DNA was used as a template to design a 5′ biotin-labeled probe for the *Rv2023c-higB2-higA2* core promoter region. The labeled probe was purified using a gel recovery kit to recover the probe, pre-mixed with 5 μg of biotin-labeled DNA and 500 μg of H37Rv total protein (Shanghai Gene-optimal Science & Technology, Shanghai, China) on the ice. Next, 100 μL of BeyoMag™ Streptavidin Magnetic Beads (Beyotime, Shanghai, China) was pretreated with cool PBS, and the mixture of DNA and protein was added to the magnetic beads. The beads were resuspended and incubated at 4 °C for 1 h. The beads were centrifuged at 5000× *g* for 30 s, the supernatant was removed, the precipitate was collected, and the beads were washed five times with cool PBS and centrifuged at 5000× *g* for 1 min. Then, as much of the supernatant as possible was removed, 200 μL of PBS was added to resuspend the magnetic beads, and the proteins were identified by LC-MS/MS (Beijing Bio-Tech Pack Technology Company, Beijing, China).

3. Results

3.1. HigA2 Exhibits Toxicity in M. smegmatis

We cloned *higA2*, *higB2*, and *higB2-higA2* into pJV53 and successfully constructed the shuttle plasmid of *M. smegmatis* pJV-*higA2*, pJV-*higB2*, and pJV-*higB2-higA2*. The plasmid was transformed into *M. smegmatis* competent cells by electroporation. It was verified by colony PCR, and the primers are shown in Table S1. Acetamide was added to induce gene expression, and it was found that growth inhibition appeared in the *M. smegmatis* induced by the expression of *higA2* (Figure 1). HigB2 did not show obvious toxicity.

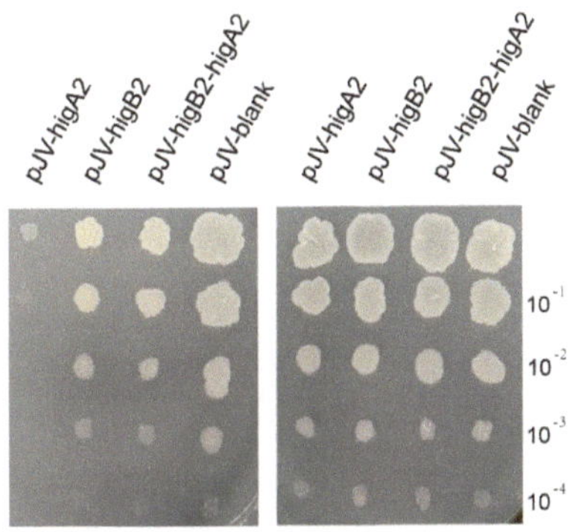

Figure 1. *M. smegmatis* growth performance check. Bacterial solution was adjusted to $OD_{600} = 0.2$. The resuspension was diluted 10^{-1}, 10^{-2}, 10^{-3}, and 10^{-4}-fold, and each diluted sample was spotted onto 7H10 solid medium with or without 0.2% Acetamide.

3.2. HigA2 Binding Site Identifications

HigA2 protein with a C-terminal 6×His-tag was successfully purified (Figure 2A). Reverse transcription PCR of the two-by-two spacer regions of *Rv2023c*, *higB2*, and *higA2* genes identified the existence of co-transcription of the three genes (Figure 2B). To determine the binding site of HigA2, the sequences 59 bp upstream of the coding frames of HigA2 and the DNA substrates *higA2*, *higB2*, and *Rv2023c* were subjected to an EMSA reaction, and the data did not show any binding of HigA2 to the three fragments. A DNA probe design was performed inside the *higB2-higA2* gene, the HigA2 protein and DNA probe were subjected to the EMSA reaction, and the DNA probe positions are shown here. Figure 2C shows that the HigA2 protein binds to the internal sequence seq2 of *higB2* (Figure 2D). The region was narrowed down to seq4. The three possible binding sites within *higB2*, namely seq5, seq6, and seq7, were further examined by EMSA, and the final binding sequence was determined to be seq6.

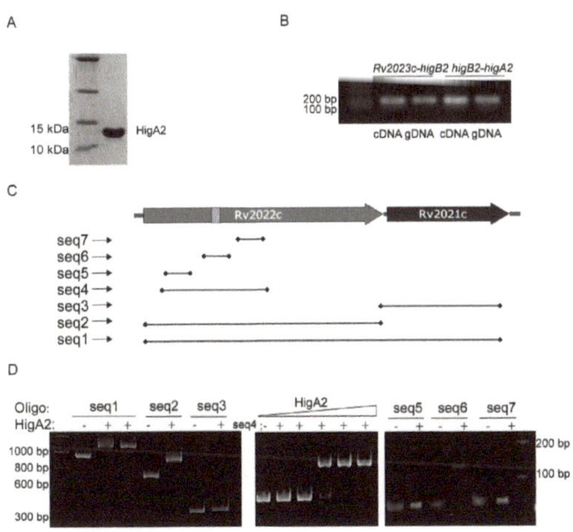

Figure 2. EMSA assay of HigA2 protein with possible binding sites. (**A**) C-terminal 6×His-tagged HigA2 was successfully expressed and purified. (**B**) PCR assay of *Rv2023c-higB2* and *higB2-higA2* intergenic regions in cDNA and gDNA of *Mtb* H37Rv. (**C**) The location of the HigA2 EMSA oligos. (**D**) EMSA detection of probe seq1-seq7 with HigA2.

3.3. Identification of Conserved HigA2 Recognition Motifs

The seq6 sequence has a palindromic motif formed by two inverted repeats (5′-ATATCAC(N)6GTGATAT-3′), and in order to confirm the importance of the specific recognition of this sequence, eight mutants originating from a DNA fragment of *higB2* promoter DNA with a length of 59 bp were used as substrates (Figure 3A). Among them, the flag region or inter-region of the m1 to m5 fragments were replaced by random sequences, respectively, or simultaneously, and the m6 fragment was reduced by two external bases. The two flag regions of the m7 fragment were reduced by one internal base, and the two flag regions of the m8 fragment were reduced by two internal bases. The results showed that HigA2 was able to bind to the DNA substrates mu4, mu5, mu6, and mu7 and weakly to DNA fragments with both binding site spacer lengths (8 bp), whereas neither a single binding site nor half of the palindromic sequence could bind to HigA2 (Figure 3B).

A

seq6 CGCTGCTCTTACCCGGAGGAATATCACCCTTGTGTGATATCGCGTATCGCGCTACTCTG
mu1 CGCTGCTCTTACCCGGAGGAATATCACCCTTGT**ggatccg**CGCGTATCGCGCTACTCTG
mu2 CGCTGCTCTTACCCGGAGGA**gctcgag**CCTTGTGTGATATCGCGTATCGCGCTACTCTG
mu3 CGCTGCTCTTACCCGGAGGA**gctcgag**CCTTGT**ggatccg**CGCGTATCGCGCTACTCTG
mu4 CGCTGCTCTTACCCGGAGGAATATCAC**aggtcc**GTGATATCGCGTATCGCGCTACTCTG
mu5 **ctgctcaaagatcgactagt**ATATCAC**aggtcc**GTGATAT**gacgtatacctttggtgca**
mu6 CGCTGCTCTTACCCGGAGGA**gc**ATCACCCTTGTGTGAT**ca**CGCGTATCGCGCTACTCTG
mu7 **ctgctcaaagatcgactagt**ATATCA**taggtcct**TGATAT**gacgtatacctttggtgca**
mu8 **ctgctcaaagatcgactagt**ATATC**ctaggtccta**GATAT**gacgtatacctttggtgca**

B

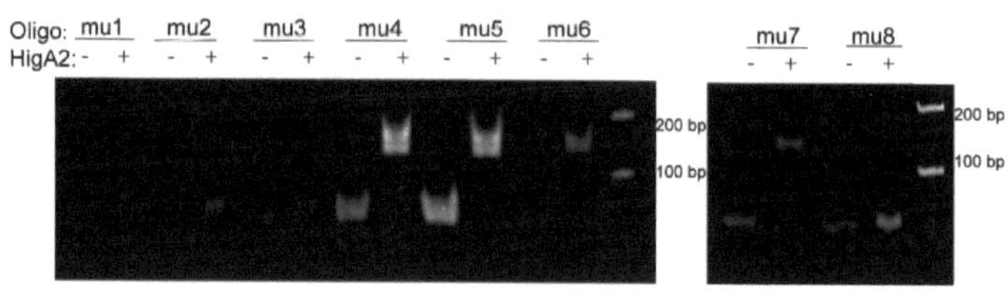

Figure 3. EMSA assay of HigA2 protein with promoter DNA mutants. (**A**) Sequences of seq6 and the eight mutants, with substitutions indicated by lowercase letters. Flag region is the inverted repeat sequence outlined and underlined, and the inter-region is the sequence between the two repeats. (**B**) EMSA experiments were performed on HigA2 and 8 mutant oligos.

3.4. HigA2 Regulating Sites Exploration on the Genome Scale

Following the above results, we searched the *Mtb* H37Rv genome based on 5′-ATCAC (N)4GTGAT-3′, 5′-ATCAC(N)5GTGAT-3′, 5′-ATCAC(N)6GTGAT-3′, and 5′-ATCAC(N)7GT GAT-3′, and 31 possible binding sites were identified (Table S2). Thirty-one oligo fragments containing the above motifs were synthesized and subjected to an EMSA assay. Finally, it was determined that HigA2 could bind to eight oligos. Different genes were tested for their binding ability, and different concentrations of the HigA2 protein with regulatory motifs were subjected to EMSA experiments. HigA2 was able to bind essentially completely to *higB2*, *Rv2044c*, and *Rv0258c* at 5 μM; HigA2 was able to bind essentially completely to *Rv0010c*, *Rv1733c*, and *Rv3396c* at 8 μM. HigA2 binds essentially completely to *Rv0086* and *Rv2434c* at 20 μM (Figure 4).

The positions of the binding sites on the 11 genes and the functions of these genes are shown in Table 1, while the remaining genes may be involved in the establishment of the holding mechanism. Of these, Rv2043c is the proazinamidase *pncA*, which is responsible for the activation of the antibiotic proazinamide [22]. *Rv0086* and *Rv0087* are located within the *Rv0081-Rv0088* operon, which is involved in hypoxic adaptation [23,24]. Rv3396c is guanosine synthase *guaA*, an essential gene for basal metabolism, and is also involved in ppGpp signaling regulation [25]. Rv1733c is a membrane protein that has been found to act as a dormancy-associated surface antigen that activates the host immune system, with the potential to be developed as a vaccine for holdout bacteria [26]. Rv2433c is a secreted protein that can be recognized by T cells as an antigen [27]. Rv0010c is an unknown protein whose mutation is involved in pyrazinamide resistance [28].

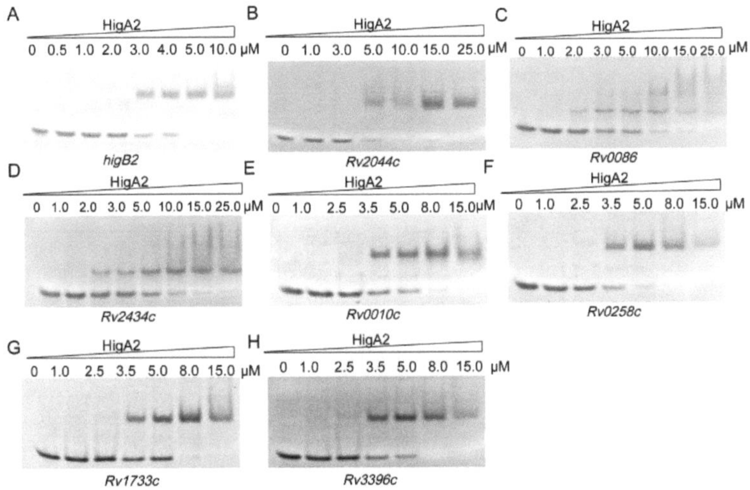

Figure 4. EMSA assay of HigA2 on the target oligos. (**A–H**) Eight oligos labeled with a length of 40 bp were all at a concentration of 1 μM and co-incubated with increasing concentrations of HigA2 protein.

Table 1. Conserved recognition motifs for HigA2.

Gene	Location	Feature	Illustration [a]
Rv2043c	5′-UTR	*pncA*: Pyrazinamidase (PZase)	
Rv2044c	CDS	relative with pyrazinamide resistance	
Rv0086	CDS	*hycQ*: hydrogenase HycQ	
Rv0087	5′-UTR	*hycE*: Possible formate hydrogenase	
Rv3396c	CDS	*guaA*: GMP synthase	
Rv1733c	CDS	transmembrane protein, surface antigen	
Rv2433c	5′-UTR	Secretory protein; T cell antigen	
Rv2434c	CDS	transmembrane protein	
Rv0010c	5′-UTR	relative with pyrazinamide resistance	
Rv0258c	5′-UTR	hypothetical protein	
higB2	CDS	*higB2*: hypothetical protein	

[a] The dark arrow indicates the gene, the light arrow is the adjacent gene, and the white rectangle indicates the binding site.

The relative expression of all eight genes showed differences in the Δ*higA2* mutant strain. *higB2*, *Rv2044c*, *Rv2434c*, *Rv0010c*, and *Rv0258c* were up-regulated, while *Rv0086*, *Rv1733c*, and *Rv3396c* showed down-regulation. *Rv0086* showed significant down-regulation. For the binding site located at the end of the gene, *Rv0086* neighboring gene *Rv0087*, *Rv2434c* neighboring gene *Rv2433c*, and *Rv2044c* neighboring gene *Rv2043c* also appeared to be upregulated (Figure 5).

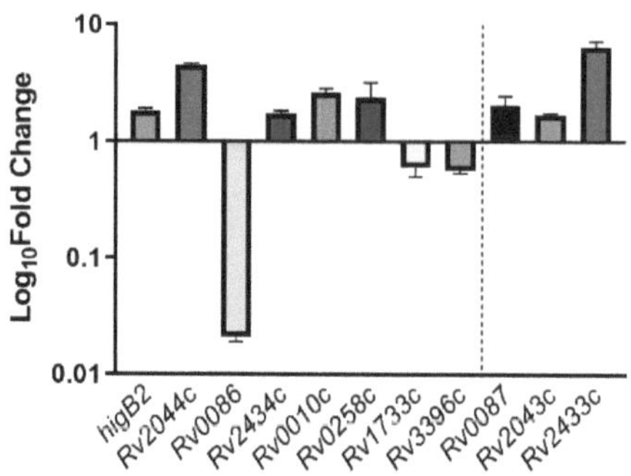

Figure 5. The expression of HigA2 downstream gene. The expression of HigA2 downstream genes was analyzed by reverse transcription PCR in the Δ*higA* mutant strain. The data are expressed as the relative fold expression of mRNA compared to *sigA*, the endogenous control.

3.5. Exploration of Upstream Regulatory Genes of HigBA2

In order to find the gene that regulates HigA2 transcription, we designed two FAM fluorescence-labeled DNA probes for the 5′-UTR region of the HigA2 gene and performed a DNA pull-down assay on the total protein extract of *Mtb* H37Rv (Figure 6).

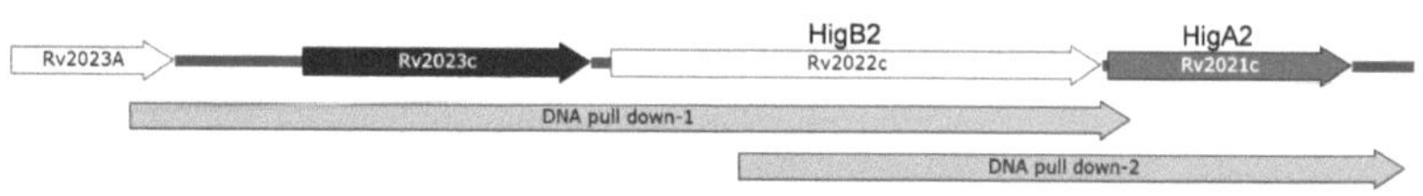

Figure 6. Illustration of DNA pull-down probes. Illustration of DNA pull-down-1 and DNA pull-down-2 in *Rv2023c-higB2-higB2*.

The samples obtained from the elution were subjected to protein profiling, and 412 proteins were identified that could bind to the above region. Among them, 12 were transcriptional regulators, and the specific information is shown in Table 2. All of the above-mentioned upstream regulatory genes may promote the formation of *Mtb*-holding bacteria by regulating the transcription of HigA2.

Table 2. Transcription factors bound by *higBA2* pull-down.

Protern	Feature
Rv0903c	essential gene *prrA*, transcriptional regulator of the two component system PrrA/PrrB
Rv0744c	Possible transcriptional regulatory protein, similar to a two-component sensor
Rv3574	KstR, probable TetR-family transcriptional regulator involved in lipid metabolism
Rv3833	Probable AraC-family transcriptional regulatory protein
Rv2488c	Probable LuxR-family transcriptional regulatory protein
Rv3676	Crp, cAMP-activated global transcriptional regulator
Rv1909c	FurA, ferric uptake regulation protein, transcriptional regulator
Rv2166c	essential gene MraZ, transcriptional regulator
Rv3124	MoaR1, transcriptional regulator
Rv0474	Cu^{2+} responsive transcriptional regulator
Rv2603c	Probable transcriptional regulator
Rv3583c	essential gene CarD, RNA polymerase-binding transcription factor

4. Discussion

Bacteria have evolved complex regulatory controls and multiple cellular transition states in response to a variety of environmental stresses. In order to survive, cells slow down their growth rate and redirect their metabolic resources until conditions are such that growth can be resumed [29,30]. The transcriptional activation mechanism of the TA system is essential for bacterial persistence, and although the mechanism of action can vary greatly and different DNA-binding domains and transcriptional regulatory mechanisms can be found even among members of the same TA family, toxin activity is diverse regardless of TA function and has been shown to interfere with basic cellular function [31].

4.1. HigBA2 Remains Uncleared for TA System Activity

Based on previous studies, the antitoxin HigA2 in *Mtb* acts as a transcriptional regulator with self-cleavage and structural flexibility and may bind DNA through HTH motifs [17]. Our study found that HigA2 binds within the *higB2* gene, and the expression of HigA2 in *M. smegmatis* resulted in growth inhibition of the bacteria, whereas the toxin HigB2 did not show toxicity.

Unfortunately, we did not successfully express HigB2 after replacing a variety of vectors, which may be due to inaccurate ORF prediction. Therefore, the binding activity of the antitoxin HigA2 and the toxin HigB2 is unknown. In *Mtb* and *E. coli*, the expression of the HigB1 toxin prevented bacterial growth and led to cell death [32–34]. After the induction of HigB1, a significant loss of viability was observed, leaving only a subset with potential acquisition persistence. tmRNA is a conserved target of HigB1 [32].

4.2. Downstream Regulatory Genes of HigA2 May Be Involved in the Establishment of Persistence

In the *Mtb* genome, HigA2 binds to and regulates a number of genes, six of which are bound to the interior of genes and five to the 5′-UTR region. TF regulates the transcription of genes in different ways by binding to CDS [35]. HigA2 regulates the transcription of these genes and may contribute to *Mtb* persistence.

pncA (*Rv2043c*) is co-transcribed into a polycistron along with *Rv2044c*, which is located 40 bp upstream of *pncA*. *pncA* encodes the pyrazinamidase enzyme, which is responsible for the conversion of an important first-line anti-TB drug, pyrazinamide (PZA), into its active form, and resistance to PZA is primarily due to mutations in *pncA* [36,37]. The detection of novel pyrazinamide-resistant mutations in clinical isolates of multidrug-resistant *Mtb* revealed novel non-synonymous mutations (Tyr70His, Ile71Asn) with effects on *PZase* activity in *Rv2044c*, all in the HigA2 binding site [38]. However, *Rv2044c* was significantly upregulated after *higA2* was knocked out and *pcnA* showed only slight upregulation. We hypothesized that *higA2* may influence *pcnA* to function under certain circumstances. The examination of intergenic distances suggests that the seven genes in *Rv0081-Rv0087* may form one operon [39]. *Rv0086* encodes proteins for the possible hydrogenase HycQ and is possibly involved in hydrogen metabolism. Based on the genetics of *M. smegmatis*, which induce the expression of their genes during starvation and hypoxia, they improve survival by scavenging atmospheric H_2 [40,41]. *M. smegmatis* increased the transcription and synthesis of a form of dehydrogenase by 50-fold in response to organic carbon limitation, a process associated with persistence [42]. Guanosine monophosphate synthetase (GMPS), encoded by the *Rv3396c* (*guaA*), is a key enzyme in the biosynthesis of guanine nucleotides in *Mtb*. *guaA* is essential for the growth of *Mtb* H37Rv, and deletion of the *guaA* gene resulted in *Mtb* lethality [25]. In *E. coli*, *guaA* is essential for the formation of persister cells, as their absence significantly enhances cell sensitivity to various antibiotics [43]. GuaA acted as an upstream reaction in the ppGpp biosynthesis pathway and may affect persistence by disrupting ppGpp regulation [44].

Many anti-TB drugs, such as INH and EMB, target biological cell membranes to break down this barrier and weaken the bacteria. These anti-TB drugs that target the cell envelope can allow other drugs to pass through by interacting with the membrane. Importantly, many of the mutations that confer resistance to anti-TB drugs occur in these cell envelope biogenesis pathways, which is critical for new anti-TB drugs to combat TB resistance [45]. *Rv0010c* encodes a conserved membrane protein of unknown function that may play a role in the cell wall and cellular processes. There was a DnaA interaction site in the intergenic region of *Rv0010c-Rv0011c*, which was the same as the HigA2 binding site. At the same time, this study showed that the *mutation* of *Rv0010c-Rv0011c* in this position led to INH susceptibility. This mutation and *dnaA* mutation regulated the cell cycle and INH resistance in the same way [46]. *Rv2434c* may encode a conserved membrane protein, possibly cyclic glycine-binding proteins, involved in cAMP signaling pathway action [47]. The structure of Rv2433c (CFP11) has been studied [48]. It was identified as a human immunodominant T-cell antigen in a previous study [49], and CFP11 significantly increased immunoglobulin levels in human serum and promoted lymphocyte proliferation and interferon production [27]. Hypoxia-associated latency antigen Rv1733c is a possible integral membrane protein that stimulates cells to produce a certain level of cellular immunity, and this antigenic protein is readily recognized by the immune system of latent tuberculosis infection (LTBI) [50].

4.3. *HigBA2 May Be Regulated by Multiple Upstream Proteins in Response to External Stresses*

The results of DNA pull-down experiments suggest a possible response mechanism for *Mtb* during external stress. There are 12 transcription factors that may regulate the expression of HigBA2. These include two two-component signaling system regulators: Rv0903c (prrA) belonging to the prrAB two-component system in response to starvation conditions such as nitrogen source limitation [51], and Rv0744c, which responds to starvation conditions such as carbon source limitation [52]. Five induced variant transcriptional regulators, namely Rv3574 (TetR-family), Rv3833 (AraC-family), Rv2488c (LuxR-family), Rv3676 (Crp) sensed cAMP, and Rv1909c (FurA), could sense Fe^{2+}. It was found that Rv2603c contributed to *Mtb* tolerance in macrophages [53]. Rv3124 (MoaR1) regulated molybdenum chitosan biosynthesis and was involved in hypoxic adaptation [54]. Rv0474 responded to Cu^{2+}, inhibiting *rpoB* transcription to put the bacterium into dormancy [55]. Rv3583c (CarD) is a global transcriptional regulator that responds to starvation conditions [56]. These transcriptional regulators have the function of sensing external regulatory signals and may be involved in the transduction and regulation of external stress signals. However, their binding activities need to be further verified.

In summary, HigA2 regulates the expression of itself and multiple genes and may be involved in regulating *Mtb* hypoxia adaptation and the ppGpp pathway to promote persistence. Although the series of base substitutions on the DNA binding site in this study were able to illustrate the characteristics of the HigA2 recognition motif, more precise detection methods such as NMR titration or X-ray crystallography are still necessary to determine the accurate structure of the bound form of HigA2 and DNA, which is in the scope of our next research study. Due to lab limitations, we could not characterize the persistence difference between the wild H37Rv strain and the mutant Δ*higA2* strain since the cultivation of *Mtb* requires a BSL-3 laboratory. We will strive to seek cooperation with BSL-3 experimental platforms to conduct subsequent research such as persistence testing, co-cultivation with macrophages, etc. In our next study, the upstream regulatory genes of HigA2 should be further identified, and the regulatory mechanism of up-stream regulatory gene expression induced by external stress conditions such as anti-tuberculosis drugs, starvation conditions, and continuous hypoxia should be verified so as to determine the complete signal pathway and mechanism of *Mtb* persistence caused by external stress.

5. Conclusions

HigA2 may interfere with protein expression by binding to these genes, thereby affecting processes associated with *Mtb* persistence. There may exist a transcriptional regulatory network centered on *higA2* that regulates *Mtb*'s response to multiple external stresses and the transcription of multiple downstream genes, driving *Mtb* into a persistent state. Further study of this regulatory network can provide ideas for understanding the establishment of the *Mtb* persistence mechanism and the development of anti-TB drugs.

Supplementary Materials: The following supporting information can be downloaded at: https://www.mdpi.com/article/10.3390/microorganisms12061244/s1, Table S1: Primers used in this study; Table S2: The distribution of target sites of HigA2 conserved recognition motifs in *Mtb* genome.

Author Contributions: Conceptualization, T.H. and L.G.; data curation, M.X. and M.L.; funding acquisition, T.H. and L.G.; methodology, M.X., T.L. and X.P.; project administration, T.H.; validation, M.L., T.L. and Q.R.; visualization, M.L.; writing—original draft, M.X.; writing—review and editing, T.H. All authors have read and agreed to the published version of the manuscript.

Funding: This research was funded by the National Natural Science Foundation of China (NSFC), grant number 31800067; and the Natural Science Foundation of Hebei Province, grant number H2020209001.

Data Availability Statement: The original contributions presented in the study are included in the article and Supplementary Materials, further inquiries can be directed to the corresponding authors.

Conflicts of Interest: The authors declare no conflicts of interest.

References

1. World Health Organization (WHO). *Global Tuberculosis Report 2023*; World Health Organization: Geneva, Switzerland, 2023.
2. Boldrin, F.; Provvedi, R.; Cioetto Mazzabò, L.; Segafreddo, G.; Manganelli, R. Tolerance and Persistence to Drugs: A Main Challenge in the Fight Against *Mycobacterium tuberculosis*. *Front. Microbiol.* **2020**, *11*, 1924. [CrossRef] [PubMed]
3. Balaban, N.Q. Persistence: Mechanisms for triggering and enhancing phenotypic variability. *Curr. Opin. Genet. Dev.* **2011**, *21*, 768–775. [CrossRef] [PubMed]
4. Zhang, Y. Persisters, persistent infections and the Yin-Yang model. *Emerg. Microbes Infect.* **2014**, *3*, e3. [CrossRef] [PubMed]
5. Fasani, R.A.; Savageau, M.A. Molecular mechanisms of multiple toxin–antitoxin systems are coordinated to govern the persister phenotype. *Proc. Natl. Acad. Sci. USA* **2013**, *110*, e2528–e2537. [CrossRef] [PubMed]
6. Tanaka, M.M.; Akarsu, H.; Bordes, P.; Mansour, M.; Bigot, D.-J.; Genevaux, P.; Falquet, L. TASmania: A bacterial Toxin-Antitoxin Systems database. *PLOS Comput. Biol.* **2019**, *15*, e1006946.
7. Ziemski, M.; Leodolter, J.; Taylor, G.; Kerschenmeyer, A.; Weber-Ban, E. Genome-wide interaction screen for *Mycobacterium tuberculosis* ClpCP protease reveals toxin-antitoxin systems as a major substrate class. *FEBS J.* **2021**, *288*, 111–126. [CrossRef] [PubMed]
8. Kamruzzaman, M.; Wu, A.Y.; Iredell, J.R. Biological Functions of Type II Toxin-Antitoxin Systems in Bacteria. *Microorganisms* **2021**, *9*, 1276. [CrossRef] [PubMed]
9. Soo, V.W.; Wood, T.K. Antitoxin MqsA represses curli formation through the master biofilm regulator CsgD. *Sci. Rep.* **2013**, *3*, 3186–3192. [CrossRef] [PubMed]
10. Sala, A.; Bordes, P.; Genevaux, P. Multiple toxin-antitoxin systems in *Mycobacterium tuberculosis*. *Toxins* **2014**, *6*, 1002–1020. [CrossRef]
11. Mansour, M.; Giudice, E.; Xu, X.; Akarsu, H.; Bordes, P.; Guillet, V.; Bigot, D.J.; Slama, N.; D'Urso, G.; Chat, S.; et al. Substrate recognition and cryo-EM structure of the ribosome-bound TAC toxin of *Mycobacterium tuberculosis*. *Nat. Commun.* **2022**, *13*, 2641–2654. [CrossRef]
12. Turkarslan, S.; Peterson, E.J.R.; Rustad, T.R.; Minch, K.J.; Reiss, D.J.; Morrison, R.; Ma, S.; Price, N.D.; Sherman, D.R.; Baliga, N.S. A comprehensive map of genome-wide gene regulation in *Mycobacterium tuberculosis*. *Sci. Data* **2015**, *2*, 150010. [CrossRef]
13. Keren, I.; Minami, S.; Rubin, E.; Lewis, K. Characterization and transcriptome analysis of *Mycobacterium tuberculosis* persisters. *mBio* **2011**, *2*, 100–111. [CrossRef]
14. Betts, J.C.; Lukey, P.T.; Robb, L.C.; McAdam, R.A.; Duncan, K. Evaluation of a nutrient starvation model of *Mycobacterium tuberculosis* persistence by gene and protein expression profiling. *Mol. Microbiol.* **2002**, *43*, 717–731. [CrossRef]
15. Bähler, J.; Rustad, T.R.; Harrell, M.I.; Liao, R.; Sherman, D.R. The Enduring Hypoxic Response of *Mycobacterium tuberculosis*. *PLoS ONE* **2008**, *3*, e1502.
16. Gupta, A.; Venkataraman, B.; Vasudevan, M.; Gopinath Bankar, K. Co-expression network analysis of toxin-antitoxin loci in *Mycobacterium tuberculosis* reveals key modulators of cellular stress. *Sci. Rep.* **2017**, *7*, 5868–5881. [CrossRef]
17. Richardson, W.; Kang, G.W.; Lee, H.J.; Kwon, K.M.; Kim, S.; Kim, H.J. Chasing the structural diversity of the transcription regulator *Mycobacterium tuberculosis* HigA2. *IUCrJ* **2021**, *8 Pt 5*, 823–832. [CrossRef]
18. Chi, X.; Chang, Y.; Li, M.; Lin, J.; Liu, Y.; Li, C.; Tang, S.; Zhang, J. Biochemical characterization of mt-PemIK, a novel toxin-antitoxin system in *Mycobacterium tuberculosis*. *FEBS Lett.* **2018**, *592*, 4039–4050. [CrossRef]
19. Agarwal, S.; Tiwari, P.; Deep, A.; Kidwai, S.; Gupta, S.; Thakur, K.G.; Singh, R. System-Wide Analysis Unravels the Differential Regulation and In Vivo Essentiality of Virulence-Associated Proteins B and C Toxin-Antitoxin Systems of *Mycobacterium tuberculosis*. *J. Infect. Dis.* **2018**, *217*, 1809–1820. [CrossRef]
20. Gou, L.; Han, T.; Wang, X.; Ge, J.; Liu, W.; Hu, F.; Wang, Z. A Novel TetR Family Transcriptional Regulator, CalR3, Negatively Controls Calcimycin Biosynthesis in Streptomyces chartreusis NRRL 3882. *Front. Microbiol.* **2017**, *8*, 2371. [CrossRef] [PubMed]

21. Gao, C.H.; Yang, M.; He, Z.G. An ArsR-like transcriptional factor recognizes a conserved sequence motif and positively regulates the expression of phoP in *Mycobacteria*. *Biochem. Biophys. Res. Commun.* **2011**, *411*, 726–731. [CrossRef] [PubMed]
22. Juréen, P.; Werngren, J.; Toro, J.-C.; Hoffner, S. Pyrazinamide Resistance and pncA Gene Mutations in *Mycobacterium tuberculosis*. *Antimicrob. Agents Chemother.* **2008**, *52*, 1852–1854. [CrossRef]
23. Sun, X.; Zhang, L.; Jiang, J.; Ng, M.; Cui, Z.; Mai, J.; Ahn, S.K.; Liu, J.; Zhang, J.; Liu, J.; et al. Transcription factors Rv0081 and Rv3334 connect the early and the enduring hypoxic response of *Mycobacterium tuberculosis*. *Virulence* **2018**, *9*, 1468–1482. [CrossRef]
24. Kumar, A.; Phulera, S.; Rizvi, A.; Sonawane, P.J.; Panwar, H.S.; Banerjee, S.; Sahu, A.; Mande, S.C. Structural basis of hypoxic gene regulation by the Rv0081 transcription factor of *Mycobacterium tuberculosis*. *FEBS Lett.* **2019**, *593*, 982–995. [CrossRef]
25. Villela, A.D.; Eichler, P.; Pinto, A.F.M.; Rodrigues-Junior, V.; Yates Iii, J.R.; Bizarro, C.V.; Basso, L.A.; Santos, D.S. Gene replacement and quantitative mass spectrometry approaches validate guanosine monophosphate synthetase as essential for *Mycobacterium tuberculosis* growth. *Biochem. Biophys. Rep.* **2015**, *4*, 277–282. [CrossRef]
26. Zhang, L.; Ma, H.; Wan, S.; Zhang, Y.; Gao, M.; Liu, X. *Mycobacterium tuberculosis* latency-associated antigen Rv1733c SLP improves the accuracy of differential diagnosis of active tuberculosis and latent tuberculosis infection. *Chin. Med. J.* **2022**, *135*, 63–69. [CrossRef]
27. Eweda, G.; Suzuki, D.; Nagata, T.; Tsujimura, K.; Koide, Y. Identification of murine T-cell epitopes on low-molecular-mass secretory proteins (CFP11, CFP17, and TB18.5) of *Mycobacterium tuberculosis*. *Vaccine* **2010**, *28*, 4616–4625. [CrossRef]
28. Shi, W.; Chen, J.; Zhang, S.; Zhang, W.; Zhang, Y. Identification of Novel Mutations in LprG (rv1411c), rv0521, rv3630, rv0010c, ppsC, and cyp128 Associated with Pyrazinoic Acid/Pyrazinamide Resistance in *Mycobacterium tuberculosis*. *Antimicrob. Agents Chemother.* **2018**, *62*, e00430-18. [CrossRef]
29. Moreno-Del Alamo, M.; Marchisone, C.; Alonso, J.C. Antitoxin ε Reverses Toxin zeta-Facilitated Ampicillin Dormants. *Toxins* **2020**, *12*, 801. [CrossRef]
30. Balaban, N.Q.; Helaine, S.; Lewis, K.; Ackermann, M.; Aldridge, B.; Andersson, D.I.; Brynildsen, M.P.; Bumann, D.; Camilli, A.; Collins, J.J.; et al. Definitions and guidelines for research on antibiotic persistence. *Nat. Rev. Microbiol.* **2019**, *17*, 441–448. [CrossRef] [PubMed]
31. De Bruyn, P.; Girardin, Y.; Loris, R. Prokaryote toxin-antitoxin modules: Complex regulation of an unclear function. *Protein Sci.* **2021**, *30*, 1103–1113. [CrossRef] [PubMed]
32. Schuessler, D.L.; Cortes, T.; Fivian-Hughes, A.S.; Lougheed, K.E.A.; Harvey, E.; Buxton, R.S.; Davis, E.O.; Young, D.B. Induced ectopic expression of HigB toxin in *Mycobacterium tuberculosis* results in growth inhibition, reduced abundance of a subset of mRNAs and cleavage of tmRNA. *Mol. Microbiol.* **2013**, *90*, 195–207. [CrossRef]
33. Bordes, P.; Cirinesi, A.-M.; Ummels, R.; Sala, A.; Sakr, S.; Bitter, W.; Genevaux, P. SecB-like chaperone controls a toxin–antitoxin stress-responsive system in *Mycobacterium tuberculosis*. *Proc. Natl. Acad. Sci. USA* **2011**, *108*, 8438–8443. [CrossRef]
34. Gupta, A. Killing activity and rescue function of genome-wide toxin-antitoxin loci of *Mycobacterium tuberculosis*. *FEMS Microbiol. Lett.* **2009**, *290*, 45–53. [CrossRef]
35. Hua, C.; Huang, J.; Wang, T.; Sun, Y.; Liu, J.; Huang, L.; Deng, X.; Chang, Y.-F. Bacterial Transcription Factors Bind to Coding Regions and Regulate Internal Cryptic Promoters. *mBio* **2022**, *13*, e0164322. [CrossRef]
36. Baddam, R.; Kumar, N.; Wieler, L.H.; Lankapalli, A.K.; Ahmed, N.; Peacock, S.J.; Semmler, T. Analysis of mutations in pncA reveals non-overlapping patterns among various lineages of *Mycobacterium tuberculosis*. *Sci. Rep.* **2018**, *8*, 4628–4636. [CrossRef]
37. Mahmood, N.; Bhatti, S.; Abbas, S.N.; Shahid, S.; Nasir, S.B. The pncA gene mutations of *Mycobacterium tuberculosis* in multidrug-resistant tuberculosis. *Biotechnol. Appl. Biochem.* **2021**, *69*, 2195–2204. [CrossRef]
38. Hameed, H.M.A.; Tan, Y.; Islam, M.M.; Lu, Z.; Chhotaray, C.; Wang, S.; Liu, Z.; Fang, C.; Tan, S.; Yew, W.W.; et al. Detection of Novel Gene Mutations Associated with Pyrazinamide Resistance in Multidrug-Resistant *Mycobacterium tuberculosis* Clinical Isolates in Southern China. *Infect. Drug Resist.* **2020**, *13*, 217–227. [CrossRef]
39. Bacon, J.; James, B.W.; Wernisch, L.; Williams, A.; Morley, K.A.; Hatch, G.J.; Mangan, J.A.; Hinds, J.; Stoker, N.G.; Butcher, P.D.; et al. The influence of reduced oxygen availability on pathogenicity and gene expression in *Mycobacterium tuberculosis*. *Tuberculosis* **2004**, *84*, 205–217. [CrossRef]
40. Berney, M.; Greening, C.; Conrad, R.; Jacobs, W.R.; Cook, G.M. An obligately aerobic soil bacterium activates fermentative hydrogen production to survive reductive stress during hypoxia. *Proc. Natl. Acad. Sci. USA* **2014**, *111*, 11479–11484. [CrossRef] [PubMed]
41. Berney, M.; Cook, G.M. Unique flexibility in energy metabolism allows *Mycobacteria* to combat starvation and hypoxia. *PLoS ONE* **2010**, *5*, e8614. [CrossRef] [PubMed]
42. Cordero, P.R.F.; Bayly, K.; Man Leung, P.; Huang, C.; Islam, Z.F.; Schittenhelm, R.B.; King, G.M.; Greening, C. Atmospheric carbon monoxide oxidation is a widespread mechanism supporting microbial survival. *ISME J.* **2019**, *13*, 2868–2881. [CrossRef] [PubMed]
43. Mohiuddin, S.G.; Massahi, A.; Orman, M.A. High-Throughput Screening of a Promoter Library Reveals New Persister Mechanisms in *Escherichia Coli*. *Microbiol. Spectr.* **2022**, *10*, e0225321. [CrossRef] [PubMed]
44. Hauryliuk, V.; Atkinson, G.C.; Murakami, K.S.; Tenson, T.; Gerdes, K. Recent functional insights into the role of (p)ppGpp in bacterial physiology. *Nat. Rev. Microbiol.* **2015**, *13*, 298–309. [CrossRef] [PubMed]

45. Batt, S.M.; Burke, C.E.; Moorey, A.R.; Besra, G.S. Antibiotics and resistance: The two-sided coin of the mycobacterial cell wall. *Cell Surf.* **2020**, *6*, 100044–100061. [CrossRef] [PubMed]
46. Hicks, N.D.; Giffen, S.R.; Culviner, P.H.; Chao, M.C.; Dulberger, C.L.; Liu, Q.; Stanley, S.; Brown, J.; Sixsmith, J.; Wolf, I.D.; et al. Mutations in dnaA and a cryptic interaction site increase drug resistance in *Mycobacterium tuberculosis*. *PLoS Pathog.* **2020**, *16*, e1009063. [CrossRef] [PubMed]
47. Johnson, R.M.; McDonough, K.A. Cyclic nucleotide signaling in *Mycobacterium tuberculosis*: An expanding repertoire. *Pathog. Dis.* **2018**, *76*, fty048. [CrossRef] [PubMed]
48. Bu, L.; Brooks, C.L., 3rd. De novo prediction of the structures of *M. tuberculosis* membrane proteins. *J. Am. Chem. Soc.* **2008**, *130*, 5384–5385.
49. Sable, S.B.; Kumar, R.; Kalra, M.; Verma, I.; Khuller, G.K.; Dobos, K.; Belisle, J.T. Peripheral Blood and Pleural Fluid Mononuclear Cell Responses to Low-Molecular-Mass Secretory Polypeptides of *Mycobacterium tuberculosis* in Human Models of Immunity to Tuberculosis. *Infect. Immun.* **2005**, *73*, 3547–3558. [CrossRef]
50. Zhang, W.; Jiang, H.; Bai, Y.L.; Kang, J.; Xu, Z.K.; Wang, L.M. Construction and Immunogenicity of the DNA Vaccine of *Mycobacterium Tuberculosis* Dormancy Antigen Rv1733c. *Scand. J. Immunol.* **2014**, *79*, 292–298. [CrossRef]
51. Haydel, S.E.; Malhotra, V.; Cornelison, G.L.; Clark-Curtiss, J.E. The prrAB Two-Component System Is Essential for *Mycobacterium tuberculosis* Viability and Is Induced under Nitrogen-Limiting Conditions. *J. Bacteriol.* **2012**, *194*, 354–361. [CrossRef]
52. Subbian, S.; Gautam, U.S.; Mehra, S.; Kaushal, D. In-Vivo Gene Signatures of *Mycobacterium tuberculosis* in C3HeB/FeJ Mice. *PLoS ONE* **2015**, *10*, e0135208.
53. Gao, L.-Y.; Groger, R.; Cox, J.S.; Beverley, S.M.; Lawson, E.H.; Brown, E.J. Transposon Mutagenesis of Mycobacterium marinumIdentifies a Locus Linking Pigmentation and Intracellular Survival. *Infect. Immun.* **2003**, *71*, 922–929. [CrossRef]
54. Mendoza Lopez, P.; Golby, P.; Wooff, E.; Garcia, J.N.; Garcia Pelayo, M.C.; Conlon, K.; Gema Camacho, A.; Hewinson, R.G.; Polaina, J.; Suárez García, A.; et al. Characterization of the transcriptional regulator Rv3124 of *Mycobacterium tuberculosis* identifies it as a positive regulator of molybdopterin biosynthesis and defines the functional consequences of a non-synonymous SNP in the Mycobacterium bovis BCG orthologue. *Microbiology* **2010**, *156*, 2112–2123.
55. Raghunandanan, S.; Ramachandran, R.; Gomez, R.L.; Devanarayanan, S.; Bommakanti, A.; Kondapi, A.K.; Varadarajan, R.; Kumar, R.A. Rv0474 is a copper-responsive transcriptional regulator that negatively regulates expression of RNA polymerase β subunit in *Mycobacterium tuberculosis*. *FEBS J.* **2018**, *285*, 3849–3869. [CrossRef]
56. Li, X.; Chen, F.; Liu, X.; Xiao, J.; Andongma, B.T.; Tang, Q.; Cao, X.; Chou, S.H.; Galperin, M.Y.; He, J. Clp protease and antisense RNA jointly regulate the global regulator CarD to mediate mycobacterial starvation response. *eLife* **2022**, *11*, e73347. [CrossRef]

microorganisms MDPI

Article

A Comparative Transcriptome Analysis Unveils the Mechanisms of Response in Feather Degradation by *Pseudomonas aeruginosa* Gxun-7

Chaodong Song [1], Rui Liu [1], Doudou Yin [1], Chenjie Xie [1], Ying Liang [1], Dengfeng Yang [2], Mingguo Jiang [1], Hongyan Zhang [1,*] and Naikun Shen [1,*]

1 Guangxi Key Laboratory of Polysaccharide Materials and Modification, School of Marine Sciences and Biotechnology, Guangxi Minzu University, Nanning 530000, China; 15036104955@163.com (C.S.); liiiiiuy@163.com (R.L.); yin13419755068@163.com (D.Y.); xiechenjie2021@163.com (C.X.); ly_ng_bjyx200112@163.com (Y.L.); mzxyjiang@163.com (M.J.)

2 Guangxi Key Laboratory of Marine Natural Products and Combinatorial Biosynthesis Chemistry, Guangxi Beibu Gulf Marine Research Center, Guangxi Academy of Sciences, No. 98, Daxue Road, Nanning 530007, China; yangdengfeng@gxas.cn

* Correspondence: hongyanzhang2008@163.com (H.Z.); shennaik05@126.com (N.S.)

Abstract: Microbial degradation of feathers offers potential for bioremediation, yet the microbial response mechanisms warrant additional investigation. In prior work, *Pseudomonas aeruginosa* Gxun-7, which demonstrated robust degradation of feathers at elevated concentrations, was isolated. However, the molecular mechanism of this degradation remains only partially understood. To investigate this, we used RNA sequencing (RNA-seq) to examine the genes that were expressed differentially in *P. aeruginosa* Gxun-7 when exposed to 25 g/L of feather substrate. The RNA-seq analysis identified 5571 differentially expressed genes; of these, 795 were upregulated and 603 were downregulated. Upregulated genes primarily participated in proteolysis, amino acid, and pyruvate metabolism. Genes encoding proteases, as well as those involved in sulfur metabolism, phenazine synthesis, and type VI secretion systems, were notably elevated, highlighting their crucial function in feather decomposition. Integration of Gene Ontology (GO) and Kyoto Encyclopedia of Genes and Genomes (KEGG) taxonomies, combined with a review of the literature, led us to propose that metabolic feather degradation involves environmental activation, reducing agent secretion, protease release, peptide/amino acid uptake, and metabolic processes. Sulfite has emerged as a critical activator of keratinase catalysis, while cysteine serves as a regulatory mediator. qRT–PCR assay results for 11 selected gene subset corroborated the RNA-seq findings. This study enhances our understanding of the transcriptomic responses of *P. aeruginosa* Gxun-7 to feather degradation and offers insights into potential degradation mechanisms, thereby aiding in the formulation of effective feather waste management strategies in poultry farming.

Keywords: *Pseudomonas aeruginosa*; feather degradation; comparative transcriptome analysis; degradation mechanism

Citation: Song, C.; Liu, R.; Yin, D.; Xie, C.; Liang, Y.; Yang, D.; Jiang, M.; Zhang, H.; Shen, N. A Comparative Transcriptome Analysis Unveils the Mechanisms of Response in Feather Degradation by *Pseudomonas aeruginosa* Gxun-7. *Microorganisms* **2024**, *12*, 841. https://doi.org/10.3390/microorganisms12040841

Academic Editor: Tomohiro Shimada

Received: 6 March 2024
Revised: 15 April 2024
Accepted: 17 April 2024
Published: 22 April 2024

1. Introduction

Feathers constitute a significant byproduct in the poultry industry, accounting for 5–7% of the total body weight of chickens [1,2]. On a global scale, around 2 million tons of feathers are produced each year as a byproduct of the poultry industry. Feathers consist of over 85% beta-keratin, along with 70% amino acids, vitamins, high-value elements, and other growth factors [3–5]. This composition underscores the potential use of feathers as animal feed and biofertilizers [6,7]. However, the structural integrity of feather keratin is notably robust, primarily due to the abundance of cysteine (Cys) residues linked by disulfide bonds, resulting in tightly packed structures that resist chemical and physical factors [8]. Consequently, a substantial portion of these feather wastes is either discarded or incinerated, posing significant environmental concerns [9]. Furthermore, feather waste can serve as a habitat for various pathogenic microorganisms, including *Vibrio* and *Salmonella*, while also emitting pollutants such as ammonia, nitrous oxide, and hydrogen sulfide, thereby posing risks to human health and the environment [10,11]. Consequently, the recycling of keratinous waste resources has emerged as an urgent crucial issue.

Methods for keratin hydrolysis typically include physical techniques such as pressurized hydrolysis, high-temperature puffing, and microwaving puffing, as well as chemical approaches involving acid and alkali treatments, which have traditionally been employed in the conversion of feathers into animal feed [7]. Nevertheless, these processes frequently involve significant energy consumption and can lead to the degradation of specific amino acids, such as lysine, methionine, and tryptophan. Additionally, these methods release sulfur and ammonia waste gases. Microbial keratinases offer an alternative solution by hydrolyzing rigid and highly cross-linked keratin substrates, directly targeting hydrophobic amino acid residues [4]. Consequently, the biodegradation of feather wastes by keratinolytic bacteria into valuable products, including free amino acids, peptides, and ammonium ions, has emerged as an efficient, cost-effective, and environmentally friendly approach in recent decades [12].

Microbial keratinase-producing organisms are widely distributed, including bacteria, actinomycetes, and fungi. Among these, bacteria have been the primary focus of research in the domain of keratin degradation, with *Bacillus licheniformis*, *Bacillus subtilis*, and *Bacillus amyloliquefaciens* demonstrating significant keratinase-producing capabilities [13,14]. The keratin hydrolysate resulting from keratinase activity is rich in ammonia and amino acids, presenting opportunities for the development of biological fertilizers [15]. For instance, the supplementation of animal feed with keratinase has been reported to enhance nutrient digestibility, palatability, and immune response. Nevertheless, the formidable cross-linking facilitated by disulfide bonds in keratins poses a major challenge to their degradation by proteases [16]. While keratinases, a class of proteases, are adept at breaking peptide bonds and degrading keratin, they do not effectively act on hydrophobic substrates or disrupt disulfide bonds [17]. The biodegradation of keratin substrates involves a complex process, and the underlying mechanisms are not yet fully understood. Current research suggests that keratinases often require additional enzymes or reducing agents to cleave disulfide bonds before initiating proteolysis [18]. These reduction reactions can be catalyzed by disulfide bond reductases or exogenous reducing agents. Studies on the mechanisms of disulfide bond disruption have proposed four primary theories: biomembrane potential, mechanical pressure, thiolysis, and enzymatic hydrolysis [19–21]. However, the precise mechanism underlying the efficient degradation of keratin by these microorganisms remains to be fully elucidated, presenting a challenge to the sustainable utilization and industrial development of keratin waste.

In our previous research, a highly efficient feather-degrading strain capable of near-complete feather degradation at 35 °C for 48 h was isolated. Nevertheless, limited knowledge exists regarding the mechanisms underlying feather degradation, particularly concerning the mechanisms involved in disulfide bond disruption [22]. Transcriptomic analysis has emerged as an indispensable and convenient tool for investigating differential gene expression. In comparison to alternative transcriptome sequencing techniques, RNA sequencing (RNA-seq) not only offers the ability to detect a broader range of expression levels with a reduced number of RNA samples but also yields highly reproducible results, allowing for both technical and biological replication.

Therefore, in the context of this study, we employed RNA-seq for transcriptomic analysis and quantitative real-time polymerase chain reaction (qRT–PCR) to assess the expression levels of key genes associated with feather degradation. Building upon the foundation of transcriptomics, we conducted a comparative analysis of genes exhibiting differential expression at crucial time points and explored the metabolic pathways implicated in feather degradation. These correlations between degradation genes and feather biodegradation pathways were established based on our prior investigations. The results of this study offer valuable insights into the mechanisms underlying feather biodegradation. These insights may prove instrumental in the engineering of strains with enhanced feather degradation capabilities and in the further development of resources for degradation enzymes.

2. Materials and Methods

2.1. Strains, Chemicals, and Media

The strain *P. aeruginosa* Gxun-7 was isolated from sludge samples obtained from a coastal duck farm in China using an enrichment culture method and was securely stored in our laboratory. It is worth noting that the GenBank accession number for this strain is MW579860.1, and it has been assigned the strain preservation number GDMCC 61615.

To activate the bacteria, they were initially cultured in LB medium for a duration of 24 h at a temperature of 35 °C. Following this incubation period, the fermentation broth was harvested at specified time intervals and subsequently subjected to centrifugation for 10 min at 13,400× *g*. The resultant crude enzyme extract obtained was utilized for subsequent experimental analyses.

2.2. Determination of Keratinase Activity and Feather Degradation Rate and Amino Acid

The methodology employed for assessing diagonal protease activity and feather degradation rate was adapted from Shen [22]. Mix 200 µL of fermentation supernatant with 300 µL of 2% (m/v) casein (Yuanye, Shanghai, China) substrate at pH 7.5, and react at 50 °C for 10 min. After the reaction is completed, immediately add 500 µL of 4 mol/L TCA, centrifuge at 13,400× *g* for 10 min, transfer 200 µL of the supernatant to a 2 mL centrifuge tube, add 1 mL of 0.5 mol/L $NaCO_3$ (aladdin, Shanghai, China), and 200 µL of Folin's phenol (aladdin, Shanghai, China). Mix well and react in a water bath at 50 °C for 10 min. Measure the absorbance at 660 nm using a spectrophotometer. In this context, a single keratinase unit was defined as the quantity of enzyme necessary to hydrolyze casein, resulting in the production of 1 µg of tyrosine per min at a temperature of 50 °C.

The determination of the feather degradation rate was carried out using the weight loss method. The fermentation medium underwent filtration through filter papers, and the remaining feather residue underwent thorough washing with distilled water, followed by drying and subsequent weighing to calculate the extent of weight loss. The results are presented as a percentage relative to the initial dry weight of the feather [23,24].

The fermentation broth was taken out and centrifuged at 13,400× *g* for 10 min, then treated with 5% (*v*/*v*) sulfosalicylic acid (aladdin, Shanghai, China) at a ratio of 1:8 at 4 °C for 12 h, passed through a 0.22 µm filter membrane, and the amino acids were determined by an automatic amino acid analyzer.

2.3. Bacterial Growth and Sample Preparation

P. aeruginosa Gxun-7 was added to 50 mL of non-feather medium (CG) and feather medium (TG) with three biological replicates per group. The medium of the CG was LB broth, and the medium of the TG was added: 40 g/L feather, 1.4 g/L K_2HPO_4, 0.7 g/L KH_2PO_4, and 0.5 g/L NaCl. Subsequently, bacterial cultures were collected during the exponential phase at 24 h and rinsed with phosphate-buffered saline.

Following this step, both the CG and TG samples underwent centrifugation (at 7100× *g* for 10 min) at 4 °C, followed by a washing step employing diethyl pyrocarbonate-treated water. Finally, these samples were promptly collected and subjected to flash freezing using liquid nitrogen, and they were then securely stored at a temperature of −80 °C.

2.4. RNA Extraction and Sequencing

Total RNA was rigorously extracted from the tissue employing TRIzol® Reagent in strict accordance with the manufacturer's instructions (Invitrogen, Waltham, MA, USA). To ensure the purity of the RNA sample, any genomic DNA present was effectively eliminated using DNase I treatment (TaKara, Shanghai, China). Subsequently, the quality of the RNA was rigorously assessed using a 2100 Bioanalyzer (Agilent Technologies, Santa

Clara, CA, USA), and its concentration was accurately determined using the ND-2000 spectrophotometer (NanoDrop Technologies, Waltham, MA, USA).

For the construction of an RNA-seq transcriptome library, we rigorously followed the protocols outlined in the TruSeqTM RNA sample preparation kit from Illumina (San Diego, CA, USA). The transformation of raw images into sequences, base-calling, and the computation of quality values were conducted through the utilization of the Illumina GA pipeline, a process expertly carried out by Shanghai Majorbio Biopharm Technology Co., Ltd (Shanghai, China). This comprehensive analysis yielded 150 base pair paired-end reads. It is pertinent to note that libraries for transcriptome analysis were established employing RNA samples obtained from both the CG and TG groups.

2.5. Analysis of RNA-Seq Data and Feather Degradation Mechanism

The initial raw data in fastq format underwent preprocessing using Trimmomatic (version 0.36) to obtain clean reads. Subsequently, these clean reads were subjected to mapping against the NCBI Rfam databases to remove any rRNA sequences, a process carried out using Bowtie2 (Version 2.4.5) for sequence alignment.

For gene expression quantification, RNA-seq by expectation–maximization was employed, and the quantitative index utilized was the Trusted Platform Module. To identify differentially expressed genes (DEGs), the data were analyzed with DESeq2, and the criteria for DEG selection were set as Padj < 0.05 and $|\log_2 FC| \geq 1$ [25].

Functional annotation was performed using various tools, including EggNOG, GO, and KEGG. GO pathway enrichment analysis was executed using Goatools software (https://github.com/tanghaibao/Goatools) (accessed on 8 July 2022), while KEGG pathway enrichment analysis was conducted using KOBAS (http://kobas.cbi.pku.edu.cn/home.do) (accessed on 8 July 2022) via Fisher's exact test. To ensure statistical rigor, p values were subjected to correction for multiple tests, with genes considered differentially expressed when the $p < 0.05$ [26].

2.6. Analysis by qRT–PCR and Determination of Sulfur-Containing Compound Contents

The RNA extracted for transcriptome sequencing was employed as a template for reverse transcription, with all reagents being mixed while maintaining a low-temperature environment. The resulting synthesized cDNA was subsequently used for qRT–PCR. The 16S rRNA gene served as the reference gene, and all reactions were conducted in triplicate. Details of the gene and primer sequences utilized for qRT–PCR can be found in Table S1 in the Supplementary Material. Data analysis was performed using the $2^{-\Delta\Delta Ct}$ method [27].

The determination of sulfite content was carried out using the pararosaniline hydrochloride method. A reaction solution was prepared by mixing 1 mL of the supernatant with 2 mL of formaldehyde–pararosaniline. After the development of a stable color, the absorbance was measured at 550 nm. The sulfite content was calculated based on a prepared Na_2SO_3 standard curve [28].

To assess sulfate content, barium chromate spectrophotometry was employed. Briefly, the reaction mixture consisted of 100 μL of culture supernatant, 400 μL of H_2O, and 250 μL of $BaCrO_4$. After incubation at room temperature for 30 min, 50 μL of calcium–ammonia miscible liquids and 500 μL of 95% anhydrous ethanol were added to initiate the reaction. The resulting mixture was then centrifuged at 13,400× *g* for 10 min, and the absorbance of the supernatant was measured at 420 nm. The sulfate content was determined based on a prepared Na_2SO_4 standard curve [29].

Furthermore, the release of sulfhydryl compounds into the fermentation medium was assessed using the method described by Ellman. Upon adding 1 mL of 10 mol/L DTNB (5,5′-Dithiobis-(2-nitrobenzoic acid)) to 500 μL of supernatant, the mixture was incubated for 10 min. Subsequently, the absorbance was measured at 412 nm after the color had stabilized, and the sulfhydryl content was calculated based on a prepared Cys standard curve [30]. The reagents used above and the configuration methods can be found in Table S5.

2.7. Statistical Analysis

The data underwent rigorous statistical analysis utilizing analysis of variance through SPSS software, version 17.0. To distinguish means, Duncan's multiple range test was employed. Statistical significance was deemed at a significance level of $p < 0.05$.

3. Results

3.1. Effects of P. aeruginosa Gxun-7 on Growth and Feather Degradation

P. aeruginosa Gxun-7 displayed a distinct transparent zone on the casein plate (Figure 1a), and its morphology, as revealed by scanning electron microscopy, exhibited a spore-free, rod-shaped structure (Figure 1b). Following incubation at 35 °C for 48 h, a noticeable degradation of feathers was evident, with a feather degradation rate reaching as high as 82.6% (Figure 1c,d). Scanning electron microscopy images further depicted the significant breakdown and disassembly of feathers (Figure 1e,f).

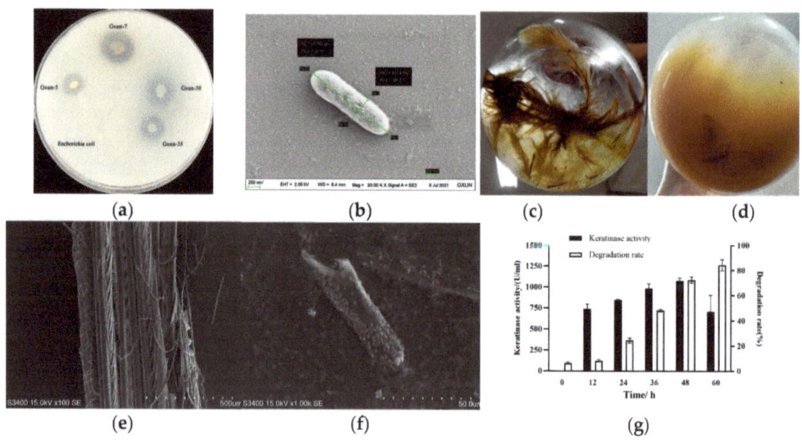

Figure 1. The effect of *P. aeruginosa* Gxun-7 on the degradation of feathers. (**a**): The transparent ring state of Gxun-7 on the casein-containing agar plate; (**b**): the SEM of Gxun-7, "*" indicates scale; (**c**,**d**): a comparison of feather degradation before and after Gxun 7 treatment; (**e**,**f**): the SEM of c and d; (**g**): measurement of protease activity and degradation rate in the degradation solution at different time intervals.

Keratinase activity exhibited a gradual increase, reaching its maximum at the 48 h mark after Gxun-7 inoculation, with the highest recorded keratinase activity being 1075.3 U/mL. As keratinase activity escalated, so did the rate of feather degradation, indicating a direct proportional relationship between feather degradation and keratinase activity (Figure 1g). These findings underscore the remarkable feather-degrading capabilities of *P. aeruginosa* Gxun-7. However, the precise mechanisms underlying feather degradation remain poorly understood, thereby impeding the widespread application of this strain on a large scale.

3.2. RNA-seq Analysis

After filtration, over 30 million high-quality reads were obtained. The error rate for each sample was found to be less than 0.0258%. Additionally, the quality metrics Q20 and Q30 exceeded 97.36 and 93.73, respectively, indicating the reliability of the transcriptome sequencing data (refer to Table S2). A total of 5013 expressed genes were identified in the CG, while 5541 were identified in the TG. Among these genes, 4989 were co-expressed in both samples. Notably, 24 genes were found to be specifically expressed in CG, while 552 genes were specific to TG (Figure 2).

Figure 2. The Venn diagram illustrates gene differential expression. The total number of genes detected as expressed for each sample is represented by the sum within each circle, while the overlap between circles signifies genes expressed in both tested samples.

To assess the degree of similarity between these samples, we calculated both correlation and variance. In this study, we utilized principal component analysis (PCA) and a heatmap to visualize this correlation. As illustrated in the heatmap and PCA plots (Figures S1 and S2), there was a high degree of correlation between the CG and TG samples, underscoring the accuracy of the sampling process. The gene cluster heat map analysis showed that the gene expression of the samples was well differentiated (Figure S3). DESeq2 was employed to assess DEGs between the samples in comparison to CG. In the TG, a total of 438 genes exhibited significant differential expression, comprising 282 genes that were significantly upregulated ($|Log_2FC| \geq 1$ and Padj ≤ 0.05) and 156 genes that were significantly downregulated ($|Log_2FC| \leq -1$ and Padj ≤ 0.05) (Figure 3).

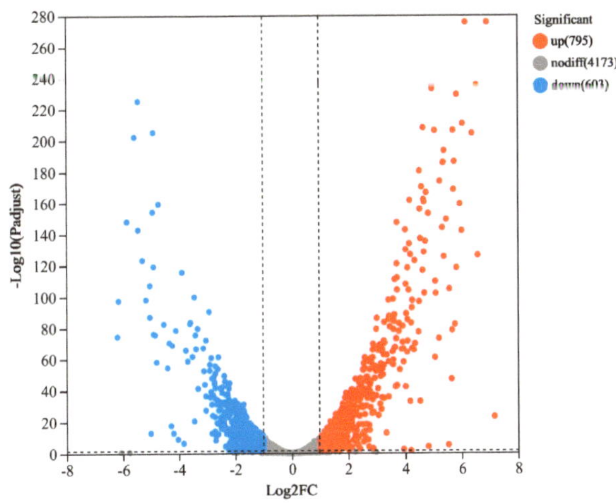

Figure 3. The volcano plot illustrates the differentially expressed genes (DEGs) between the treated group (TG) and the control group (CG). The y-axis represents the mean expression value of log10 (p-adjust), while the x-axis displays the log2FC value. Gray dots indicate non-differentially expressed genes, red dots represent upregulated genes, and blue dots denote downregulated genes.

3.3. Functional Enrichment Analysis of DEGs in *P. aeruginosa* Gxun-7

The Gene Ontology (GO) classification system serves as an international framework for standardizing and describing the attributes of genes and gene products across various organisms. In this analysis, all DEGs were classified into three categories based on their respective biological processes, cellular components, and molecular functions, accounting for 102 (6.72%), 753 (49.64%), and 662 (43.64%) genes, respectively (Figure 4). In terms of "biological process", 58 genes were found to be related to DNA template and transcriptional regulation (GO: 0006355), while 44 genes were associated with translation (GO: 0006412). The highest number of DEGs in the "cellular component" were observed in the membrane (GO: 0016021), with a total of 285 genes. Within the "molecular process" category, the top three classifications were metal ion binding (GO: 0046872), ATP binding (GO: 0005524), and DNA binding (GO: 0003677), with 94, 92, and 89 DEGs, respectively. Collectively, these data suggests that the metabolic functions of *P. aeruginosa* Gxun-7 were influenced following exposure to feathers, and the DEGs associated with these GO terms may play a role in the degradation process of feathers by the Gxun-7.

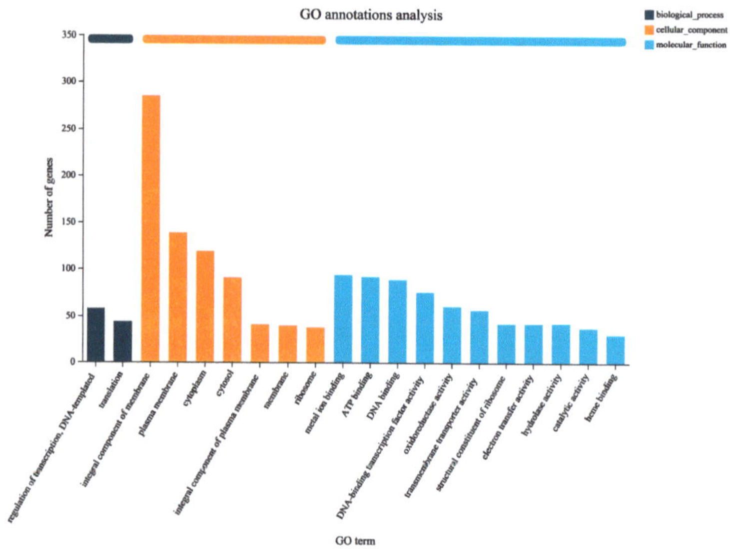

Figure 4. Gene Ontology (GO) function classification of DEGs. The abscissa represents the primary functional type, and the ordinate represents the number of DEGs.

Pathway-based analysis plays a vital role in revealing the biological functions of genes. To further explore the potential roles of DEGs, KEGG pathway analysis was conducted to identify the metabolic pathways enriched in the process of feather degradation. The results indicate that out of the 272 distinct metabolic pathways identified between the CG and TG, they can be categorized into six major groups: metabolism, genetic information processing, environmental information processing, cellular processes, organismal systems, and human diseases. These categories comprised 123, 4, 59, 76, 5, and 5 different metabolic pathways, respectively. A total of 446 DEGs (57.18%) were associated with metabolic pathways, with environmental information processing involving 139 genes (17.82%). The amino acid metabolism pathway exhibited the greatest number of upregulated and downregulated genes, with 47 and 44 DEGs, respectively, followed by the cellular community metabolism and carbohydrate metabolism pathways with 89 and 80 DEGs (Figure 5a). A total of 16 of the top 20 differential pathways were related to metabolism, suggesting that Gxun-7 adopts an alternative metabolic strategy when utilizing feathers as its exclusive carbon and nitrogen source compared to a medium without feathers. The remaining 4 pathways in the top 20 included flagellar assembly, quorum sensing, oxidative phosphorylation, and the bacterial secretion system, indicating the functions of Gxun-7 in environmental sensing, effective oxidative respiration, and material transport during feather degradation (Figure 5b).

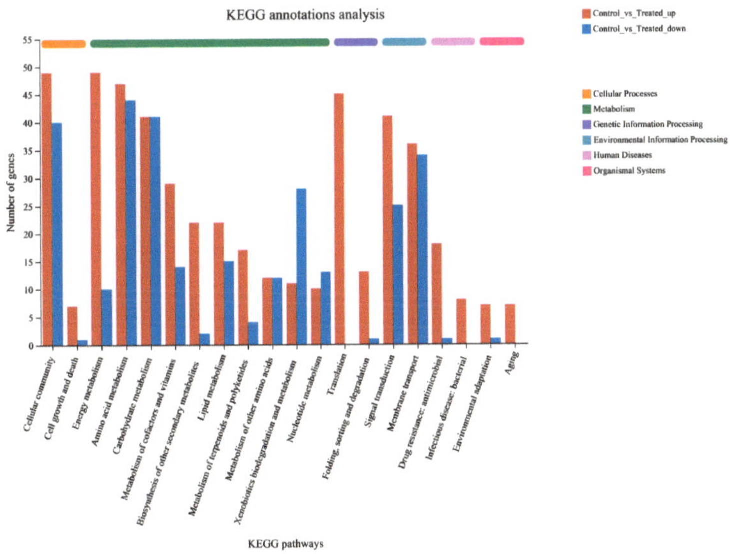

(**a**)

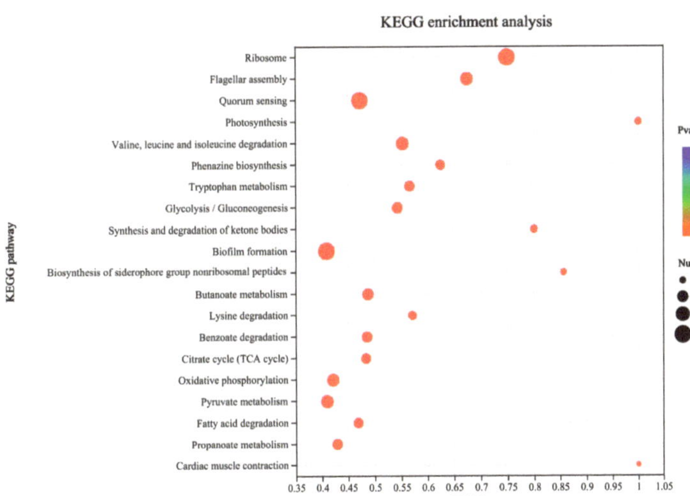

(**b**)

Figure 5. Pathway classification map of significantly enriched DEGs. (**a**): KEGG annotation results. The abscissa represents the pathway type, and the ordinate represents the number of DEGs. Red indicates upregulated genes and blue indicates downregulated genes. (**b**): KEGG enrichment results. The abscissa represents the enrichment rate, and the ordinate represents the pathway type. Bubble size indicates the number of differentially expressed genes, and bubble color indicates significance.

Furthermore, 141 DEGs were annotated into the amino acid anabolic pathway, encompassing various pathways such as "Valine, leucine and isoleucine degradation" (map00280) with 26 DEGs, "Tryptophan metabolism" (map00380) with 17 DEGs, "Glycine, serine and threonine metabolism" (map00260) with 14 DEGs, "Lysine degradation" (map00310) with 12 DEGs, "Glutathione metabolism" (map00480) with 7 DEGs, "Cysteine and methionine metabolism" (map00270) with 7 DEGs, and more. This indicates that the synthesis and metabolism of amino acids play a crucial role throughout the entire biological process. Moreover, 40 DEGs were annotated into "ABC transporters" (map02010). ABC transporters are essential for detoxification as they excrete toxic substances. Notably, the transporter genes of sulfate/thiosulfate, zinc, phosphate, phosphonate, nucleoside, and amino acid exhibited significant changes. The activity of these transporters highlights the importance of these substances. It is worth mentioning that the pathways "sulfur metabolism" (map00920) and "sulfur relay system" (map04122) appear to be potentially related to sulfite formation, warranting further investigation. Relatively fewer DEGs were annotated in organismal systems and human disease pathways, with 16 and 27 genes, respectively.

3.4. Transcriptome Analysis of the Gene Expression Differences in Cells of CG and TG

After performing GO and KEGG correlation analyses, we identified 20 key genes that could potentially be associated with feather degradation (Table 1). These genes were found to be associated with protease synthesis, pyocyanine biosynthesis, chitinase and catalase synthesis, amino acid metabolism, sulfite synthesis, and metabolism. The upregulated genes were mainly extracellular proteases, metabolism-related genes, transmembrane transport-related genes, and membrane proteins. Type secretion system protein and keratinase coding gene (*kp2*, NCBI accession number: OM992359) were found to be the two most upregulated with Log_2FC 7.15 and 6.91, respectively. Furthermore, the genes of reductase and phenozine metabolism related were also significantly upregulated (Table 1). These suggest that the addition of feathers to the medium acts as an inducer, stimulating the expression of the *tssA*, *kp2*, *katA*, and ribosomal gene in *P. aeruginosa* Gxun-7, ultimately promoting the synthesis of keratinase and facilitating feather hydrolysis. However, their specific roles in feather degradation is still unknown and requires further investigation. The results of the analysis on differential gene expression show a significant upregulation of the keratinase expression gene, *kp2*, indicating that the bacterium synthesizes a large amount of keratinase to degrade keratin in feathers. Meanwhile, the upregulation of genes related to the type secretion systems is also significant ($Log_2FC = 7.15$), which assists in the secretion of keratins. Therefore, the enhanced expression of this efflux pump could transport more phenazines out of the cell.

Table 1. Table of key gene screening.

Gene	Gene Description	Log_2FC
tssA	Type secretion system protein	7.15
kp2	Keratinase KP2	6.91
ahpF	Alkyl hydroperoxide reductase subunit F	6.15
phzB1	Phenazine biosynthesis protein	5.76
dhcB	Dehydrocarnitine CoA transferase C subunit B	5.71
phzA1	Phenazine biosynthesis protein	5.63
phzC2	Phenazine biosynthesis protein	5.52
dhcA	Dehydrocarnitine CoA transferase C subunit A	5.34
pchD	Pyochelin biosynthesis protein	4.97
sodM	Superoxide dismutase	4.82
phzF1	Phenazine biosynthesis protein	4.81
phzM	Phenazine-specific methyltransferase	4.72
katA	Catalase	4.68
pchE	Dihydroaeruginoic acid synthetase	4.65
phzD1	Phenazine biosynthesis protein	4.53
katB	Catalase	4.21
chiC	Chitinase	4.12
lasA	Protease precursor	2.87
sir2	Sulfite reductase	1.61
gor	Glutathione reductase	1.18

Moreover, the gene expression levels of catalase and superoxide dismutase (SOD) increased along with the increase in phenazines production, which has been demonstrated to increase the levels of ROS [31]. Extracellular proteases degrade degradable proteins into short peptides and partial amino acids outside the cell, which are then transported into the cell for further degradation into amino acids. Disulfide bonds are critical for keratin degradation, so breaking the disulfide bonds is a key step in keratin degradation [19,21].

The experimental results presented above suggest that the degradation of feathers by *P. aeruginosa* Gxun-7 is likely a complex mechanism involving several factors. This includes the enzymatic action of various proteases, mainly keratinases, as well as the action of sulfite and the reducing power provided by the bacterium. The degradation pathways of feather by Gxun-7 obtained from transcriptome analysis and pathway enrichment are illustrated in Figure 6.

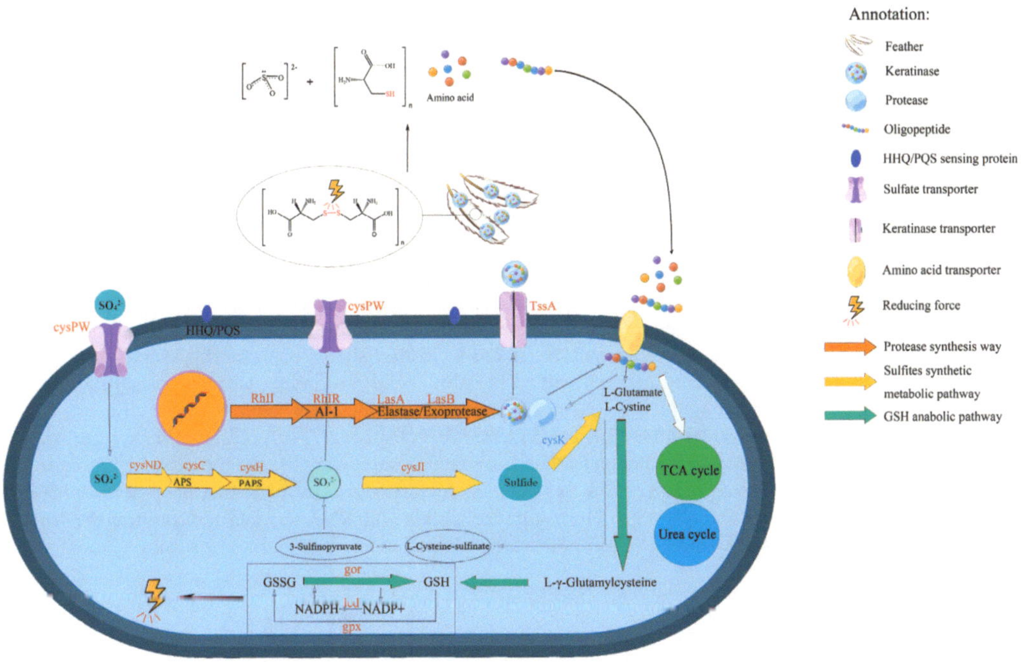

Figure 6. Hypothetical model of *P. aeruginosa* Gxun-7 feather utilization mechanisms. The color red indicates that the gene is upregulated and the color blue is down-regulated. (GSH: reduced glutathione; GSSG: oxidized glutathione; APS: Adenosine 5′-phosphosulfate; PAPS: 3′-Phosphoadenosine 5′-phosphosulfate); the relevant genetic information is provided in Supplementary Table S3.

3.5. Results of Sulfate, Sulfite, and Sulfur-Containing Compounds

Generally, the degradation process of feather keratins causes changes in the content of sulfur-containing compounds. In the present study, we analyzed the levels of sulfur-containing compounds, specifically sulfate and sulfite, at various time intervals (12, 24, 36, 48, 60 h) during the fermentation of feathers (Figure 7). Initially, the medium contained only sulfate but no sulfite. With the proceeding of feather degradation, the content of sulfate and sulfur-containing compounds concentration increased continuously in the whole degradation period. However, the content of sulfite reached a maximum of 0.03 g/L at 36 h, and then began to decrease with the extension of fermentation time. This suggests that *P. aeruginosa* Gxun-7 undergoes a transformation among sulfate, sulfite, and sulfur-containing compounds during the degradation of feathers.

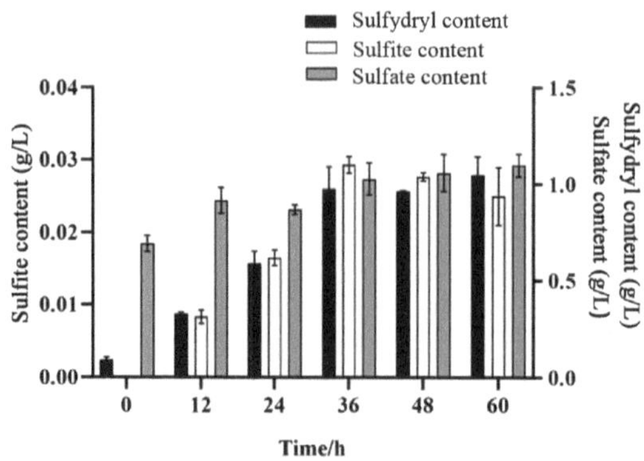

Figure 7. Changes in sulfates, sulfites, and sulfydryl-containing compounds during degradation at different times.

3.6. Validation by qRT–PCR

Eleven genes involved in the feather degradation pathway were selected for further verification using qRT–PCR analysis. One of the prominently upregulated genes is *kp2*, which plays a pivotal role in the synthesis of keratinase. This observation suggests a significant increase in the synthesis of keratinase in the TG. The *phz* genes are associated with the biosynthesis of pyocyanine, which is a specific pathway in *P. aeruginosa*. Moreover, this pathway plays a crucial role in the metabolic fate of amino acids [31]. The upregulation of chitinase genes may be linked to feather shafts, contributing to feather keratinization. Additionally, genes involved in sulfite synthesis and metabolism exhibited upregulation, indicating the possible presence of sulfite in the TG. The presence of sulfite has a profound effect on the disulfide bond formation in feathers. The thiolysis theory, recognized by most scholars in studies on feather degradation, emphasizes the significance of sulfite as an essential substance. Figure 8 shows the relative transcription level of selected genes in both RNA-seq data and qRT–PCR. The result is consistent with RNA-seq data and confirms the upregulation and downregulation of the feather degradation genes in the *P. aeruginosa* Gxun-7.

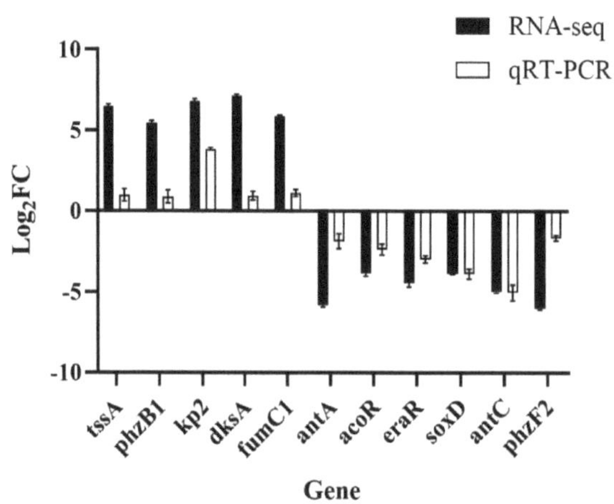

Figure 8. Validation of RNA-seq data by qRT–PCR. Eleven DEGs involved in the feather degradation by Gxun-7 were selected for verification. The data are expressed as the mean fold change relative to the control samples.

4. Discussion

The feature that renders keratins highly resistant to degradation by proteases is their extensive cross-linking through disulfide bonds. The biodegradation of keratin substrates is a complex process, and its mechanism remains incompletely understood. According to current research, effective keratin degradation not only involves the action of keratinases but also necessitates the disruption of disulfide bonds through specific redox mechanisms involving reductases or reductants. [32]. Our previous study demonstrates that *P. aeruginosa* Gxun-7 is a highly effective feather-degrading strain [22]. Therefore, our research employed transcriptome analysis to investigate how feather exposure influences gene expression and cellular metabolism.

The denaturation of keratin by the reduction of its disulfide bonds is the first step in feather keratin degradation. Sulfite (SO_3^{2-}) is a potent reducing agent and is capable of breaking the disulfide bonds present in keratin, making it more susceptible to degradation by keratinases [1]. Microorganisms release sulfites into the environment while degrading keratin, as seen in dermatophytes and *Streptomyces sp.* In this process, cysteine undergoes sulfonation through cysteine dioxygenases (Cdos), and the reducing agent sulfite disrupts the S–S bond in keratin, yielding L-cysteine and thiocysteine. This, in turn, leads to the denaturation of keratin, thereby facilitating its degradation [33]. Sulfite and free cysteinyl groups are secreted to reduce the disulfide bonds in the feather keratin, allowing proteases to access the peptide bonds of keratin. This process hydrolyzes the bonds, generating peptides and amino acids. These breakdown products are then imported into the cell through upregulated peptide and amino acid transporters. Within the cell, peptides may undergo further hydrolysis by intracellular proteases. Cysteine dioxygenases (CDOs), enzymes that catalyze the sulfoxidation of cysteine to cysteine sulfinic acid, have been extensively researched in eukaryotes due to their involvement in various diseases [34]. Contrastingly, only a limited number of prokaryotic enzymes of this category have undergone investigation. In *Pseudomonas aeruginosa*, two homologues of cysteine dioxygenase (Cdo and p3MDO) have been previously identified. Nevertheless, no alterations in genes associated with L-cysteine–sulfinate were discerned from our transcriptome data. Consequently, we posit the existence of alternative pathways in strain Gxun-7 for sulfite production.

The genes (cysNC, cysC, csyH, and cysJI) involved in sulfate transport and metabolism to sulfite are upregulated in the transcriptome of TG. In the previous study, we also found that adding 0.1g/L $ZnSO_4$ in the medium could significantly improve the feather degradation effect. Peng also found that the external addition of sulfites could further promote feather degradation [35]. It may be that a high concentration of sulfate in the environment inhibited metabolic pathways related to Cdo of Gxun-7. Therefore, the sulfite formation mechanism of this strain may be as follows: extracellular sulfates are transferred into the cells and converted to adenylyl sulfate (APS) through the action of sulfate transfer protein (CysFUWA), followed by further conversion to 3′-Phospho-5′-adenylyl sulfate (PAPS) by APS synthase and reduction to sulfite by phosphoadenosine phosphosulfate reductase. Sulfite is then excreted to the outside, further contributing to the reduction of disulfide bonds in feathers. In order to prevent damage to cells caused by excessive sulfites, the strain reduced sulfites to sulfides by upregulating the sulfite reductase gene (*Sir*) [36]. The above-mentioned genes showed upregulated expression in the comparative transcriptome, suggesting that these pathways may play important roles in disulfide bond reduction (Figure 6). Furthermore, sulfite was also detected in the fermentation broth, indicating the presence of thiolysis in the feather degradation by *P. aeruginosa* Gxun-7 (Figure 7).

Previous study has also identified the role of γ-glutamyl transpeptidase-glutathione (GGT-GSH) serving as a redox principle during degradation of feather keratin by recombinant *Escherichia coli* harboring keratinase from *P. aeruginosa* KS-1 [37]. The γ-glutamyl transpeptidase (GGT) in the periplasm generates a free cysteinyl group, acting as a potent reductant that enhances the sulfitolysis of resistant proteins, rendering them more susceptible to degradation by keratinases. GGT catalyzes the hydrolysis of glutathione (GSH) to form cysteinyl–glycine, a significantly more robust reductant for feather breakdown compared to glutathione alone [38]. GGT is also an important component of the γ-glutamyl cycle. The genes (gor, gpx, and icd) involved in the GSH pathway, which can form a conversion cycle with oxidized glutathione (GSSG) were discovered to be significantly upregulation in the transcriptome of TG. Furthermore, by analyzing the amino acid content in the fermentation broth, the concentrations of Cys, Glu, and Gly, which are critical substances for GSH formation, were unexpectedly low in the fermentation broth (Table S4). Meanwhile, during the GSH and GSSG conversion process, glutathione reductase (GR), which possesses disulfide bond reductase activity, may also play an important role in disulfide bond cleavage. GR has also been reported to have disulfide bond reductase activity, which may also play a major role in disulfide bond cleavage [33,39]. The related mechanisms of glutathione synthesis and feather metabolism may be as follows: Glu and Cys are first converted to γ-glutamylcysteine by the action of glutamate–cysteine ligase (GCLC), next, γ-glutamylcysteine and Gly are reduced to GSH under the action of glutathione synthase (GSS). GSH has strong reducing power and can be interconverted with its oxidized state glutathione disulfide (GSSG) through the actions of glutathione peroxidase (gpx) and glutathione reductase (gor) (Figure S5). Meanwhile, GSH also is hydrolyzed by GGT to produce cysteinyl–glycine and γ-glutamylcysteine [40]. Therefore, another possible mechanism by which the strain Gxun-7 disrupt disulfide bonds of feather keratins is secreting reducing agents like cysteinyl–glycine.

The gene cluster *phzABCDEFG* encodes enzymes that catalyze the conversion of chorismate, the end product of the shikimate pathway, into PCA (phenazine-1-carboxylic acid), which serves as the precursor for synthesizing phenazine derivatives [41]. Transcriptome analysis indicated that the gene expression level of phenazine biosynthesis cluster (*phzABCDEFG*) and resistance to reactive oxygen species (ROS) (*sodM* and *katAB*) increased significantly by the addition of feather comparing to the TG group (Table 1). In addition, the phenazine content in feather degradation solution increased with the increase in fermentation time. The possible mechanism may be after a long-period reduction of disulfide bonds in feathers causes an increase in ROS levels in the cell. In order to avoid the damage of ROS, cells reduce the level of ROS by upregulating related scavenger enzyme genes such as superoxide dismutase (SOD), catalases, peroxidases, and dismutases. It has been reported that SOD, catalases, and peroxidases, which can convert ROS into less harmful products to reduce the damaging effect of hydrogen peroxide on cells, play critical roles in cell resistance to ROS [42]. SOD is a crucial component of the microbial redox system, serving as a vital scavenger of reactive oxygen species (ROS) within the cell. It plays a pivotal role in the cellular antioxidant system, safeguarding the cell from damage [43]. Catalase catalyzes the breakdown of hydrogen peroxide into water and oxygen, thereby safeguarding cells against oxidative damage [44]. Therefore, the addition of feather increased the expression of genes involved in ROS resistance and resulted in lower level of ROS. Concurrently, there was a notable decrease in the levels of tyrosine, phenylalanine, and tryptophan, synthesized from chorismate, upon the addition of feather. These findings indicate that the inclusion of feather heightened the carbon flux directed towards the biosynthesis of phenazines, thereby facilitating an increase in phenazine production.

Bacterial type VI secretion systems (T6SS) are widely distributed among Gram-negative bacteria and serve as a versatile protein export mechanism that transports effectors into eukaryotic or prokaryotic target cells. Additionally, T6SS plays a role in metal ion uptake, including iron, manganese, and zin [45–48], conferring an advantage during bacteria–bacteria competition. *P. aeruginosa* possesses three T6SSs, designated H1-, H2-, and H3-T6SS. Thirteen essential genes are conserved in all T6SSs [49]. Generally, T6SS are composed of 13–15 structural proteins divided into three interlocking structures: the intermembrane anchor, the baseplate, and the needle/sheath. The T6SS secretes two categories of proteins, valine–glycine repeat protein G (VgrG) and hemolysin coregulated protein (Hcp). The Hcp protein shares structural similarities with a phage tail tube component, while VgrG proteins exhibit resemblance to the puncturing device located at the tip of the phage tube [50]. It has been suggested that the primary role of Hcp is the formation of nanotubes on the bacterial surface, which could facilitate the transportation of other effector proteins dependent on the T6SS. Conversely, VgrG proteins assemble into trimeric complexes that potentially function as puncturing devices, enabling the perforation of membranes to facilitate the passage of proteins or macromolecular complexes [48]. Notably, Hcp is considered as the hallmark of T6SS and can be used as a biomarker for activation of T6SS. The expression of T6SS in *P. aeruginosa* is regulated by the quorum sensing (QS) system [51]. QS is crucial for collective adaptive responses and regulates both bacterial virulence and biofilm formation. *Pseudomonas aeruginosa* possesses multiple QS systems, including two N-acyl-homoserine lactone-based systems (las and rhl systems) and one quinolone PQS system (pqs) [47]. Transcriptome analysis indicated that the genes expression level of T6SS structural proteins and QS systems (*sodM* and *katA*) increased significantly with the addition of feather comparing to the TG group. Thus, the synthesized keratinase, phenazine, and reducing agents, such as sulfite and cysteinyl–glycine, can be quickly removed from the cell.

5. Conclusions

In summary, this study provides deep insights into the molecular mechanisms of response in feather degradation by *P. aeruginosa* Gxun-7 through comprehensive biochemical and transcriptomics analyses. The RNA-seq analysis identified 5571 differentially expressed genes; of these, 795 were upregulated and 603 were downregulated. Upregulated genes primarily participated in proteolysis, amino acid, and pyruvate metabolism. This investigation culminated in the synthesis of a hypothetical degradation mechanism: under the activation of QS signals, Gxun-7 produces a large amount of proteases, including keratin, which first degrade keratins with broken disulfide bonds into peptides and amino acids. Extracellular sulfates, peptides, and amino acids are transported into the cells through T6SS. The sulfate is converted into sulfite inside the cell, which is further transported outside the cell to break the disulfide bond of keratin. The abundant Cys, Glu, and Gly participate in cellular metabolism, and synthesis of GSH. Next, the reductive cleavage of disulfide bonds in feather by the interaction of oxidation, sulfite, GSH, and cysteinyl–glycine, which loosens the structure of keratin, enhances its solubility, and exposes more protease attack sites. Finally, the remaining loose structure is completely degraded by keratinase. This study enhances our understanding of the transcriptomic responses of *P. aeruginosa* Gxun-7 to feather degradation and offers insights into potential degradation mechanisms, thereby aiding in the formulation of effective feather waste management strategies in poultry farming.

Supplementary Materials: The following supporting information can be downloaded at: https://www.mdpi.com/article/10.3390/microorganisms12040841/s1, Figure S1: An analysis of the sample correlation heat map; Figure S2: Principal component analysis (PCA) correlation analysis was conducted to examine the correlation levels between samples; Figure S3: Gene cluster heatmap analysis; Figure S4: Schematic representation of sulfite metabolism; Figure S5: Schematic representation of GSH production; Table S1: Primer sequences; Table S2: Statistical table for quality control of sequencing; Table S3: Relevant major genetic information in Figure 6; Table S4: Analysis of amino acid species and contents before and after degradation; Table S5: Preparation table of main reagents.

Author Contributions: Conceptualization, C.S. and N.S.; methodology, C.X.; software, R.L.; formal analysis, D.Y. (Doudou Yin); investigation, Y.L.; resources, H.Z. and D.Y. (Dengfeng Yang); data curation, C.S.; writing—original draft preparation, C.S.; writing—review and editing, N.S. and H.Z.; supervision, M.J.; project administration, M.J. and H.Z.; funding acquisition, N.S. and H.Z. All authors have read and agreed to the published version of the manuscript.

Funding: This research was funded by the Natural Science Foundation of China (funding code: 32160017), Science and Technology Major Project of Guangxi Province (funding code: AB21220020), The Scientific Research Project for Introducing High-level Talents of Guangxi Minzu University (funding code: 2018KJQD17), and the Opening Project of Guangxi Key Laboratory of Marine Natural Products and Combinatorial Biosynthesis Chemistry (funding code: GXMNPCBC-2022-01).

Data Availability Statement: Raw transcriptome data have been uploaded to National Center for Biotechnology Information (NCBI) (Accession SRA number: PRJNA1005561). All the data presented in this study are available on request from the corresponding author.

Conflicts of Interest: The authors declare no conflicts of interest.

References

1. Yamamura, S.; Morita, Y.; Hasan, Q.; Yokoyama, K.; Tamiya, E. Keratin Degradation: A Cooperative Action of Two Enzymes from *Stenotrophomonas* Sp. *Biochem. Biophys. Res. Commun.* **2002**, *294*, 1138–1143. [CrossRef]
2. Fraser, R.D.B.; Parry, D.A.D. The Structural Basis of the Filament-Matrix Texture in the Avian/Reptilian Group of Hard β-Keratins. *J. Struct. Biol.* **2011**, *173*, 391–405. [CrossRef] [PubMed]
3. Yusuf, I.; Ahmad, S.A.; Phang, L.Y.; Syed, M.A.; Shamaan, N.A.; Abdul Khalil, K.; Dahalan, F.A.; Shukor, M.Y. Keratinase Production and Biodegradation of Polluted Secondary Chicken Feather Wastes by a Newly Isolated Multi Heavy Metal Tolerant *Bacterium-Alcaligenes* Sp. AQ05-001. *J. Environ. Manag.* **2016**, *183*, 182–195. [CrossRef]
4. Akram, F.; Aqeel, A.; Shoaib, M.; Haq, I.U.; Shah, F.I. Multifarious Revolutionary Aspects of Microbial Keratinases: An Efficient Green Technology for Future Generation with Prospective Applications. *Environ. Sci. Pollut. Res.* **2022**, *29*, 86913–86932. [CrossRef] [PubMed]
5. Chukwunonso Ossai, I.; Shahul Hamid, F.; Hassan, A. Valorisation of Keratinous Wastes: A Sustainable Approach towards a Circular Economy. *Waste Manag.* **2022**, *151*, 81–104. [CrossRef] [PubMed]
6. Callegaro, K.; Welter, N.; Daroit, D.J. Feathers as Bioresource: Microbial Conversion into Bioactive Protein Hydrolysates. *Process Biochem.* **2018**, *75*, 1–9. [CrossRef]
7. Li, Q. Progress in Microbial Degradation of Feather Waste. *Front. Microbiol.* **2019**, *10*, 2717. [CrossRef] [PubMed]
8. Lai, Y.; Wu, X.; Zheng, X.; Li, W.; Wang, L. Insights into the Keratin Efficient Degradation Mechanism Mediated by *Bacillus* Sp. CN2 Based on Integrating Functional Degradomics. *Biotechnol. Biofuels* **2023**, *16*, 59. [CrossRef] [PubMed]
9. Yahaya, R.S.R.; Phang, L.Y.; Normi, Y.M.; Abdullah, J.O.; Ahmad, S.A.; Sabri, S. Feather-Degrading *Bacillus Cereus* HD1: Genomic Analysis and Its Optimization for Keratinase Production and Feather Degradation. *Curr. Microbiol.* **2022**, *79*, 166. [CrossRef] [PubMed]
10. Brandelli, A. Bacterial Keratinases: Useful Enzymes for Bioprocessing Agroindustrial Wastes and Beyond. *Food Bioprocess Technol.* **2008**, *1*, 105–116. [CrossRef]
11. Daroit, D.J.; Brandelli, A. A Current Assessment on the Production of Bacterial Keratinases. *Crit. Rev. Biotechnol.* **2014**, *34*, 372–384. [CrossRef] [PubMed]
12. Peng, Z.; Mao, X.; Zhang, J.; Du, G.; Chen, J. Biotransformation of Keratin Waste to Amino Acids and Active Peptides Based on Cell-Free Catalysis. *Biotechnol. Biofuels* **2020**, *13*, 61. [CrossRef] [PubMed]
13. Ramnani, P.; Singh, R.; Gupta, R. Keratinolytic Potential of *Bacillus Licheniformis* RG1: Structural and Biochemical Mechanism of Feather Degradation. *Can. J. Microbiol.* **2005**, *51*, 191–196. [CrossRef] [PubMed]
14. Nnolim, N.E.; Okoh, A.I.; Nwodo, U.U. *Bacillus* Sp. FPF-1 Produced Keratinase with High Potential for Chicken Feather Degradation. *Molecules* **2020**, *25*, 1505. [CrossRef] [PubMed]
15. Hamiche, S.; Mechri, S.; Khelouia, L.; Annane, R.; El Hattab, M.; Badis, A.; Jaouadi, B. Purification and Biochemical Characterization of Two Keratinases from *Bacillus Amyloliquefaciens* S13 Isolated from Marine Brown Alga Zonaria Tournefortii with Potential Keratin-Biodegradation and Hide-Unhairing Activities. *Int. J. Biol. Macromol.* **2019**, *122*, 758–769. [CrossRef] [PubMed]
16. Li, Q. Structure, Application, and Biochemistry of Microbial Keratinases. *Front. Microbiol.* **2021**, *12*, 674345. [CrossRef] [PubMed]
17. Abd El-Aziz, N.M.; Khalil, B.E.; Ibrahim, H.F. Enhancement of Feather Degrading Keratinase of *Streptomyces Swerraensis* KN23, Applying Mutagenesis and Statistical Optimization to Improve Keratinase Activity. *BMC Microbiol.* **2023**, *23*, 158. [CrossRef] [PubMed]
18. Garg, S.K.; Alam, M.S.; Soni, V.; Radha Kishan, K.V.; Agrawal, P. Characterization of Mycobacterium Tuberculosis WhiB1/Rv3219 as a Protein Disulfide Reductase. *Protein Expr. Purif.* **2007**, *52*, 422–432. [CrossRef] [PubMed]
19. Peng, Z.; Zhang, J.; Du, G.; Chen, J. Keratin Waste Recycling Based on Microbial Degradation: Mechanisms and Prospects. *ACS Sustain. Chem. Eng.* **2019**, *7*, 9727–9736. [CrossRef]
20. González, V.; Vargas-Straube, M.J.; Beys-da-Silva, W.O.; Santi, L.; Valencia, P.; Beltrametti, F.; Cámara, B. Enzyme Bioprospection of Marine-Derived Actinobacteria from the Chilean Coast and New Insight in the Mechanism of Keratin Degradation in *Streptomyces* Sp. G11C. *Mar. Drugs* **2020**, *18*, 537. [CrossRef]
21. Wang, Z.; Chen, Y.; Yan, M.; Li, K.; Okoye, C.O.; Fang, Z.; Ni, Z.; Chen, H. Research Progress on the Degradation Mechanism and Modification of Keratinase. *Appl. Microbiol. Biotechnol.* **2023**, *107*, 1003–1017. [CrossRef] [PubMed]
22. Shen, N.; Yang, M.; Xie, C.; Pan, J.; Pang, K.; Zhang, H.; Wang, Y.; Jiang, M. Isolation and Identification of a Feather Degrading *Bacillus Tropicus Strain* Gxun-17 from Marine Environment and Its Enzyme Characteristics. *BMC Biotechnol.* **2022**, *22*, 11. [CrossRef] [PubMed]

23. Aktayeva, S.; Baltin, K.; Kiribayeva, A.; Akishev, Z.; Silayev, D.; Ramankulov, Y.; Khassenov, B. Isolation of *Bacillus* Sp. A5.3 Strain with Keratinolytic Activity. *Biology* **2022**, *11*, 244. [CrossRef] [PubMed]
24. Peng, S.; Li, H.; Zhang, S.; Zhang, R.; Cheng, X.; Li, K. Isolation of a Novel Feather-Degrading *Ectobacillus* Sp. JY-23 Strain and Characterization of a New Keratinase in the M4 Metalloprotease Family. *Microbiol. Res.* **2023**, *274*, 127439. [CrossRef] [PubMed]
25. Love, M.I.; Huber, W.; Anders, S. Moderated Estimation of Fold Change and Dispersion for RNA-Seq Data with DESeq2. *Genome Biol.* **2014**, *15*, 550. [CrossRef] [PubMed]
26. Li, S.; Ma, J.; Li, S.; Chen, F.; Song, C.; Zhang, H.; Jiang, M.; Shen, N. Comparative Transcriptome Analysis Unravels the Response Mechanisms of *Fusarium oxysporum* f.sp. *cubense* to a Biocontrol Agent, *Pseudomonas aeruginosa* Gxun-2. *Int. J. Mol. Sci.* **2022**, *23*, 15432. [CrossRef] [PubMed]
27. Arocho, A.; Chen, B.; Ladanyi, M.; Pan, Q. Validation of the 2-DeltaDeltaCt Calculation as an Alternate Method of Data Analysis for Quantitative PCR of BCR-ABL P210 Transcripts. *Diagn. Mol. Pathol.* **2006**, *15*, 56–61. [CrossRef] [PubMed]
28. Li, Y.; Zhao, M. Simple Methods for Rapid Determination of Sulfite in Food Products. *Food Control* **2006**, *17*, 975–980. [CrossRef]
29. Sakuragawa, A.; Nakayama, S.; Okutani, T. Flow-Injection Spectrophotometric Determination of Micro Amounts of Sulfate Ion in Surface- and Sea-Water Samples with a Barium Chromate Reaction Column. *Anal. Sci.* **1994**, *10*, 77–81. [CrossRef]
30. Ellman, G. Tissue Sulfhydryl Groups. *Arch. Biochem. Biophys.* **1959**, *82*, 70–77. [CrossRef] [PubMed]
31. Veremeenko, E.G.; Maksimova, N.P. Activation of the Antioxidant Complex in *Pseudomonas Aurantiaca*—Producer of Phenazine Antibiotics. *Microbiology* **2010**, *79*, 439–444. [CrossRef]
32. Ramakrishna Reddy, M.; Sathi Reddy, K.; Ranjita Chouhan, Y.; Bee, H.; Reddy, G. Effective Feather Degradation and Keratinase Production by *Bacillus Pumilus* GRK for Its Application as Bio-Detergent Additive. *Bioresour. Technol.* **2017**, *243*, 254–263. [CrossRef] [PubMed]
33. Grumbt, M.; Monod, M.; Yamada, T.; Hertweck, C.; Kunert, J.; Staib, P. Keratin Degradation by Dermatophytes Relies on Cysteine Dioxygenase and a Sulfite Efflux Pump. *J. Investig. Dermatol.* **2013**, *133*, 1550–1555. [CrossRef] [PubMed]
34. Hennicke, F.; Grumbt, M.; Lermann, U.; Ueberschaar, N.; Palige, K.; Böttcher, B.; Jacobsen, I.D.; Staib, C.; Morschhäuser, J.; Monod, M.; et al. Factors Supporting Cysteine Tolerance and Sulfite Production in Candida Albicans. *Eukaryot. Cell* **2013**, *12*, 604–613. [CrossRef] [PubMed]
35. Peng, Z.; Xu, P.; Song, Y.; Du, G.; Zhang, J.; Chen, J. Cysteine-Mediated Cyclic Metabolism Drives the Microbial Degradation of Keratin. *ACS Sustain. Chem. Eng.* **2021**, *9*, 9861–9870. [CrossRef]
36. Kshetri, P.; Roy, S.S.; Sharma, S.K.; Singh, T.S.; Ansari, M.A.; Prakash, N.; Ngachan, S.V. Transforming Chicken Feather Waste into Feather Protein Hydrolysate Using a Newly Isolated Multifaceted Keratinolytic Bacterium *Chryseobacterium Sediminis* RCM-SSR-7. *Waste Biomass Valoriz.* **2019**, *10*, 1–11. [CrossRef]
37. Sharma, R.; Gupta, R. Extracellular Expression of Keratinase *Ker P* from *Pseudomonas Aeruginosa* in E. Coli. *Biotechnol. Lett.* **2010**, *32*, 1863–1868. [CrossRef] [PubMed]
38. Kunutsor, S.K. Gamma-Glutamyltransferase-Friend or Foe Within? *Liver Int.* **2016**, *36*, 1723–1734. [CrossRef]
39. Li, Z.-W.; Liang, S.; Ke, Y.; Deng, J.-J.; Zhang, M.-S.; Lu, D.-L.; Li, J.-Z.; Luo, X.-C. The Feather Degradation Mechanisms of a New *Streptomyces* Sp. Isolate SCUT-3. *Commun. Biol.* **2020**, *3*, 191. [CrossRef] [PubMed]
40. Inada, S.; Watanabe, K. Draft Genome Sequence of *Meiothermus Ruber* H328, Which Degrades Chicken Feathers, and Identification of Proteases and Peptidases Responsible for Degradation. *Genome Announc.* **2013**, *1*, e00176-13. [CrossRef] [PubMed]
41. Mentel, M.; Ahuja, E.G.; Mavrodi, D.V.; Breinbauer, R.; Thomashow, L.S.; Blankenfeldt, W. Of Two Make One: The Biosynthesis of Phenazines. *ChemBioChem* **2009**, *10*, 2295–2304. [CrossRef] [PubMed]
42. Mishra, S.; Imlay, J. Why Do Bacteria Use so Many Enzymes to Scavenge Hydrogen Peroxide? *Arch. Biochem. Biophys.* **2012**, *525*, 145–160. [CrossRef]
43. Lalaouna, D.; Baude, J.; Wu, Z.; Tomasini, A.; Chicher, J.; Marzi, S.; Vandenesch, F.; Romby, P.; Caldelari, I.; Moreau, K. RsaC sRNA Modulates the Oxidative Stress Response of *Staphylococcus Aureus* during Manganese Starvation. *Nucleic Acids Res.* **2019**, *47*, 9871–9887. [CrossRef] [PubMed]
44. Queiroux, C.; Bonnet, M.; Saraoui, T.; Delpech, P.; Veisseire, P.; Rifa, E.; Moussard, C.; Gagne, G.; Delbès, C.; Bornes, S. Dialogue between *Staphylococcus Aureus* SA15 and *Lactococcus Garvieae* Strains Experiencing Oxidative Stress. *BMC Microbiol.* **2018**, *18*, 193. [CrossRef] [PubMed]
45. Wang, T.; Si, M.; Song, Y.; Zhu, W.; Gao, F.; Wang, Y.; Zhang, L.; Zhang, W.; Wei, G.; Luo, Z.-Q.; et al. Type VI Secretion System Transports Zn^{2+} to Combat Multiple Stresses and Host Immunity. *PLoS Pathog.* **2015**, *11*, e1005020. [CrossRef] [PubMed]
46. Chen, W.-J.; Kuo, T.-Y.; Hsieh, F.-C.; Chen, P.-Y.; Wang, C.-S.; Shih, Y.-L.; Lai, Y.-M.; Liu, J.-R.; Yang, Y.-L.; Shih, M.-C. Involvement of Type VI Secretion System in Secretion of Iron Chelator Pyoverdine in *Pseudomonas Taiwanensis*. *Sci. Rep.* **2016**, *6*, 32950. [CrossRef] [PubMed]
47. Lin, J.; Zhang, W.; Cheng, J.; Yang, X.; Zhu, K.; Wang, Y.; Wei, G.; Qian, P.-Y.; Luo, Z.-Q.; Shen, X. A *Pseudomonas* T6SS Effector Recruits PQS-Containing Outer Membrane Vesicles for Iron Acquisition. *Nat. Commun.* **2017**, *8*, 14888. [CrossRef]
48. Si, M.; Zhao, C.; Burkinshaw, B.; Zhang, B.; Wei, D.; Wang, Y.; Dong, T.G.; Shen, X. Manganese Scavenging and Oxidative Stress Response Mediated by Type VI Secretion System in *Burkholderia Thailandensis*. *Proc. Natl. Acad. Sci. USA* **2017**, *114*, E2233–E2242. [CrossRef] [PubMed]
49. Filloux, A. The Type VI Secretion System: A Tubular Story. *EMBO J.* **2009**, *28*, 309–310. [CrossRef] [PubMed]

50. Chen, L.; Zou, Y.; She, P.; Wu, Y. Composition, Function, and Regulation of T6SS in *Pseudomonas Aeruginosa*. *Microbiol. Res.* **2015**, *172*, 19–25. [CrossRef] [PubMed]
51. Miranda, S.W.; Asfahl, K.L.; Dandekar, A.A.; Greenberg, E.P. *Pseudomonas Aeruginosa* Quorum Sensing. *Adv. Exp. Med. Biol.* **2022**, *1386*, 95–115. [CrossRef]

microorganisms

Article

It Takes Two to Make a Thing Go Right: Epistasis, Two-Component Response Systems, and Bacterial Adaptation

Brittany R. Sanders, Lauren S. Thomas, Naya M. Lewis, Zaria A. Ferguson, Joseph L. Graves, Jr. and Misty D. Thomas *

Department of Biology, North Carolina Agricultural and Technical State University, Greensboro, NC 27411, USA; brsander@ncat.edu (B.R.S.); lauren_thomas@brown.edu (L.S.T.); naya.m.lewis@vanderbilt.edu (N.M.L.); zferguson5@student.gsu.edu (Z.A.F.); gravesjl@ncat.edu (J.L.G.J.)

* Correspondence: mthomas1@ncat.edu; Tel.: +1-336-285-2178

Abstract: Understanding the interplay between genotype and fitness is a core question in evolutionary biology. Here, we address this challenge in the context of microbial adaptation to environmental stressors. This study explores the role of epistasis in bacterial adaptation by examining genetic and phenotypic changes in silver-adapted *Escherichia coli* populations, focusing on the role of beneficial mutations in two-component response systems (TCRS). To do this, we measured 24-hour growth assays and conducted whole-genome DNA and RNA sequencing on *E. coli* mutants that confer resistance to ionic silver. We showed recently that the R15L *cusS* mutation is central to silver resistance, primarily through upregulation of the *cus* efflux system. However, here we show that this mutation's effectiveness is significantly enhanced by epistatic interactions with additional mutations in regulatory genes such as *ompR*, *rho*, and *fur*. These interactions reconfigure global stress response networks, resulting in robust and varied resistance strategies across different populations. This study underscores the critical role of epistasis in bacterial adaptation, illustrating how interactions between multiple mutations and how genetic backgrounds shape the resistance phenotypes of *E. coli* populations. This work also allowed for refinement of our model describing the role TCRS genes play in bacterial adaptation by now emphasizing that adaptation to environmental stressors is a complex, context-dependent process, driven by the dynamic interplay between genetic and environmental factors. These findings have broader implications for understanding microbial evolution and developing strategies to combat antimicrobial resistance.

Keywords: epistasis; bacterial adaptation; two-component response systems; silver resistance; gene-by-environment interactions

Citation: Sanders, B.R.; Thomas, L.S.; Lewis, N.M.; Ferguson, Z.A.; Graves, J.L., Jr.; Thomas, M.D. It Takes Two to Make a Thing Go Right: Epistasis, Two-Component Response Systems, and Bacterial Adaptation. *Microorganisms* **2024**, *12*, 2000. https://doi.org/10.3390/microorganisms12102000

Academic Editor: Tomohiro Shimada

Received: 27 August 2024
Revised: 27 September 2024
Accepted: 30 September 2024
Published: 30 September 2024

1. Introduction

Understanding the relationships between genotype and fitness is a fundamental question in evolutionary biology, especially in the context of microbial adaptation to environmental stressors [1–3]. A key aspect of this understanding is the study of epistasis, which refers to interactions between different genetic loci and their combined effect on an organism's fitness. Epistatic interactions are crucial for revealing the complexity of adaptive evolution, as they can significantly influence the trajectory of evolutionary changes in microbial populations [4,5]. Fitness epistasis highlights different types of genetic interactions—whether positive, negative, or sign epistasis—can alter the expected outcomes of adaptive mutations, making it a vital consideration in the study of complex traits [4,6–8].

Historically, silver has been used for centuries in antimicrobial applications due to its broad-spectrum antibacterial activity. Silver has been shown to impact disruption of the cell wall and membrane; interact with the thiol groups of respiratory enzymes; disrupt respiration due to binding to the cell membrane; disrupt the uptake of phosphorus and cause release of phosphates, mannitol, succinate, proline, and glutamine; disrupt metabolism, cell signaling, DNA replication, transcription, translation, and cell division, either directly or through the generation of reactive oxygen species (ROS). This makes it a valuable tool in preventing infections and preserving materials [9–11]. However, the emergence of silver-resistant bacteria poses a significant challenge to its effectiveness. The rising prevalence of silver resistance is not only a concern in clinical settings but also has broader implications for environmental health and the efficacy of silver-based antimicrobials [12–15].

The adaptive response of bacteria to such environmental stressors is often mediated by sophisticated regulatory networks, including two-component response systems (TCRS) [16–18]. TCRS are widespread in bacteria and play a pivotal role in sensing and responding to environmental changes. These systems typically consist of a sensor kinase and a response regulator that work together to modulate gene expression in response to specific stimuli [19–21]. In particular, TCRS like the CusS/CusR system are essential for metal ion homeostasis, regulating the efflux of toxic ions such as copper and silver, and thus play a central role in bacterial survival under metal stress [22–24].

The interplay between two-component response systems (TCRS) and epistasis is particularly significant in the context of bacterial resistance to antimicrobial agents [25–27]. When adaptive mutations occur within TCRS genes, these mutations do not act in isolation; rather, their phenotypic effects are often modulated by the genetic background of the organism and the specific environmental conditions in which the bacteria find themselves. This interaction can lead to complex and sometimes unpredictable adaptive responses [25]. Furthermore, epistasis can further complicate the outcome of adaptive mutations within TCRS. A mutation that increases resistance in one genetic background may have a different effect, or even an opposite effect, in another, depending on the presence of other interacting mutations [4,8,28,29]. Moreover, epistasis can also influence protein–protein interactions within TCRS and can result in varying degrees of resistance, depending on the combination of mutations and environmental pressures [27]. This suggests that the adaptive landscape of TCRS is highly contingent on both the internal genetic network and external environmental factors, making the evolution of resistance a dynamic process, and this complexity underscores the challenges in predicting bacterial resistance [30–32].

Experimental evolution has become a powerful approach to studying these processes, allowing researchers to observe evolutionary changes in real-time and identify adaptive mutations through techniques such as whole-genome resequencing [2,33,34]. Several studies have employed these methods to evaluate the potential for silver resistance in *Escherichia coli* and found that several mutations, namely, *cusS*, *ompR*, *rpoB*, and *purL*, were associated with the adaptive response [35–37]. We then recently demonstrated that specific mutations in the *cusS* gene could drive significant changes in gene expression and bacterial fitness, particularly in response to silver stress [38]. However, these mutations alone were insufficient to fully explain the observed resistance in the past studies, suggesting that additional genetic interactions were at play. We therefore hypothesized that these *cusS* mutations interact with other genetic changes via positive epistasis to enhance the overall resistance phenotype.

Given this context, our current study aims to further explore the role of epistasis in bacterial adaptation by examining the interplay between *cusS* mutations and other genetic changes within silver-adapted *Escherichia coli* populations. Specifically, we seek to answer the following key questions:

1. How do *cusS* mutations, in combination with other genetic changes, influence the overall fitness and resistance of *E. coli* to silver?
2. What are the specific gene expression patterns associated with these mutations, and how do they contribute to the observed phenotypes?
3. How do epistatic interactions among these mutations shape the evolutionary trajectory of the bacteria in response to silver stress?
4. How does our model of TCRS-driven bacterial adaptation need to be refined in light of these findings?

To address these questions, we employed a combination of 24-hour growth assays along with whole-genome DNA and RNA sequencing to systematically analyze the phenotypic and genetic characteristics of the silver-adapted mutants. This approach allowed us to dissect the complex genetic interactions that underpin silver resistance in these bacterial populations and to better describe the epistatic interactions that complement adaptive mutations in TCRS genes.

2. Materials and Methods

2.1. Bacterial Strains and Culturing

Escherichia coli K-12 MG1655 (ATCC #10798D-5, Manassas, VA, USA) was used as our wild-type (WT) strain, as it served as the ancestral strain in the experimental evolution (EE) studies where our mutants were initially identified [36,37]. The archived stocks of the silver-adapted mutants (SAMs) were revived from storage at −80 °C. Of the original 16 strains, only 7 were successfully recovered during the revival process, and these populations have been designated as SAM1-7. We previously published the design, construction, and whole genome sequencing of our single *cusS* variants (R15L, T14P, T17P, N279H) [39], which were utilized again in this study. All growth experiments were conducted in Davis Minimal Broth (DMB—7 g/L dipotassium phosphate, 2 g/L monopotassium phosphate, 0.5 g/L sodium citrate, 1 g/L ammonium sulfate, and 0.1 g/L magnesium sulfate, ThermoFisher, Waltham, MA, USA) supplemented with 10% dextrose as a carbon source, incubated overnight at 37 °C with continuous shaking at 200 rpm, unless otherwise noted. DMB was selected as it was the original medium used during the silver adaptation study, where these mutants had initially evolved.

2.2. Whole Genome Illumina Sequencing of the SAM Populations

Bacterial stocks from the silver-adapted mutant (SAM) populations in the Tajkarimi et al., 2017 [37] EE study were grown overnight in DMB, pelleted, and sent for whole genome Illumina sequencing at the Sequencing Center (SeqCenter.com) in Pittsburgh, PA on an Illumina NovaSeq X Plus (San Diego, CA, USA). Sequence alignment and variant calling were performed using the breseq 0.30.0 pipeline [40], with alignment to the *E. coli* K-12 MG1655 reference sequence (NC_000913). The original WT strain was sequenced in one of our previous studies [36]. It is important to note that after DNA sequencing, the only *cusS* mutation detected in the SAM populations was R15L. As a result, this was the only single adaptive mutant used for the remainder of the study and will be referred to simply as R15L from this point forward. Whole genome sequencing fastq files are available in the NCBI SRA database using the BioProject ID: PRJNA1160277 and BioSamples: SAMN43760522-SAMN43760542.

2.3. RNAseq and Differential Gene Expression

Overnight cultures of the WT and each mutant population (R15L and SAM1-7) were grown and then subcultured at a 1:100 ratio in DMB (6 replicates). These cultures were grown to an OD_{600} of 0.2, as this was where our populations reach mid-log phase when grown in DMB. Once the desired optical density was reached, three replicates of each culture were incubated for 1 h at 37 °C, then pooled, pelleted, and stored at −80 °C. The remaining three replicates were exposed to 50 ng/mL silver nitrate (the original selection concentration) for 1 h at 37 °C, then pooled, pelleted, and stored at −80 °C. The frozen bacterial pellets were shipped to SeqCenter (seqcenter.com). At SeqCenter, samples were treated with Invitrogen DNase (RNase-free) (Waltham, MA, USA) and RNA extractions were performed. Library preparation was conducted using Illumina's Stranded Total RNA Prep Ligation with Ribo-Zero Plus kit and 10 bp IDT for Illumina indices. Sequencing was performed on a NextSeq 2000, generating 2 × 51 bp reads. Demultiplexing, quality control, and adapter trimming were performed using bcl-convert (v3.9.3) [41]. The average reads per sample exceeded 20 million, with over 94% of bases having a quality score greater than Q30. Read mapping was carried out with HISAT2 (v2.2.0) [42], and read quantification was performed using Subread's featureCounts (v2.0.1) functionality. The read counts were loaded into R (4.0.2) and normalized using edgeR's (v1.14.5) Trimmed Mean of M values (TMM) algorithm [43]. Subsequent values were converted to counts per million (CPM). Differential expression analysis was performed using edgeR's exact test to compare differences between two groups of negative-binomial counts, with an estimated dispersion value of 0.1. All populations were compared to the WT in the absence of silver nitrate for this analysis. Individual genes were then grouped into categories based on their biological function using a combination of literature searchers and databases such as UniProt, KEGG, and Gene Ontology. RNAseq fastq files are available in the NCBI SRA database using the BioProject ID: PRJNA1160277 and BioSamples: SAMN43760522-SAMN43760542.

2.4. *Twenty-Four-Hour Growth Assays*

The 24-hour growth assays were conducted using the WT, R15L, and SAM1-6 (SAM7 was lost after sequencing and could not be included) to evaluate changes in bacterial growth and determine minimum inhibitory concentrations (MICs) in the presence of increasing concentrations of silver nitrate (0–750 ng/mL). To begin, archived glycerol stocks of each population were inoculated in DMB and incubated overnight at 37 °C with continuous shaking at 200 rpm. The following day, overnight cultures were diluted to an OD_{600} of 0.05 for normalization and added (in triplicate) to a 96-well plate containing a gradient of silver nitrate concentrations. The 96-well plates were then covered with optically clear MicroAmp Adhesive film (ThermoFisher, Waltham, MA, USA) and absorbance readings were taken every hour from 0 to 24 h using a pre-programmed protocol on the Varioskan Lux 96-well plate reader (ThermoFisher, Waltham, MA, USA) at 37 °C with shaking. To normalize the data, blank readings (in triplicate) were averaged, and the mean values were subtracted from each replicate population at each concentration. The resulting data were plotted using GraphPad Prism version 8.0.0 for Mac OS X. We then used the Growthcurver package (v0.3.1) in R to fit our growth curve data for each population at each concentration of silver nitrate to a standard form of the logistic equation [44]. This enabled us to calculate the initial population size (n0), the carrying capacity (k), the growth rate (r), the time to midpoint (t_mid), the generation time (t_gen), the area under the curve—log phase (auc_l) and the goodness-of-fit (sigma). The (k), (r), (t_mid) and (t-gen) were plotted in GraphPad Prism. MICs were determined by identifying the lowest concentration at which no growth was observed for each population and are reported on their corresponding curve. Growth curves were also used as the basis for calculating the relative fitness. The calculation for relative fitness (ω) was performed by taking the OD value for each population at a specific time point and dividing it by the OD value of the fittest population at that same time point. These data were then plotted in Prism, and one-way ANOVAs with multiple pairwise comparisons were used to determine statistical differences.

3. Results

3.1. *DNA Sequencing and Mutation Analysis*

Populations carried variation in the number of total mutations, but most notably the R15L *cusS* mutation was found in SAM2, SAM3, SAM6, and SAM7, all to fixation (f = 1.00). All moderate and high frequency mutations are given in Table 1 and a complete list of all mutations can be found in Table S1.This mutation was the most common in the original evolution study and the only one revivable from storage [37]. Notably, the R15L mutation was consistently associated with additional mutations in other key regulatory genes, such as *rho* and *ompR* in all four of these populations. SAM2 and SAM3 fixed an R66C mutation in *rho* and a 1 bp deletion in *ompR*, while SAM6 and SAM7 fixed G63V in *rho* and R182C in *ompR*. These R15L populations then diverged, with SAM2 and SAM3 sharing additional mutations in *dnaK*, *yggN*, *yghS*, and *yicO*, in addition to a 103 bp deletion in *yicC*, and SAM6 and SAM7 both carrying K9N in *fur* and V282L in *rpo*. SAM7 also uniquely fixed three intergenic mutations. Overall, SAM2 had 39 mutations, SAM3 had 24, SAM6 had 35, and SAM7 had 17. SAM1, SAM4, and SAM5 lacked *cusS* mutations or did not reach fixation

(SAM4 had a *cusS* mutation with an *f* = 0.055). SAM1 fixed A17V in *glnH*, while SAM4 and SAM5 had no fixed mutations. SAM4's highest was an 82 bp deletion in *rph* (*f* = 0.533), and SAM5's was E783V in *glnE* (*f* = 0.408). In total, SAM1 carried 17 mutations, SAM4 had 12, and SAM5 had 17—generally fewer overall than the R15L populations. This suggests that the *cusS* mutation may be crucial for driving the selection and fixation of other genes in response to silver nitrate, though *cusS* alone is not enough for full resistance [38].

Table 1. BreSeq-detected moderate to high-frequency mutations (>0.25) in SAM populations following DNA sequencing: mutations highlighted in the table were detected to fixation.

	Location	Mutation	Frequency	Annotation	Gene	Description
SAM1	847,955	G→A	1.000	A17V (GCG→GTG)	*glnH* ←	glutamine transporter subunit
	4,188,510	C→A	0.877	T1054N (ACC→AAC)	*rpoC* →	RNA polymerase, beta prime subunit
	1,868,984	C→A	0.849	N10K (AAC→AAA)	*yeaH* →	UPF0229 family protein
	3,815,809	Δ1 bp	0.807	intergenic (−41/+25)	*pyrE* ←/← *rph*	orotate phosphoribosyltransferase/ ribonuclease PH (defective);enzyme; Degradation of RNA; RNase PH
	3,831,168	C→T	0.732	L238L (CTA→TTA)	*yicH* →	putative inner membrane-anchored periplasmic AsmA family protein
SAM2	12,661	C→G	1.000	R167G (CGT→GGT)	*dnaK* →	chaperone Hsp70, with co-chaperone DnaJ
	594,727	C→A	1.000	R15L (CGC→CTC)	*cusS* ←	sensory histidine kinase in two-component regulatory system with CusR, senses copper ions
	3,101,306	G→T	1.000	I106I (ATC→ATA)	*yggN* ←	DUF2884 family putative periplasmic protein
	3,133,461	G→A	1.000	A166V (GCA→GTA)	*yghS* ←	putative ATP-binding protein
	3,536,061	Δ1 bp	1.000	coding (524/720 nt)	*ompR* ←	response regulator in two-component regulatory system with EnvZ
	3,816,605	Δ103 bp	1.000		*[yicC]*	*[yicC]*
	3,843,548	A→C	1.000	I81R (ATA→AGA)	*yicO* ←	putative adenine permease
	3,966,612	C→T	1.000	R66C (CGT→TGT)	*rho* →	transcription termination factor
	3,992,735	C→A	0.930	S528 * (TCG→TAG)	*cyaA* →	adenylate cyclase
	3,177,973	IS*1* (+) +9 bp	0.552	intergenic (−63/−134)	*nudF* ←/→ *tolC*	ADP-ribose pyrophosphatase/ transport channel
	915,226	T→C	0.446	intergenic (−369/+126)	*ybjE* ←/← *aqpZ*	putative transporter/aquaporin Z
	2,229,177	A→G	0.376	intergenic (+112/+261)	*yohP* →/← *dusC*	uncharacterized protein/ tRNA-dihydrouridine synthase C
	4,296,060	C→T	0.282	intergenic (+266/+376)	*gltP* →/← *yjcO*	glutamate/aspartate:proton symporter/Sel1 family TPR-like repeat protein

Table 1. *Cont.*

	Location	Mutation	Frequency	Annotation	Gene	Description
SAM3	12,661	C→G	1.000	R167G (CGT→GGT)	*dnaK* →	chaperone Hsp70, with co-chaperone DnaJ
	594,727	C→A	1.000	R15L (CGC→CTC)	*cusS* ←	sensory histidine kinase in two-component regulatory system with CusR, senses copper ions
	3,101,306	G→T	1.000	I106I (ATC→ATA)	*yggN* ←	DUF2884 family putative periplasmic protein
	3,133,461	G→A	1.000	A166V (GCA→GTA)	*yghS* ←	putative ATP-binding protein
	3,536,061	Δ1 bp	1.000	coding (524/720 nt)	*ompR* ←	response regulator in two-component regulatory system with EnvZ
	3,816,605	Δ103 bp	1.000		*[yicC]*	*[yicC]*
	3,843,548	A→C	1.000	I81R (ATA→AGA)	*yicO* ←	putative adenine permease
	3,966,612	C→T	1.000	R66C (CGT→TGT)	*rho* →	transcription termination factor
	3,992,735	C→A	0.865	S528 * (TCG→TAG)	*cyaA* →	adenylate cyclase
	3,177,973	IS*1* (+) +9 bp	0.648	intergenic (−63/−134)	*nudF* ←/→ *tolC*	ADP-ribose pyrophosphatase/transport channel
SAM4	3,992,735	C→A	0.865	S528 * (TCG→TAG)	*cyaA* →	adenylate cyclase
	3,177,973	IS*1* (+) +9 bp	0.648	intergenic (−63/−134)	*nudF* ←/→ *tolC*	ADP-ribose pyrophosphatase/transport channel
	3,360,120	G→A	0.163	R308H (CGC→CAC)	*gltD* →	glutamate synthase, 4Fe-4S protein, small subunit
	4,296,060	C→T	0.159	intergenic (+266/+376)	*gltP* →/← *yjcO*	glutamate/aspartate:proton symporter/Sel1 family TPR-like repeat protein
	2,725,169	Δ1 bp	0.147	coding (578/1299 nt)	*kgtP* ←	alpha-ketoglutarate transporter
	2,229,205	G→C	0.116	intergenic (+140/+233)	*yohP* →/← *dusC*	uncharacterized protein/tRNA-dihydrouridine synthase C
	3,815,859	Δ82 bp	0.533		*[rph]–[rph]*	*[rph], [rph]*
	3,815,824	G→T	0.277	intergenic (−56/+10)	*pyrE* ←/← *rph*	orotate phosphoribosyltransferase/ribonuclease PH (defective);enzyme; Degradation of RNA; RNase PH
	3,198,033	IS*186* (+) +6 bp :: Δ2 bp	0.258	coding (1604–1609/2841 nt)	*glnE* ←	fused deadenylyltransferase/adenylyltransferase for glutamine synthetase
	4,296,060	C→T	0.235	intergenic (+266/+376)	*gltP* →/← *yjcO*	glutamate/aspartate:proton symporter/Sel1 family TPR-like repeat protein
	3,815,809	Δ1 bp	0.138	intergenic (−41/+25)	*pyrE* ←/← *rph*	orotate phosphoribosyltransferase/ribonuclease PH (defective);enzyme; Degradation of RNA; RNase PH
	915,226	T→C	0.112	intergenic (−369/+126)	*ybjE* ←/← *aqpZ*	putative transporter/aquaporin Z
	594,727	C→A	0.055	R15L (CGC→CTC)	*cusS* ←	sensory histidine kinase in two-component regulatory system with CusR, senses copper ions

Table 1. *Cont.*

	Location	Mutation	Frequency	Annotation	Gene	Description
SAM5	3,197,294	T→A	0.408	E783V (GAA→GTA)	*glnE* ←	fused deadenylyltransferase/ adenylyltransferase for glutamine synthetase
	3,815,809	Δ1 bp	0.373	intergenic (−41/+25)	*pyrE* ←/← *rph*	orotate phosphoribosyltransferase/ ribonuclease PH (defective);enzyme; Degradation of RNA; RNase PH
	3,815,859	Δ82 bp	0.367		*[rph]*–*[rph]*	*[rph]*, *[rph]*
	4,296,060	C→T	0.236	intergenic (+266/+376)	*gltP* →/← *yjcO*	glutamate/aspartate:proton symporter/Sel1 family TPR-like repeat protein
	3,992,588	$(ATCAGCC)_{2\to1}$	0.235	coding (1436–1442/2547 nt)	*cyaA* →	adenylate cyclase
	1,907,503	IS3 (–) +5 bp :: +T	0.178	coding (85–89/144 nt)	*yobF* ←	DUF2527 family heat-induced protein
	4,181,669	A→G	0.164	E142G (GAG→GGG)	*rpoB* →	RNA polymerase, beta subunit
	3,485,966	IS2 (+) +5 bp	0.129	intergenic (−148/−150)	*yhfA* ←/→ *crp*	OsmC family protein/cAMP-activated global transcription factor, mediator of catabolite repression
	3,897,059	C→A	0.129	L96I (CTC→ATC)	*yieH* →	phosphoenolpyruvate and 6-phosphogluconate phosphatase
SAM6	594,727	C→A	1.000	R15L (CGC→CTC)	*cusS* ←	sensory histidine kinase in two-component regulatory system with CusR, senses copper ions
	710,620	C→A	1.000	K9N (AAG→AAT)	*fur* ←	ferric iron uptake regulon transcriptional repressor; autorepressor
	3,440,186	C→A	1.000	V282L (GTA→TTA)	*rpoA* ←	RNA polymerase, alpha subunit
	3,536,041	G→A	1.000	R182C (CGC→TGC)	*ompR* ←	response regulator in two-component regulatory system with EnvZ
	3,966,604	G→T	1.000	G63V (GGT→GTT)	*rho* →	transcription termination factor
	4,232,641	C→A	0.318	R198L (CGT→CTT)	*lysC* ←	lysine-sensitive aspartokinase 3
	1,213,820	G→C	0.279	D80E (GAC→GAG)	*bluR* ←	repressor of blue light-responsive genes
	3,359,461	Δ1 bp	0.226	coding (264/1419 nt)	*gltD* →	glutamate synthase, 4Fe-4S protein, small subunit
SAM7	594,727	C→A	1.000	R15L (CGC→CTC)	*cusS* ←	sensory histidine kinase in two-component regulatory system with CusR, senses copper ions
	710,620	C→A	1.000	K9N (AAG→AAT)	*fur* ←	ferric iron uptake regulon transcriptional repressor; autorepressor
	1,428,765	T→C	1.000	intergenic (−39/−30)	*insH1* ←/→ *lomR*	IS5 transposase and trans-activator; IS, phage, Tn; Transposon-related functions; extrachromosomal; transposon related/pseudogene, Rac prophage lom homolog; Phage or Prophage Related; interrupted by IS5 and N-ter deletion

Table 1. *Cont.*

	Location	Mutation	Frequency	Annotation	Gene	Description
	3,440,186	C→A	1.000	V282L (GTA→TTA)	*rpoA* ←	RNA polymerase, alpha subunit
	3,536,041	G→A	1.000	R182C (CGC→TGC)	*ompR* ←	response regulator in two-component regulatory system with EnvZ
	3,815,801	Δ1 bp	1.000	intergenic (−33/+33)	*pyrE* ←/← *rph*	orotate phosphoribosyltrans-ferase/ribonuclease PH (defective);enzyme; Degradation of RNA; RNase PH
SAM7	3,966,604	G→T	1.000	G63V (GGT→GTT)	*rho* →	transcription termination factor
	1,212,080:1	+C	1.000	intergenic (−77/+623)	*iraM* ←/← *ycgX*	RpoS stabilizer during Mg starvation, anti-RssB factor/DUF1398 family protein
	1,213,820	G→C	0.627	D80E (GAC→GAG)	*bluR* ←	repressor of blue light-responsive genes
	3,359,461	Δ1 bp	0.590	coding (264/1419 nt)	*gltD* →	glutamate synthase, 4Fe-4S protein, small subunit
	4,296,060	C→T	0.248	intergenic (+266/+376)	*gltP* →/← *yjcO*	glutamate/aspartate:proton symporter/Sel1 family TPR-like repeat protein

* Indicates that the mutation resulted in the incorporation of a stop codon in place of the original amino acid.

3.2. Gene Expression and Functional Implications

We previously associated the R15L mutation in *cusS* with significant upregulation and constitutive expression of the *cusCFBA* operon, which is central to the bacterial response to silver stress [38]. CusC is a component of the outer membrane porin channel, CusB serves as a membrane fusion protein, CusA is the inner membrane component that actively exports silver and copper ions, and CusF is a periplasmic metallochaperone [45,46]. The increased expression enhances the efflux system's efficiency, contributing to a robust silver resistance phenotype. RNAseq analysis here revealed distinct gene expression profiles in these efflux genes across the SAM mutants compared to the WT strain and single R15L mutant (Figure 1).

The non-R15L-carrying mutants like SAM1 showed only moderate upregulation in metal homeostasis genes (Figure 2) such as *zntA* and *cueO* and all down regulated expression of all of the *cus* efflux pump genes in both the presence and absence of silver nitrate (Figure 1). While all SAM mutants carrying the R15L mutation demonstrated upregulation in metal homeostasis genes, the extent of this response varied (Figure 2). Specifically, they all exhibited constitutive upregulation in both *cusS/R* and *cusCFBA* (Figure 1). SAM7 exhibited the highest overall upregulation in the *cusCFBA* efflux genes, suggesting that additional mutations in its genome, particularly those in *rho* and *ompR*, might enhance its ability to manage metal ion stress. Clearly, the R15L mutation in *cusS* does not act in isolation, as the single mutant (R15L) displays lower expression levels than the SAMs. This supports the idea that expression of the *cusCFBA* efflux pump genes is enhanced by the other mutations observed in the genetic background of the SAM populations. Providing evidence of positive epistasis suggests that the regulatory disruptions caused by the additional mutations may facilitate a more robust and sustained activation of the *cus* operon, leading to heightened silver resistance.

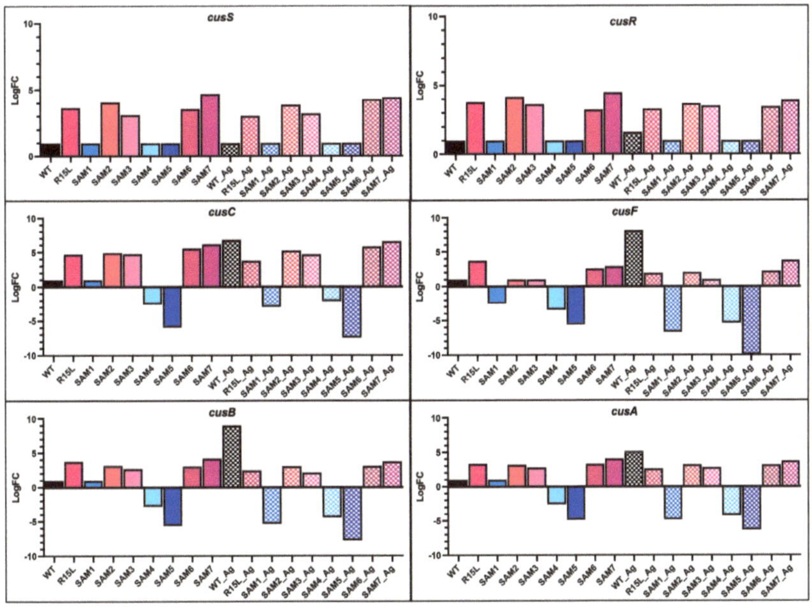

Figure 1. Differential expression of *cus* genes across silver-adapted *E. coli* populations: This figure illustrates the expression levels of the *cusS/R* two-component response system (TCRS) genes and the *cusCFBA* efflux genes, which are essential for silver and copper ion efflux, across various silver-adapted *E. coli* populations in both the presence (checkered bars) and absence of silver nitrate (solid bars). All expression levels are normalized to the wild-type (WT) in the absence of silver nitrate, where WT is assigned a log fold change (logFC) of 1. LogFC values were plotted using GraphPad Prism. The upregulation observed in populations with the R15L mutation in *cusS* highlights its role in enhancing the efficiency of the efflux system, thereby increasing silver resistance. This figure also demonstrates the variation in gene expression across different populations, reflecting their differing capacities to manage metal ion toxicity. Notably, SAM populations that do not carry the R15L mutation exhibit no expression in the *cusS/R* genes and downregulate the efflux pump genes, indicating a distinct response mechanism.

Beyond the *cus* operon, differential gene expression analysis revealed several other biological pathways significantly altered in response to silver adaptation across the SAM mutants (Figure 2):

1. Oxidative Stress Response Genes: Many SAM mutants, particularly those with *cusS* mutations, showed upregulation in genes associated with the oxidative stress response. For instance, *msrP* (methionine sulfoxide reductase) and *huiH* (hypothetical protein involved in stress response) were consistently upregulated. Also, *msrP* is known for its role in reducing methionine sulfoxide residues in proteins, which helps protect cells from oxidative damage caused by reactive oxygen species (ROS) [47]. The upregulation of these genes suggests that the mutants have developed a preemptive defense mechanism to mitigate oxidative damage associated with silver ion exposure, contributing to their enhanced survival under stress.
2. Transport Systems Genes: The expression of various transport-related genes, beyond the *cus* operon, was significantly altered. SAM2 and SAM5, for instance, showed strong upregulation of *zntA*, a gene encoding a P-type ATPase that functions as a zinc efflux pump. This gene plays a dual role in protecting the cell from toxic levels of zinc and potentially providing cross-resistance to other heavy metals, including silver [48]. Additionally, *copA*, a gene encoding a copper-transporting ATPase, was variably expressed among the mutants. In SAM7, *copA* was downregulated, which might

indicate a shift in the cell's strategy to manage copper and silver ion homeostasis more efficiently under stress.

3. Amino Acid Biosynthesis and Metabolism Genes: SAM3 and SAM6 exhibited notable upregulation in genes involved in amino acid biosynthesis and metabolism, such as *gltB* (glutamate synthase large subunit) and *asd* (aspartate-semialdehyde dehydrogenase). *GltB* plays a crucial role in the synthesis of glutamate, a key amino acid involved in nitrogen metabolism and as a precursor for other amino acids [49]. *Asd* is involved in the biosynthesis of lysine and other essential metabolites. The upregulation of these genes suggests an increased demand for amino acid synthesis, likely to support protein repair and synthesis during stress conditions. The prioritization of these metabolic pathways may help sustain cellular function and growth during silver exposure, providing a buffer against the detrimental effects of stress.
4. Cell Envelope Integrity Genes: Several mutants, particularly SAM7, showed upregulation of genes involved in maintaining cell envelope integrity. For instance, *murein* (peptidoglycan synthesis genes) such as *mrcA* (penicillin-binding protein 1A) and *lpoA* (lipoprotein involved in peptidoglycan synthesis) were upregulated. These genes are critical for maintaining the structural integrity of the bacterial cell wall, especially under conditions where membrane integrity might be compromised by silver ions.
5. Regulatory Functions Genes: The SAM mutants exhibited differential expression in several genes associated with regulatory functions. For example, *rpoA* (RNA polymerase alpha subunit) and *rpoS* (sigma factor S) were upregulated in SAM6, suggesting a heightened global stress response. *RpoS* is particularly important for the bacterial stress response, regulating the expression of numerous genes involved in survival during the stationary phase and under various environmental stresses. The upregulation of *rpoS* indicates that SAM6 might have an enhanced ability to manage multiple stressors, contributing to its superior fitness and resistance profile.
6. Metabolic Suppression Genes: Interestingly, several mutants, especially SAM7, exhibited downregulation of genes involved in central metabolism, including those associated with glycolysis (*pfkA*, phosphofructokinase) and the TCA cycle (*sdhA*, succinate dehydrogenase). This downregulation may reflect a strategic metabolic shift to conserve energy and resources, redirecting them toward essential stress responses and repair mechanisms. This metabolic suppression likely serves as a trade-off, allowing the cells to prioritize survival overgrowth in the presence of silver.

These widespread changes in gene expression across the SAM mutants highlight the complex and multifaceted nature of bacterial adaptation to silver stress. The upregulation of stress response genes, such as *msrP* and *huiH*, coupled with alterations in metal ion transport systems (*zntA*, *copA*), amino acid metabolism (*gltB*, *asd*), and regulatory networks (*rpoA*, *rpoS*), underscores the importance of coordinated genetic responses in developing robust resistance. The downregulation of metabolic genes, particularly in SAM7, suggests a strategic shift in cellular priorities, where survival mechanisms are favored over growth and energy production.

The differential expression of genes involved in cell envelope integrity and global stress response further emphasizes the role of epistasis in fine-tuning these adaptive responses. By modifying multiple pathways simultaneously, the SAM mutants demonstrate the intricate interplay between genetic mutations and environmental pressures, leading to the emergence of distinct resistance phenotypes. This comprehensive adaptation strategy, involving both upregulation of protective mechanisms and downregulation of non-essential pathways, highlights the sophisticated nature of bacterial evolution in response to antimicrobial agents.

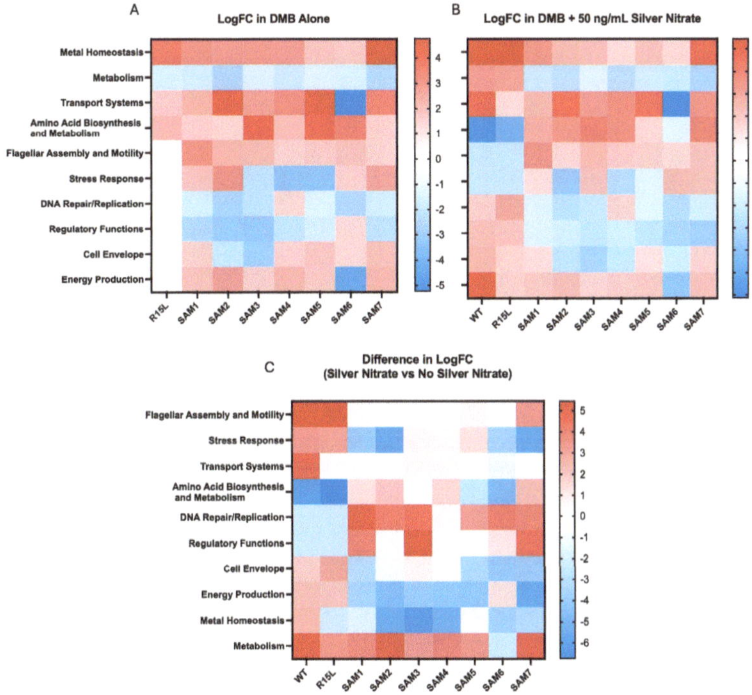

Figure 2. Comparative gene expression across 10 biological categories in silver-adapted *E. coli* populations: Heatmaps illustrate differential expression of genes across 10 key biological categories in various silver-adapted *E. coli* populations, normalized to the wild-type (WT) in the absence of silver nitrate. Differentially expressed genes were categorized based on their biological functions, and the heatmaps display the averaged differential expression for all genes within each category. Warmer colors indicate higher expression levels, while cooler colors indicate lower expression levels. The heatmaps were generated using GraphPad Prism. Subfigure (**A**) shows gene expression in the absence of silver nitrate, serving as a baseline to display natural variations in gene regulation and adaptation strategies across different populations. Subfigure (**B**) highlights gene expression in the presence of silver nitrate. Subfigure (**C**) is a difference map illustrating the changes in gene expression between conditions with and without silver nitrate, providing a direct comparison of how silver exposure affects gene expression across different biological categories. These heatmaps collectively emphasize how specific genetic backgrounds modulate these adaptive responses to silver exposure.

3.3. Comparative Fitness Analysis

Growth assays were conducted over a range of silver nitrate concentrations (0–750 ng/mL) to evaluate the fitness and adaptability of the SAM mutants (Supplemental Figure S1 for full growth curves at all silver nitrate concentrations and Figure 3). The Growthcurver package was used to extract detailed growth metrics for each population, including the growth rate (r), generation time (t_gen), midpoint time (t_mid), and carrying capacity (k) (Figure 4). These metrics provided insights into how each strain adapts to varying levels of silver stress. Data from the 24-hour growth curves were also used to calculate relative fitness across silver concentrations and time points (Figure 5). It is important to note that relative fitness, as calculated in this study, reflects the performance of each population at a given time point in relation to the fittest population at that same time point, rather than indicating an absolute increase in fitness or resilience over time.

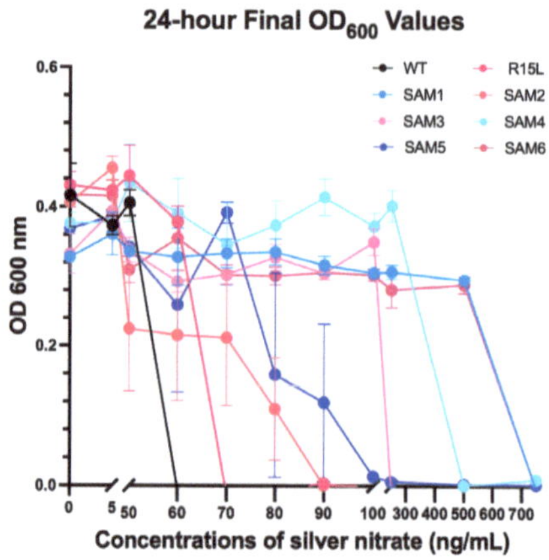

Figure 3. 24-hour final OD600 values from growth response assays of *E. coli* populations to increasing silver nitrate concentrations. The 24-hour time point from growth response assays of WT, R15L, and SAM1-6 populations in Davis Minimal Broth (DMB) is shown under increasing concentrations of silver nitrate (0–750 ng/mL). OD600 values were measured hourly over 24 h, matching the selection time point used in the original experimental evolution study where the SAM populations evolved. Data were collected in triplicate, with means and standard errors of the mean (SEMs) plotted using GraphPad Prism. To calculate statistical variation between each time point and the WT, we performed a two-way ANOVA with multiple comparisons. Detailed statistical results are provided in Table S2. These growth curves were also used to determine the minimum inhibitory concentration (MIC), defined as the lowest silver concentration at which no growth was observed for a population.

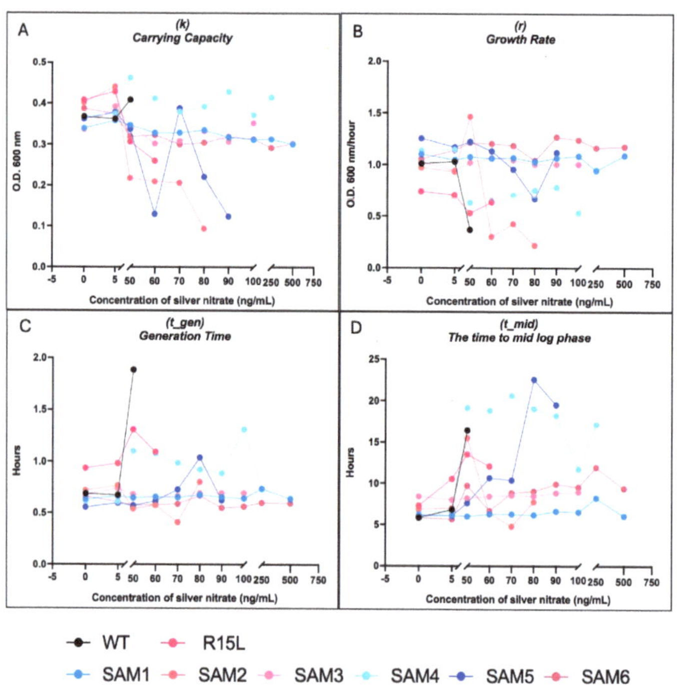

Figure 4. Detailed growth metrics across silver nitrate concentrations. This figure presents detailed growth metrics for each *E. coli* population under varying silver nitrate concentrations, analyzed

using the R package Growthcurver (v0.3.1). Growthcurver fits the growth data to a logistic model, providing key metrics that describe the dynamics of each population. These metrics include growth rate (r), reflecting how quickly the population grows; generation time (t_gen), indicating the time required for the population to double; midpoint time (t_mid), representing the time at which the population reaches half its carrying capacity; and carrying capacity (k), denoting the maximum population size supported under the given conditions. The data generated by Growthcurver were plotted using GraphPad Prism. (**A**) (k) carrying capacity, (**B**) (r) on growth rate, (**C**) on generation time, and (**D**) on time to mid-log phase. These results, derived from Growthcurver's comprehensive analysis, highlight the impact of silver stress on growth dynamics and reveal the differing adaptive responses among the populations.

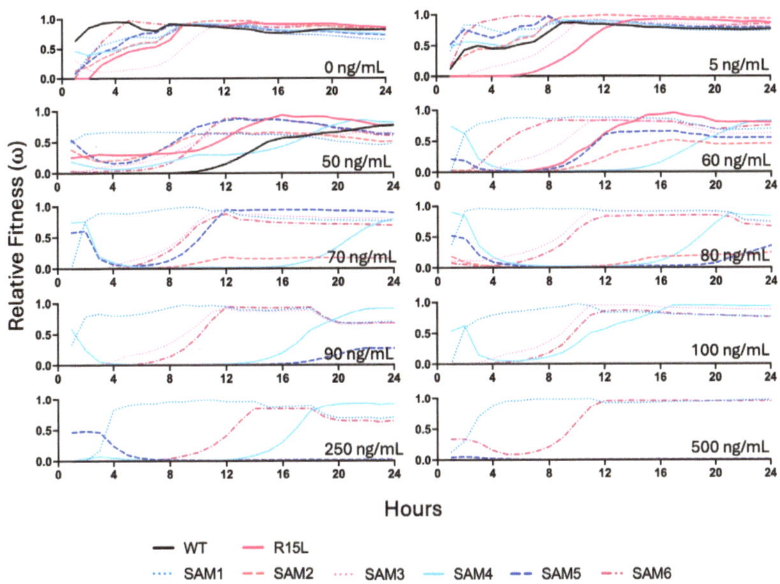

Figure 5. Comparative fitness of *E. coli* populations under silver nitrate stress. Relative fitness (ω) of each *E. coli* population under varying concentrations of silver nitrate (0–750 ng/mL) is shown. To calculate relative fitness, the OD_{600} of each population was divided by the maximum OD_{600} observed among other genotypes in the population at the same time points. Prior to this calculation, all negative growth values were set to 0 to ensure accurate comparisons, and the data were plotted in GraphPad Prism. Statistical analyses were conducted using a one-way ANOVA with pairwise multiple comparisons (Table S3) in GraphPad Prism. These results underscore the impact of silver stress on the competitive dynamics among the populations.

Growth curves showed that WT had an MIC of 60 ng/mL (Figure 3), which is consistent with previous studies that used a sub-MIC of 50 ng/mL silver nitrate for selection of silver resistance (37). At 50 ng/mL, WT displayed a drop in growth rate from 1 at 0 ng/mL to 0.36 at 50 ng/mL in addition to an increase from a time to mid-log from 6 h to 16 h, albeit while still maintaining its carrying capacity (Figure 4A–D).

The R15L single mutant has a slightly higher MIC of 70 ng/mL (Figure 3), albeit did not show much of a decline in growth rate across silver concentrations and only an increase from 7 to 12 h in its time to mid-log accounting for its increase in fitness over the WT at 50 ng/mL (Figure 5).

SAM1 consistently demonstrated the highest relative fitness (Figure 5) across most silver concentrations, with a growth rate (r) of 1.2 at 50 ng/mL and 1.15 at 750 ng/mL and stable carrying capacity (k), indicating strong adaptability and resistance, even under high silver stress (Figures 3 and 4A,B). SAM6 also showed significant resilience, particularly at higher concentrations, with a slightly lower but still robust growth rate (r) of 1.18 at 50 ng/mL and 1.1 at 750 ng/mL. Both SAM1 and SAM6 had an MIC of 750 ng/mL, with short generation times (t_gen) and minimal lag phase extensions, highlighting their superior resistance (Figures 3 and 4C).

In contrast, WT and R15L-carrying populations, such as SAM2 and SAM3, struggled as silver concentration increased. SAM2 showed an MIC of 90 ng/mL. The growth rate dropped sharply from 1.0 at 50 ng/mL to 0.2 at 80 ng/mL, with a corresponding decline in carrying capacity and extended lag phases, suggesting reduced adaptability at higher concentrations. At lower silver concentrations, SAM2 exhibited statistically lower fitness than WT at 0 µg/mL, highlighting its inherent difficulty in thriving even without stress. SAM3, with an MIC of 250 ng/mL, showed steady metrics across concentrations of silver nitrate, including carrying capacity, growth rate, generation time, and time to mid-log phase.

SAM4, despite lacking a fixed *cusS* mutation, displayed strong resistance with an MIC of 500 ng/mL, though its growth rate and carrying capacity varied more significantly, reflecting a slower adaptive response. SAM5 showed moderate resistance with an MIC of 100 ng/mL, maintaining stable growth but struggling with higher silver stress. SAM4 and SAM5 both showed a large increase in their time to mid-log phase, at almost 20 h from SAM4 above silver concentrations of 50 ng/mL, and for SAM5 it remained around 10 h until 80 ng/mL and rose to 20 h at 90 ng/mL, where both are at ~6 h in absence of silver nitrate.

The differential fitness among the SAM mutants likely results from a variation in epistatic interactions between mutations. In particular, this is evident in the SAM populations that carry R15L mutations while all showing drastically different relative fitness levels, MIC, and growth metrics.

4. Discussion

The R15L *cusS* mutation has been predicted to be central to silver resistance through the upregulation of the *cus* efflux system [37]. However, as demonstrated in this study, its efficacy is not uniform across all silver-adapted populations. This variability can be attributed to the distinct genetic compositions of each population, which modulate how the *cusS* mutation interacts with other mutations in regulatory genes such as *ompR*, *rho*, and *fur* [5]. These interactions lead to different epistatic effects that either enhance or diminish the silver resistance phenotype in each population.

For instance, populations such as SAM6 and SAM7, which share the R15L mutation, exhibit superior resistance due to additional mutations in other regulatory genes, whereas populations lacking these additional mutations show lower resistance. This highlights that while the R15L *cusS* mutation plays a significant role, the broader genetic background is crucial in determining the overall efficacy of silver resistance.

Consequently, our findings suggest that the R15L *cusS* mutation alone is not universally applicable as a predictor of resistance across all bacterial strains. Instead, its impact is highly context-dependent, shaped by the specific genetic and environmental conditions of each population. This context dependency complicates the development of universal models of bacterial resistance and suggests that evolutionary trajectories may vary widely among populations, even when key adaptive mutations are shared.

4.1. Role of Epistasis in Silver Resistance

The variations in gene expression and fitness among the SAM mutants underscore the critical role of epistatic interactions in shaping bacterial adaptation to silver stress. The combination of mutations in *cusS*, *ompR*, *rho*, and other regulatory genes creates a more robust and adaptive response to silver stress than would be expected from single mutations alone.

Comparative analysis reveals that SAM mutants with *cusS* mutations, particularly those combined with *ompR* and *rho* mutations, generally exhibit a more integrated and potent resistance phenotype. This suggests a synergistic effect, where the combined impact of these mutations enhances the overall fitness and resistance of the bacterium beyond the additive effects of individual mutations suggesting positive epistasis. The epistatic interactions observed in SAM6, for instance, likely contribute to its exceptional resistance and fitness, as evidenced by its ability to maintain growth under high silver concentrations. In contrast, SAM1, which lacks a fixed *cusS* mutation, may rely on alternative epistatic interactions involving different regulatory pathways, resulting in a distinct but equally effective adaptive strategy.

Our findings highlight that the adaptive advantage conferred by the *cusS* mutations is significantly modulated by their interaction with mutations in regulatory genes such as *ompR* and *rho*. These interactions do not merely enhance silver resistance but also reconfigure global stress response networks, leading to a more integrated and resilient phenotype.

4.2. New Adaptive Traits That Are Not Predictable from the Individual Effects of Each Mutation Alone

We further assessed potential epistatic effects in these populations by identifying the commonly differentially expressed genes between the R15L strain and the SAM populations that carry the R15L mutation (SAM2, SAM3, SAM6, and SAM7). We then compared their differential gene expression profiles under conditions with and without silver exposure (Table 2). Specifically, in the presence of silver nitrate, the differential expression data strongly suggest that the enhanced silver resistance in SAM mutants is a result of positive epistasis, where the interaction between the R15L mutation in the *cusS* gene and additional mutations in the SAM strains leads to a phenotype that is greater than the sum of its parts.

Table 2. Evidence of epistasis: The table lists biological function and genes that are differentially regulated in R15L and all R15L-carrying SAM populations (SAM2, SAM3, SAM6, and SAM7) under both the absence and presence of silver nitrate.

In DMB Alone											
Biological Function	**Gene**	**R15L_logFC**		**SAM2_logFC**		**SAM3_logFC**		**SAM6_logFC**		**SAM7_logFC**	
Metal Homeostasis (Zinc Transport)	*znuA*	1.694	Up	−5.107	Down	−4.516	Down	−3.553	Down	−4.875	Down
Metal Homeostasis (Copper Export)	*copA*	−1.723	Down	−7.115	Down	−6.099	Down	−6.739	Down	−6.873	Down
Regulatory Functions (Redox Stress Response)	*hprR*	3.078	Up	2.780	Up	4.052	Up	1.609	Up	3.787	Up
Metal Homeostasis (Copper Detoxification)	*cueO*	−1.560	Down	−5.212	Down	−4.669	Down	−6.070	Down	−4.945	Down
Membrane Proteins (Potential Stress Response or Transport)	*shoB*	1.941	Up	3.827	Up	3.525	Up	4.512	Up	3.934	Up
Energy Production (Anaerobic Respiration)	*napH*	2.023	Up	4.341	Up	4.789	Up	4.199	Up	4.403	Up
Bacteriophage Interaction (Phage Entry)	*nfrB*	1.617	Up	2.174	Up	1.619	Up	2.803	Up	1.790	Up
Cell Envelope (Peptidoglycan Remodeling)	*mepM*	1.372	Up	−3.552	Down	−3.977	Down	−3.465	Down	−4.218	Down
Metal Homeostasis (Copper/Silver Efflux)	*cusA*	3.285	Up	3.167	Up	2.753	Up	3.237	Up	4.018	Up
Metal Homeostasis (Iron Transport)	*yfhH*	1.594	Up	2.597	Up	2.571	Up	3.292	Up	2.572	Up
Metal Homeostasis (Copper/Silver Efflux)	*cusB*	3.746	Up	3.164	Up	2.676	Up	3.026	Up	4.154	Up
Transport Systems (Amino Acid Transport)	*pheP*	2.550	Up	3.148	Up	2.715	Up	3.146	Up	3.517	Up
Nucleotide Metabolism (Purine Salvage Pathway)	*ghxP*	2.450	Up	5.917	Up	3.650	Up	2.340	Up	5.340	Up

Table 2. *Cont.*

In Presence of 50 ng/mL Silver Nitrate											
Biological Function	**Gene**	**R15L_logFC**		**SAM2_logFC**		**SAM3_logFC**		**SAM6_logFC**		**SAM7_logFC**	
Metal Homeostasis (Zinc Transport)	*znuA*	5.344	Up	−4.877	Down	−4.602	Down	−4.134	Down	−5.168	Down
Metal Homeostasis (Copper Export)	*copA*	5.769	Up	−6.581	Down	−6.517	Down	−5.676	Down	−7.027	Down
Regulatory Functions (Redox Stress Response)	*hprR*	−2.073	Down	2.971	Up	3.221	Up	2.544	Up	3.570	Up
Metal Homeostasis (Copper Detoxification)	*cueO*	5.807	Up	−5.137	Down	−4.975	Down	−5.726	Down	−5.079	Down
Membrane Proteins (Potential Stress Response or Transport)	*shoB*	−1.593	Down	3.410	Up	3.196	Up	3.972	Up	3.797	Up
Energy Production (Anaerobic Respiration)	*napH*	−3.244	Down	4.158	Up	4.843	Up	2.773	Up	4.864	Up
Bacteriophage Interaction (Phage Entry)	*nfrB*	−2.437	Down	2.269	Up	1.867	Up	3.743	Up	2.115	Up
Cell Envelope (Peptidoglycan Remodeling)	*mepM*	2.915	Up	−4.029	Down	−3.910	Down	−3.861	Down	−4.885	Down
Metal Homeostasis (Copper/Silver Efflux)	*cusA*	2.293	Up	3.184	Up	2.735	Up	3.129	Up	3.674	Up
Metal Homeostasis (Iron Transport)	*yfhH*	−2.223	Down	2.936	Up	2.577	Up	4.065	Up	2.348	Up
Metal Homeostasis (Copper/Silver Efflux)	*cusB*	2.827	Up	3.013	Up	2.097	Up	3.055	Up	3.690	Up
Transport Systems (Amino Acid Transport)	*pheP*	2.377	Up	2.968	Up	2.503	Up	2.918	Up	2.879	Up
Nucleotide Metabolism (Purine Salvage Pathway)	*ghxP*	−1.647	Down	5.197	Up	4.370	Up	3.033	Up	4.724	Up

Green-shaded rows indicate genes that are oppositely regulated between R15L and the SAM populations.

This positive epistasis is evident in the opposing gene expression patterns observed between R15L and the SAM strains. For example, genes like *znuA* and *copA*, which are upregulated in the R15L strain, are downregulated in the SAM mutants. This suggests that the additional mutations in SAM strains do not simply add to the effects of the *cusS* mutation but instead modify the overall regulatory network, leading to a more effective response to silver. The SAM strains are not just following the same pathway as R15L but have developed alternative or enhanced pathways due to the combined effect of multiple mutations.

The concept of positive epistasis is important here as it demonstrates how the interaction between mutations can lead to new adaptive traits that are not predictable from the individual effects of each mutation alone. In the case of the SAM mutants, the combined genetic changes result in a bacterium that is significantly more resistant to silver than the R15L strain, highlighting the power of epistatic interactions in driving evolutionary adaptation. This phenomenon explains why the SAM mutants, with their complex genetic backgrounds, exhibit such a robust resistance phenotype, far surpassing what would be expected from the *cusS* mutation in isolation.

4.3. Mechanisms of Silver Resistance

The analysis of the SAM populations reveals distinct but overlapping mechanisms of silver resistance, each shaped by unique genetic modifications. While there is some clear overlap in these mechanisms, each population follows its own evolutionary trajectory, leading to variations in silver tolerance. These data have been summarized in Table 3.

Table 3. Summary of growth metrics, genetic mutations, and mechanisms of silver resistance in silver-adapted *E. coli* populations.

Population	MIC (ng/mL)	Growth Metrics	Notable Mutations	Differential Expression	Mechanism of Silver Resistance	Notable Observations
SAM1	750	r: 1.2 (50 ng/mL), 1.15 (750 ng/mL); t_gen: Short; t_mid: Stable; k: Strong resistance	*glnH*, *rpoC*	Moderate upregulation of *zntA* and metal transporters, downregulation of *cus* efflux system	Alternative metal transport, downregulates *cus* system	Highest relative fitness, strong adaptability, short lag phase
SAM2	90	r: 1.0 (50 ng/mL), 0.2 (80 ng/mL); t_gen: Extended; t_mid: Lower fitness; k: Declines at high concentrations	*cusS*, *rho*, *ompR*	Downregulation of *cus* efflux genes, upregulation of *zntA* and *dnaK*	Moderate *cus* efflux activity, zinc efflux upregulation	Struggles at higher silver concentrations, fitness lower than WT at low silver
SAM3	250	r: Steady; t_gen: Steady; t_mid: Moderate; k: Varies	*cusS*, *rho*, *ompR*	Upregulation of *zntA*, moderate *cus* efflux expression	Balanced *cus* efflux and metal transport systems	Moderate resistance, extended lag phases under high silver stress
SAM4	500	r: Slower growth; t_gen: Significant increase; t_mid: ~20 h (above 50 ng/mL); k: Moderate resistance	*glnE*, *rph*	No *cus* efflux upregulation, upregulation of RNA processing genes	RNA processing, alternative metabolic adaptations	Adapts slowly, extended lag phase, lacks fixed *cusS* mutation
SAM5	100	r: Stable; t_gen: ~10 h (80 ng/mL), ~20 h (90 ng/mL); t_mid: ~10–20 h; k: Moderate resistance	*rph*, *glnE*	Downregulation of central metabolism, upregulation of *zntA*	Metabolic adjustments, zinc efflux	Moderate fitness, struggles with adaptation under higher silver stress
SAM6	750	r: 1.18 (50 ng/mL), 1.1 (750 ng/mL); t_gen: Short; t_mid: Fast; k: High carrying capacity	*cusS*, *fur*, *rpoA*, *ompR*, *rho*	Strong upregulation of *cus* efflux genes, *msrQ* and *zinT* for stress protection	High-efficiency *cus* efflux, stress defense upregulation	Superior resistance, robust response to high silver concentrations, short lag phase
SAM7	750	r: 1.15 (50 ng/mL), 1.1 (750 ng/mL); t_gen: Short; t_mid: Fast; k: High carrying capacity	*cusS*, *fur*, *rpoA*, *ompR*, *rho*	Highest upregulation of *cus* efflux genes, *rho* and *ompR* for membrane stability	Robust *cus* efflux, membrane integrity	Similar to SAM6, strong genetic flexibility due to transposable elements

SAM1 demonstrates the highest relative fitness and an MIC of 750 ng/mL, showing strong adaptability even under high silver stress. This resistance is achieved through a robust efflux system and a key mutation in *rpoC* that enhances transcriptional regulation. SAM1 adapts to silver stress by downregulating traditional metal detoxification pathways (*copA* and *cusF*) while upregulating alternative metal transporters like *zntA*, reflecting a complex and effective resistance strategy.

SAM6 also exhibits significant resilience, with an MIC of 750 ng/mL. The combination of mutations in *cusS*, *fur*, *rpoA*, *ompR*, and *rho* supports metal ion efflux, transcriptional control, and outer membrane integrity. The R15L *cusS* mutation enhances the activity of the *cus* efflux system, while the upregulation of *msrQ* and *zinT* provides additional protection against oxidative stress and metal sequestration. SAM6's broad resistance strategy includes the use of other efflux pumps like *emrD* and potential reductions in membrane permeability, ensuring efficient neutralization of silver ions.

SAM2 exhibits moderate resistance (MIC of 90 ng/mL) but struggles at higher silver concentrations. It relies on upregulation of the *cus* operon, driven by the R15L mutation, to export toxic metal ions. A mutation in *dnaK* enhances protein damage management, while a deletion in *ompR* may reduce silver ion influx. The upregulation of *zntA* suggests potential cross-resistance to various metals, although SAM2's overall fitness declines significantly at higher concentrations.

SAM3 shares similar mechanisms with SAM2 but shows a broader resistance strategy through more pronounced upregulation of *zntA*. This contributes to its moderate resistance (MIC of 250 ng/mL), although SAM3 also struggles under high silver stress, with extended lag phases indicating difficulty in adaptation.

SAM4 displays strong resistance (MIC of 500 ng/mL) despite lacking a fixed *cusS* mutation. Its unique resistance mechanisms center around RNA processing, nucleotide biosynthesis, and nitrogen metabolism. The mutations in *rph* and *pyrE* suggest a reprogramming of these processes, enhancing survival under stress. SAM4's ability to adapt appears slower, with significant lag phases at high silver concentrations, potentially due to ongoing selection pressures.

SAM5 shows moderate resistance (MIC of 100 ng/mL) and employs several core resistance mechanisms similar to SAM4. However, it is distinguished by a mutation in *cyaA*, which suggests alterations in cAMP signaling pathways, allowing SAM5 to finely regulate stress responses and metabolism.

SAM7, while sharing many resistance mechanisms with SAM6, is characterized by mutations near transposable elements, providing greater genetic flexibility and adaptability. This could allow SAM7 to dynamically alter gene expression in response to environmental changes, offering a potential advantage in fluctuating environments.

These findings underscore the role of epistasis and genetic background in shaping the resistance strategies of each SAM population. While the R15L *cusS* mutation is central to the resistance mechanisms observed, its effectiveness varies depending on the presence of additional mutations that modulate stress responses and metabolic processes.

4.4. Crosstalk and Adaptation in TCRS: Evaluating a Three-Step Model

In our previous work, we proposed a three-step model for how adaptive mutations in TCRS genes drive bacterial adaptation, focusing on genotype-environment interactions (38). Initially based on single mutations in the *cusS* TCRS gene, to this model we can now add four SAM mutants (SAM2, SAM3, SAM6, and SAM7) that carry fixed mutations in *cusS* but with distinct genetic backgrounds due to additional mutations, allowing us to include the role of epistasis in the model.

Primary Response: The first step in our model involves the upregulation of TCRS genes with adaptive mutations, leading to constitutive expression of response genes, even without stimuli. As with the single adaptive mutants such as R15L, SAM mutants carrying *cusS* mutations (SAM2, SAM3, SAM6, and SAM7) also exhibit upregulation and constitutive expression of the *cus* genes, aligning with our previous findings. However, differences in *cus* gene expression suggest varying resistance profiles. SAM6 shows the highest *cus* expression, correlating with its superior resistance (MIC of 750 ng/mL), suggesting additional mutations amplify its efflux capacity. SAM7, similar to SAM6, may benefit from genetic flexibility due to transposable elements, aiding its adaptation to environmental changes.

Epistasis plays a crucial role in shaping these outcomes. For example, while SAM2 and SAM3 share similar mechanisms, their different genetic backgrounds lead to variations in fitness and expression levels. SAM6, with its combination of mutations, exhibits the most refined response, highlighting how epistasis can enhance or modify the effectiveness of resistance genes like those in the *cus* operon.

Secondary Response: The second step in the model involves the differential expression of CusR-regulated genes, often through cross-talk with other TCRS, broadening the cell's adaptive response. SAM mutants with *cusS* mutations show differential expression in CusR regulated genes, such as a significant downregulation of *cueO* and *copA*, though this varies among the mutants. SAM2 and SAM3 exhibit strong downregulation of *cueR*, *phoP*, and *rpoS*, amplifying the repression of these genes. SAM6, however, does not downregulate *cueO* or *copA*, possibly due to epistatic interactions that redirect its secondary response to other pathways, such as osmotic stress.

Cross-talk between CusR and other systems like HprR/HprS also illustrates the complexity of this response [50] SAM populations with the R15L mutation show varied upregulation of redox and oxidative stress-related genes, with differences in expression indicating that epistasis and genetic background are modifying how each strain manages cross-talk. Additionally, *ompR* mutations in some SAM populations further modify this response, affecting outer membrane integrity and stress responses. SAM6 and SAM7, for instance, focus on osmotic stress and metal ion toxicity while deprioritizing other stress responses.

Tertiary Response: The final step in the model involves the "fitness tuning" through additional gene expression changes that optimize the cell's fitness in its original environment. SAM2, SAM3, SAM6, and SAM7 show upregulation in metal homeostasis, indicating a strong adaptation to resist metal toxicity, with SAM7 exhibiting the highest upregulation. Metabolic activity is generally downregulated to conserve energy under stress, with SAM7 showing the most significant suppression. Transport system activity varies, with SAM2 showing strong upregulation while SAM6 exhibits substantial downregulation, suggesting different adaptive strategies. SAM3 and SAM6 prioritize amino acid biosynthesis and metabolism, while SAM2 and SAM7 focus on other responses like metal homeostasis. Motility-related genes are moderately upregulated in SAM3, SAM6, and SAM2, indicating an adaptive strategy involving increased motility. The differential expression across these mutants underscores the role of epistasis and genetic background in shaping their tertiary responses, leading to different survival strategies.

These findings illustrate the complexity of bacterial adaptation, where cross-talk between regulatory systems and epistatic interactions create finely tuned survival mechanisms. The R15L *cusS* mutation primarily confers silver resistance through the upregulation of the *cus* efflux system, but its effectiveness varies among SAM populations due to their unique genetic backgrounds. This adaptability is crucial for these mutants to thrive under diverse environmental conditions.

4.5. Refinements and Modifications to the Model

Our findings provide key insights that refine our existing model of bacterial adaptation, particularly in the context of genotype-by-environment (GxE) interactions. These refinements are directly informed by the new data (Figure 6).

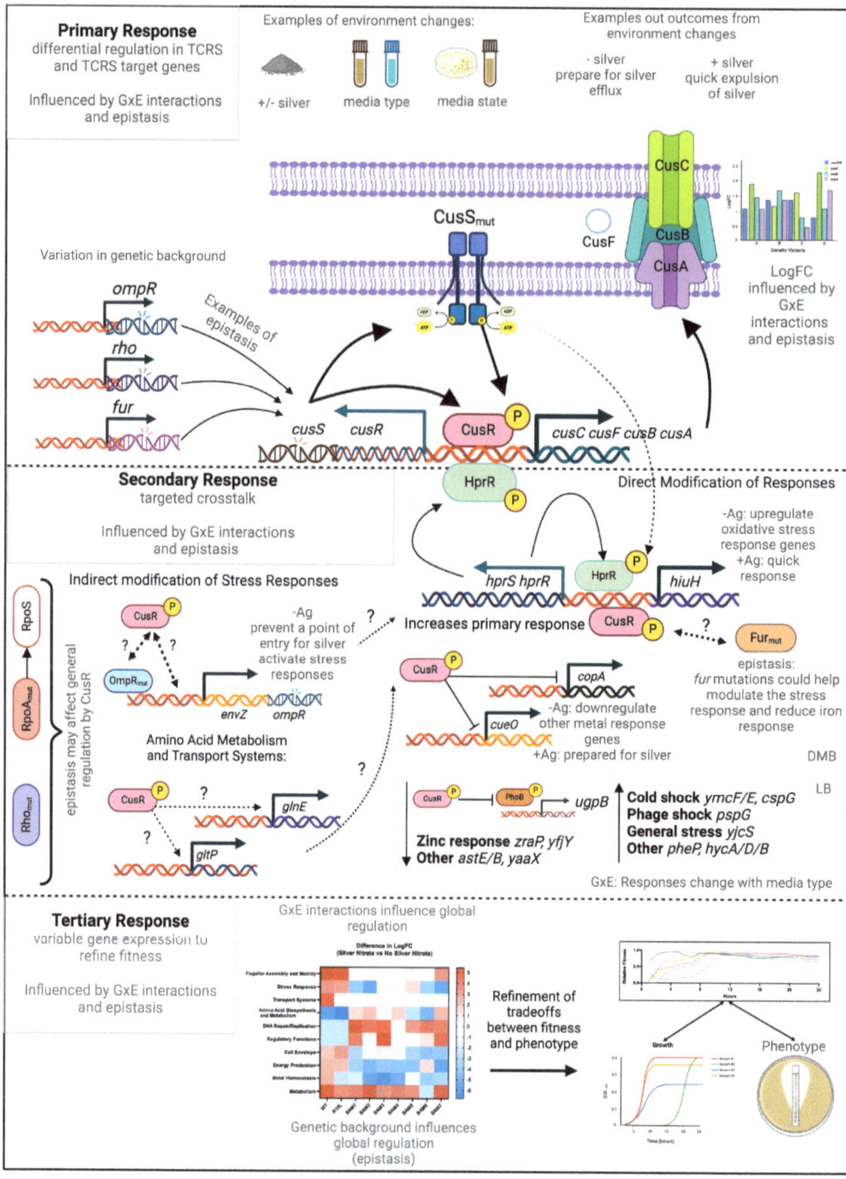

Figure 6. Refined model of adaptive responses in silver-resistant *E. coli* Mutants: A refined three-step adaptive response mechanism in silver-resistant *E. coli* mutants is illustrated, emphasizing the roles of epistasis and genotype-by-environment (GxE) dynamics throughout the adaptive process. Dotted arrows indicate interactions that have not yet been characterized, while thick, bold arrows represent the most critical pathways in the response. The model begins with the Primary Response, where epistatic interactions among mutations in the *cusS* gene and other regulatory genes lead to the enhanced expression of the CusSR two-component regulatory system (TCRS) and its downstream efflux pump genes, *cusCFBA*. These interactions, influenced by specific environmental conditions, establish a baseline level of silver resistance that varies across SAM populations due to GxE dynamics. The combined effects of these mutations create a context-dependent expression of resistance traits.

4.5.1. Epistasis in GxE Interactions

Findings: The varying gene expression and fitness profiles among the SAM mutants, especially those with the R15L *cusS* mutation, underscore the critical role of epistatic interactions. For instance, SAM6, with additional mutations in *ompR*, *rho*, and *fur*, exhibits enhanced resistance compared to SAM2 and SAM3, despite sharing the same *cusS* mutation.

In the Secondary Response, epistasis and GxE interactions further shape the crosstalk between CusR and other regulatory pathways, such as those governing oxidative stress and osmotic stress responses. These pathways are differentially activated depending on the specific genetic background and environmental conditions of each mutant. For instance, SAM6, carrying additional mutations in ompR, rho, and fur, exhibits heightened resistance through enhanced modulation of these secondary pathways. Conversely, SAM2 and SAM3, with their unique epistatic backgrounds, show varied pathway activation and resistance levels.

The Tertiary Response, or "fitness tuning", involves the fine-tuning of gene expression to optimize survival and resistance under the specific environmental conditions experienced by each population. This stage is driven by epistatic interactions that shape unique expression patterns, maximizing fitness in a context-specific manner. GxE interactions further refine these adaptations, with populations like SAM6 and SAM7 displaying distinct adaptive strategies. SAM7, for example, exhibits greater adaptability due to mutations near transposable elements, while SAM4 relies on shifts in RNA processing and metabolism to survive in silver-rich environments. This refined model underscores the integral role of epistasis and GxE interactions in shaping the evolutionary trajectories and adaptive changes driven by mutations in TCRS genes, highlighting why these genes are often selected for adaptation. This figure was created using BioRender.

Model Refinement: Epistasis is now emphasized as a key factor in GxE interactions, shaping how genetic changes influence adaptive traits across different environments. The superior resistance in SAM6 results from these interactions, demonstrating that adaptation is driven by both genetic interactions and environmental pressures.

4.5.2. Impact of Genetic Background

Findings: The study highlights the influence of genetic background on adaptation. SAM7, despite sharing key mutations with SAM6, shows greater adaptability due to mutations near transposable elements.

Model Refinement: The model now reinforces the idea that genetic background plays a crucial role in GxE interactions. Mutations in regulatory elements, such as those found in SAM6, can significantly alter adaptive outcomes under varying environmental conditions, further illustrating the context-dependent nature of bacterial adaptation.

4.5.3. Multifunctional Pathways in GxE Dynamics

Findings: Adaptive responses in SAM mutants involve not only metal ion efflux but also broader pathways like oxidative stress response and RNA processing. SAM4 compensates for the lack of a fixed *cusS* mutation by upregulating RNA processing genes, aiding in survival under silver stress.

4.5.4. Adaptive Strategy Prioritization

Findings: Different SAM populations prioritize specific adaptive strategies depending on their genetic makeup. For instance, SAM6 focuses on metal ion efflux, while SAM4 emphasizes RNA processing and metabolic shifts.

Model Refinement: The revised model recognizes that bacterial populations adopt varied adaptive strategies within the GxE framework. Adaptation is not uniform but involves the selective prioritization of pathways that best address the environmental challenges faced by each population.

4.6. Revised Model Implications

In light of these findings, our revised model now emphasizes the critical role of epistasis in not only amplifying resistance traits but also in reprogramming broader regulatory networks that enable the bacteria to thrive under varying environmental stresses. This revised perspective suggests that the evolutionary pathways leading to robust resistance are more complex and involve a greater degree of interaction between multiple genetic and environmental factors than previously understood.

5. Conclusions

This study underscores the complexity of bacterial resistance mechanisms, particularly in the context of evolving resistance to silver. Our findings reveal that identifying a single mutation associated with resistance, such as the R15L mutation, does not guarantee that the organism will exhibit resistance. The role of genetic background is critical, as evidenced by the dramatically different resistance profiles among populations carrying the same R15L mutation. Moreover, SAM1, which does not harbor any distinctive silver resistance mutations yet demonstrates the greatest fitness and highest resistance, highlights the intricate interplay of genetic background and epistatic interactions.

These insights have significant clinical implications. As sequencing technology becomes more affordable and widely used for evaluating antibiotic resistance, it is crucial to consider the broader genetic context rather than relying solely on single-gene assessments. The variability in resistance profiles among the SAM populations challenges the predictive power of identifying single mutations and suggests that bacterial adaptation is driven by a dynamic network of genetic interactions that extend beyond individual mutations.

The unique resistance strategies observed across the SAM mutants exemplify how bacteria harness multiple interacting mutations to reshape cellular processes, leading to more robust and adaptable phenotypes. This adaptability not only allows bacteria to survive in hostile environments but also to exploit new ecological niches, emphasizing the importance of genetic diversity and the potential for rapid evolution in microbial populations.

In a broader evolutionary context, these findings highlight the potential for rapid and complex evolution in response to strong selective pressures, such as antimicrobial agents. This suggests that strategies targeting single pathways may be insufficient to curb the development of resistance. Instead, a more comprehensive approach, considering epistasis and the broader genetic landscape, is necessary to combat bacterial resistance effectively.

The evolutionary trajectories observed in the SAM mutants provide valuable insights into the dynamic and multifaceted nature of microbial adaptation. These insights contribute to a deeper understanding of how bacteria evolve complex resistance mechanisms and underscore the need for innovative strategies to address the growing challenge of antimicrobial resistance in both clinical and environmental settings.

Supplementary Materials: The following supporting information can be downloaded at: https://www.mdpi.com/article/10.3390/microorganisms12102000/s1, Figure S1: Complete 24-hour growth response assays of *E. coli* populations to increasing silver nitrate concentrations. Table S1: All detected mutations from DNA sequencing analysis of all SAM populations. Table S2: Two-way ANOVA with pairwise comparisons of 24-hour growth data. Table S3: One-way ANOVA with multiple pairwise comparisons of relative fitness data.

Author Contributions: Conceptualization, M.D.T., J.L.G.J. and B.R.S.; methodology, M.D.T., J.L.G.J. and B.R.S.; validation, M.D.T. and B.R.S.; formal analysis, M.D.T. and B.R.S.; investigation, B.R.S., L.S.T., N.M.L. and Z.A.F.; resources, M.D.T.; data curation, B.R.S., L.S.T., N.M.L. and Z.A.F.; writing—original draft preparation, M.D.T. and B.R.S.; writing—review and editing, M.D.T., J.L.G.J., B.R.S., L.S.T., N.M.L. and Z.A.F.; visualization, M.D.T.; supervision, M.D.T. and B.R.S.; project administration, M.D.T. and J.L.G.J.; funding acquisition, M.D.T. and J.L.G.J. All authors have read and agreed to the published version of the manuscript.

Funding: This work was funded via support of the National Science Foundation: Excellence in Research Award (NSF Award Number 1900220) to M.D.T. and J.L.G.J. B.R.S. was supported by funding provided by the Department of Education's HBGI Grant. L.S.T. and N.M.L. were funded by a National Institutes of Health NIGMS MARC Undergraduate NRSA Institutional Grant (Award number 5T34GM083980).

Data Availability Statement: Public access code for the entire Bioproject is accessible through PRJNA1160277 and the individual datasets are also accessible directly through the SRA database using SAMN43760522-SAMN43760542.

Conflicts of Interest: The authors declare no conflicts of interest. The funders had no role in the design of the study; in the collection, analyses, or interpretation of data; in the writing of the manuscript; or in the decision to publish the results.

References

1. Kassen, R. Distribution of fitness effects among beneficial mutations before selection in experimental populations of Bacteria. *Nat. Genet.* **2006**, *38*, 484. [CrossRef] [PubMed]
2. Barrick, J.E.; Lenski, R.E. Genome dynamics during experimental evolution. *Nat. Rev. Genet.* **2013**, *14*, 827. [CrossRef]
3. Elena, S.F.; Lenski, R.E. Evolution experiments with microorganisms: The dynamics and genetic bases of adaptation. *Nat. Rev. Genet.* **2003**, *4*, 457–469. [CrossRef]
4. Phillips, P.C. Epistasis—The essential role of gene interactions in the structure and evolution of genetic systems. *Nat. Rev. Genet.* **2008**, *9*, 855–867. [CrossRef]
5. Weinreich, D.M.; Lan, Y.; Wylie, C.S.; Heckendorn, R.B. Should evolutionary geneticists worry about higher-order epistasis? *Curr. Opin. Genet. Dev.* **2013**, *23*, 700–707. [CrossRef]
6. Desai, M.M.; Fisher, D.S. Beneficial mutation–selection balance and the effect of linkage on positive selection. *Genetics* **2007**, *176*, 1759–1798. [CrossRef]
7. Otto, S.P.; Feldman, M.W. Deleterious mutations, variable epistatic interactions, and the evolution of recombination. *Theor. Popul. Biol.* **1997**, *51*, 134–147. [CrossRef]
8. Lehner, B. Molecular mechanisms of epistasis within and between genes. *Trends Genet.* **2011**, *27*, 323–331. [CrossRef]
9. Barillo, D.J.; Marx, D.E. Silver in medicine: A brief history BC 335 to present. *Burns* **2014**, *40* (Suppl. S1), S3. [CrossRef]
10. Rai, M.; Yadav, A.; Gade, A. Silver nanoparticles as a new generation of antimicrobials. *Biotechnol. Adv.* **2009**, *27*, 76–83. [CrossRef]
11. Russell, A.D.; Hugo, W.B. 7 Antimicrobial activity and action of silver. *Prog. Med. Chem.* **1994**, *31*, 351–370.
12. Haefeli, C.; Franklin, C.; Hardy, K. Plasmid-determined silver resistance in *Pseudomonas stutzeri* isolated from a silver mine. *J. Bacteriol.* **1984**, *158*, 389–392. [CrossRef] [PubMed]
13. Lemire, J.A.; Harrison, J.J.; Turner, R.J. Antimicrobial activity of metals: Mechanisms, molecular targets and applications. *Nat. Rev. Microbiol.* **2013**, *11*, 371–384. [CrossRef] [PubMed]
14. McHugh, G.L.; Moellering, R.; Hopkins, C.; Swartz, M. *Salmonella typhimurium* resistant to silver nitrate, chloramphenicol, and ampicillin: A new threat in burn units? *Lancet* **1975**, *305*, 235–240. [CrossRef] [PubMed]
15. McNeilly, O.; Mann, R.; Hamidian, M.; Gunawan, C. Emerging concern for silver nanoparticle resistance in *Acinetobacter baumannii* and other bacteria. *Front. Microbiol.* **2021**, *12*, 652863.
16. Stock, A.M.; Robinson, V.L.; Goudreau, P.N. Two-component signal transduction. *Annu. Rev. Biochem.* **2000**, *69*, 183–215. [CrossRef]
17. Hoch, J.A. Two-component and phosphorelay signal transduction. *Curr. Opin. Microbiol.* **2000**, *3*, 165–170. [CrossRef]
18. Mascher, T.; Helmann, J.D.; Unden, G. Stimulus perception in bacterial signal-transducing histidine kinases. *Microbiol. Mol. Biol. Rev.* **2006**, *70*, 910–938. [CrossRef]
19. Gupta, R.; Gupta, N. Two-component systems. In *Fundamentals of Bacterial Physiology and Metabolism*; Springer: Singapore, 2021; pp. 557–573.
20. Koretke, K.K.; Lupas, A.N.; Warren, P.V.; Rosenberg, M.; Brown, J.R. Evolution of two-component signal transduction. *Mol. Biol. Evol.* **2000**, *17*, 1956–1970. [CrossRef]
21. West, A.H.; Stock, A.M. Histidine kinases and response regulator proteins in two-component signaling systems. *Trends Biochem. Sci.* **2001**, *26*, 369–376. [CrossRef]
22. Munson, G.P.; Lam, D.L.; Outten, F.W.; O'Halloran, T.V. Identification of a copper-responsive two-component system on the chromosome of *Escherichia coli* K-12. *J. Bacteriol.* **2000**, *182*, 5864–5871. [CrossRef] [PubMed]
23. Yamamoto, K.; Ishihama, A. Transcriptional response of *Escherichia coli* to external copper. *Mol. Microbiol.* **2005**, *56*, 215–227. [CrossRef] [PubMed]
24. Gudipaty, S.A.; McEvoy, M.M. The histidine kinase CusS senses silver ions through direct binding by its sensor domain. *Biochim. Biophys. Acta. Proteins. Proteom.* **2014**, *1844*, 1656–1661. [CrossRef]
25. Capra, E.J.; Laub, M.T. Evolution of two-component signal transduction systems. *Annu. Rev. Microbiol.* **2012**, *66*, 325–347. [CrossRef] [PubMed]

26. Gotoh, Y.; Eguchi, Y.; Watanabe, T.; Okamoto, S.; Doi, A.; Utsumi, R. Two-component signal transduction as potential drug targets in pathogenic bacteria. *Curr. Opin. Microbiol.* **2010**, *13*, 232–239. [CrossRef]
27. Podgornaia, A.I.; Laub, M.T. Pervasive degeneracy and epistasis in a protein-protein interface. *Science* **2015**, *347*, 673–677. [CrossRef]
28. Costanzo, M.; Baryshnikova, A.; Bellay, J.; Kim, Y.; Spear, E.D.; Sevier, C.S.; Ding, H.; Koh, J.L.Y.; Toufighi, K.; Mostafavi, S.; et al. The genetic landscape of a cell. *Science* **2010**, *327*, 425–431. [CrossRef]
29. Chou, H.; Marx, C.J. Optimization of gene expression through divergent mutational paths. *Cell Rep.* **2012**, *1*, 133–140. [CrossRef]
30. Andersson, D.I.; Hughes, D. Antibiotic resistance and its cost: Is it possible to reverse resistance? *Nat. Rev. Microbiol.* **2010**, *8*, 260–271. [CrossRef]
31. Toprak, E.; Veres, A.; Michel, J.B.; Chait, R.; Hartl, D.L.; Kishony, R. Evolutionary paths to antibiotic resistance under dynamically sustained drug selection. *Nat. Genet.* **2012**, *44*, 101–105. [CrossRef]
32. Baquero, F.; Coque, T.M. Multilevel population genetics in antibiotic resistance. *FEMS. Microbiol. Rev.* **2011**, *35*, 705–706. [CrossRef] [PubMed]
33. Barrick, J.E.; Lenski, R.E. Genome evolution and adaptation in a long-term experiment with *Escherichia coli*. *Nat. Biotechnol.* **2009**, *461*, 1243–1247. [CrossRef] [PubMed]
34. Swallow, J.G.; Hayes, J.P.; Koteja, P.; Garland, T., Jr. Selection experiments and experimental evolution of performance and physiology. In *Evolutionary and Genetic Biology*; Regents of the University of California Press: Oakland, CA, USA, 2009; Chapter 12.
35. Randall, C.P.; Gupta, A.; Jackson, N.; Busse, D.; O'Neill, A.J. Silver resistance in Gram-negative bacteria: A dissection of endogenous and exogenous mechanisms. *J. Antimicrob. Chemother.* **2015**, *70*, 1037–1046. [CrossRef] [PubMed]
36. Graves, J.L., Jr.; Tajkarimi, M.; Cunningham, Q.; Campbell, A.; Nonga, H.; Harrison, S.H.; Barrick, J.E. Rapid evolution of silver nanoparticle resistance in *Escherichia coli*. *Front. Genet.* **2015**, *6*, 42. [CrossRef] [PubMed]
37. Tajkarimi, M.; Campbell, A.; Rhinehardt, K.; Thomas, M.; Akamu, E.J.; Boyd, S.; Turner, D.; Harrison, S.H.; Graves, J.L., Jr. Selection for ionic-silver confers silver nanoparticle resistance in *Escherichia coli*. *JSM Nanotechnol. Nanomed.* **2017**, *5*, 1047.
38. Sanders, B.R.; Miller, J.E.; Ahmidouch, N.; Graves, J.L., Jr.; Thomas, M.D. Genotype-by-environment interactions govern fitness changes associated with adaptive mutations in two-component response systems. *Front. Genet.* **2024**, *15*, 1349507. [CrossRef]
39. Sanders, B.R.; Townsend, S.E.; Ford, M.L.; Graves, J.L., Jr.; Thomas, M.D. Reporting off-target effects of recombinant engineering using the pORTMAGE system. *J. Microbiol. Methods* **2023**, *204*, 106627. [CrossRef] [PubMed]
40. Deatherage, D.E.; Barrick, J.E. Identification of mutations in laboratory-evolved microbes from next-generation sequencing data using breseq. In *Engineering and Analyzing Multicellular Systems: Methods and Protocols*; Humana Press: New York, NY, USA, 2014; pp. 165–188.
41. Illumina. BCL Convert: A proprietary Illumina Software for the Conversion of BCL Files to Basecalls. 2021. (v3.9.3). Available online: https://support-docs.illumina.com/SW/BCL_Convert/Content/SW/FrontPages/BCL_Convert.htm (accessed on 25 January 2024).
42. Zhang, Y.; Park, C.; Bennett, C.; Thornton, M.; Kim, D. Rapid and accurate alignment of nucleotide conversion sequencing reads with HISAT-3N. *Genome Res.* **2021**, *31*, 1290–1295. [CrossRef]
43. Robinson, M.D.; McCarthy, D.J.; Smyth, G.K. edgeR: A Bioconductor package for differential expression analysis of digital gene expression data. *Bioinformatics* **2010**, *26*, 139–140. [CrossRef]
44. Sprouffske, K.; Wagner, A. Growthcurver: An R package for obtaining interpretable metrics from microbial growth curves. *BMC Bioinform.* **2016**, *17*, 172. [CrossRef]
45. Franke, S.; Grass, G.; Nies, D.H. The product of the *ybdE* gene of the *Escherichia coli* chromosome is involved in detoxification of silver ions. *Microbiology* **2001**, *147*, 965–972. [CrossRef] [PubMed]
46. Chacón, K.N.; Mealman, T.D.; McEvoy, M.M.; Blackburn, N.J. Tracking metal ions through a Cu/Ag efflux pump assigns the functional roles of the periplasmic proteins. *Proc. Natl. Acad. Sci. USA* **2014**, *111*, 15373–15378. [CrossRef] [PubMed]
47. Gennaris, A.; Ezraty, B.; Henry, C.; Agrebi, R.; Vergnes, A.; Oheix, E.; Bos, J.; Leverrier, P.; Espinosa, L.; Szewczyk, J.; et al. Repairing oxidized proteins in the bacterial envelope using respiratory chain electrons. *Nature* **2015**, *528*, 409–412. [CrossRef] [PubMed]
48. Rensing, C.; Mitra, B.; Rosen, B.P. The *zntA* gene of *Escherichia coli* encodes a Zn (II)-translocating P-type ATPase. *Proc. Natl. Acad. Sci. USA* **1997**, *94*, 14326–14331. [CrossRef]
49. Goss, T.J.; Perez-Matos, A.; Bender, R.A. Roles of glutamate synthase gltBD and gltF in nitrogen metabolism of *Escherichia coli* and *Klebsiella. aerogenes*. *J. Bacteriol.* **2001**, *183*, 6607–6619. [CrossRef]
50. Urano, H.; Yoshida, M.; Ogawa, A.; Yamamoto, K.; Ishihama, A.; Ogasawara, H. Cross-regulation between two common ancestral response regulators HprR and CusR in *Escherichia coli*. *Microbiology* **2017**, *163*, 243–252. [CrossRef]

microorganisms

Article

Sigma Factor Engineering in *Actinoplanes* sp. SE50/110: Expression of the Alternative Sigma Factor Gene *ACSP50_0507* (σH^{As}) Enhances Acarbose Yield and Alters Cell Morphology

Laura Schlüter [1], Tobias Busche [2,3], Laila Bondzio [4], Andreas Hütten [4], Karsten Niehaus [5], Susanne Schneiker-Bekel [1,6], Alfred Pühler [6] and Jörn Kalinowski [1,2,*]

1 Microbial Genomics and Biotechnology, Center for Biotechnology, Bielefeld University, 33594 Bielefeld, Germany; lschluet@cebitec.uni-bielefeld.de (L.S.); schneike@cebitec.uni-bielefeld.de (S.S.-B.)
2 Technology Platform Genomics, Center for Biotechnology, Bielefeld University, 33594 Bielefeld, Germany; tbusche@cebitec.uni-bielefeld.de
3 Medical School East Westphalia-Lippe, Bielefeld University, 33594 Bielefeld, Germany
4 Faculty of Physics, Bielefeld University, 33594 Bielefeld, Germany; laila.bondzio@uni-bielefeld.de (L.B.); andreas.huetten@uni-bielefeld.de (A.H.)
5 Proteome and Metabolome Research, Faculty of Biology, Bielefeld University, 33594 Bielefeld, Germany; kniehaus@cebitec.uni-bielefeld.de
6 Genome Research of Industrial Microorganisms, Center for Biotechnology (CeBiTec), Bielefeld University, 33594 Bielefeld, Germany; puehler@cebitec.uni-bielefeld.de
* Correspondence: joern@cebitec.uni-bielefeld.de

Abstract: Sigma factors are transcriptional regulators that are part of complex regulatory networks for major cellular processes, as well as for growth phase-dependent regulation and stress response. *Actinoplanes* sp. SE50/110 is the natural producer of acarbose, an α-glucosidase inhibitor that is used in diabetes type 2 treatment. Acarbose biosynthesis is dependent on growth, making sigma factor engineering a promising tool for metabolic engineering. ACSP50_0507 is a homolog of the developmental and osmotic-stress-regulating *Streptomyces coelicolor* σH^{Sc}. Therefore, the protein encoded by *ACSP50_0507* was named σH^{As}. Here, an *Actinoplanes* sp. SE50/110 expression strain for the alternative sigma factor gene *ACSP50_0507* (*sigH*^{As}) achieved a two-fold increased acarbose yield with acarbose production extending into the stationary growth phase. Transcriptome sequencing revealed upregulation of acarbose biosynthesis genes during growth and at the late stationary growth phase. Genes that are transcriptionally activated by σH^{As} frequently code for secreted or membrane-associated proteins. This is also mirrored by the severely affected cell morphology, with hyperbranching, deformed and compartmentalized hyphae. The dehydrated cell morphology and upregulation of further genes point to a putative involvement in osmotic stress response, similar to its *S. coelicolor* homolog. The DNA-binding motif of σH^{As} was determined based on transcriptome sequencing data and shows high motif similarity to that of its homolog. The motif was confirmed by in vitro binding of recombinantly expressed σH^{As} to the upstream sequence of a strongly upregulated gene. Autoregulation of σH^{As} was observed, and binding to its own gene promoter region was also confirmed.

Keywords: σ factor; acarbose; *Actinoplanes*; transcription; regulation; cell morphology; osmotic stress response; actinobacteria

Citation: Schlüter, L.; Busche, T.; Bondzio, L.; Hütten, A.; Niehaus, K.; Schneiker-Bekel, S.; Pühler, A.; Kalinowski, J. Sigma Factor Engineering in *Actinoplanes* sp. SE50/110: Expression of the Alternative Sigma Factor Gene *ACSP50_0507* (σH^{As}) Enhances Acarbose Yield and Alters Cell Morphology. *Microorganisms* **2024**, *12*, 1241. https://doi.org/10.3390/microorganisms12061241

Academic Editor: Tomohiro Shimada

Received: 7 May 2024
Revised: 11 June 2024
Accepted: 17 June 2024
Published: 20 June 2024

1. Introduction

Actinoplanes sp. SE50/110 is a high G + C (~71%) Gram-positive aerobic soil bacterium, featuring filamentous growth and a complex life cycle. The bacterium grows in a dense mycelium, forms motile spores [1] and has high industrial relevance as the natural producer of acarbose. Acarbose is an α-glucosidase inhibitor, used as treatment for diabetes mellitus type 2. The pseudo-tetrasaccharide acarbose consists of two parts: a maltose residue, which is α-1,4-linked to a pseudodisaccharide, consisting of aminocyclitol and deoxyribose. Its inhibitory effect is based on the complex of the unsaturated C7-aminocyclitol, composed of C7-cyclitol and aminohexose with N-glycosidic linkage [2,3]. The genome harbors 22 *acb* genes involved in acarbose biosynthesis (16) as well as its molecular modification (3) and export (3) [4–9]. The C7-aminocyclitol and aminohexose moieties are synthesized trough two different biosynthesis branches, performed by the enzymes AcbCMOLNUJR and AcbABV, respectively. Further genes are essential for extracellular modifications and transport [4,10]. Within the *acb* gene cluster, regulatory elements are scarcely known. Determination of transcription start sites (TSSs) and operon structures [11] as well as transcription dynamics and protein abundancies [12] were major steps for the broader understanding of *acb* gene regulation. In prior works, the MalR type regulator AcrC was identified to bind within the *acb* gene cluster in the intergenic region between the genes *acbE* and *acbD*, but gene deletion did not result in an increased acarbose yield [13]. To date, no transcriptional regulator capable of influence acarbose biosynthesis with an increased acarbose yield has been identified.

Since acarbose biosynthesis is growth-dependent, showing no production within the stationary growth phase [14], a focus on global transcriptional regulators with a putative influence on growth is a promising approach. For other species, it was shown that the manipulation of global transcriptional regulators can be used to achieve increased metabolite production. The manipulation can be performed using different approaches, e.g., with gene deletion, using evolved transcriptional regulators or overexpression [15–17]. Sigma factors were shown to be superior targets to manipulate metabolite production, as they control larger transcriptional networks. SigA overexpression successfully led to an increase in carotenoid production in *Corynebacterium glutamicum* [17], and sigma factor HrdB evolvement enabled antibiotic production increase in *Streptomyces avermitilis* [16]. Sigma factors specifically recognize promoter elements and direct core RNA polymerase (RNAP) enzyme to facilitate transcription initiation. Housekeeping sigma factors confer transcription for primary metabolism, while alternative sigma factors are active under different conditions, e.g., stress response [18–22], growth phase [21,23–25], or during secondary metabolite production [15–17,26,27], and these redirect RNAP to specific promoters to confer alternative gene transcription. In addition, all sigma factors compete for a limited amount of core RNAP.

Alternative sigma factors are often inactivated by binding to anti-sigma factors. Bound sigma factors can be released by direct sensing, resulting in a conformational change and release of the sigma factor [28,29], proteolysis (anti-sigma factor degeneration) [30,31] or partner-switching (inactivation by binding through anti-sigma factor antagonist) [32]. Anti-sigma factor antagonists, also named anti-anti-sigma factors, can bind anti-sigma factors, preventing the inhibition of its target sigma factor. The complex hierarchical regulatory network comprises several more steps, since anti-anti-sigma factors are activated upon diverse cellular signals [24]. Alternative sigma factors enable differential transcription, facilitating response to intra- and extracellular conditions. They play a key role in diverse stress-associated processes like heat or cold-shock response, pH shock, osmotic stress response or oxidative stress response [18,19,21,33–36]. Transcriptional-sigma-factor-mediated regulation is diverse and can differ in complexity among species. *Escherichia coli* has only seven sigma factors, one housekeeping sigma factor for essential cellular processes and six alternative sigma factors for selective differential transcription [33,34]. Other species have an even more complex regulation [37,38] with several more sigma factors involved: *Bacillus subtilis* carries 18 sigma factor genes [39,40] and *S. coelicolor* carries 65 sigma factor genes [41]. This makes the investigation challenging. Several studies elucidate σ factors networks in *S. coelicolor*, especially the stress response σB^{SC} homologs with its overlapping stress responses, connected regulatory networks and promoter cross-recognition [42,43]. The regulatory network of σB^{SC} homolog σH^{SC} is extensively characterized. σH^{SC} plays a role in osmotic stress response and morphological development [21]. Genes of the σH^{SC} regulon are associated with osmotic stress response, morphological differentiation, sporulation and aerial hyphae septation. The gene deletion strain shows impaired septation of aerial hyphae and a bald phenotype due to a decreased spore titer and reduced growth under osmotic stress [21]. σH^{SC} is regulated on a transcriptional, post-translational and protein level. Transcription of the σH^{SC} operon, coding σH^{SC} and its anti-σ factor UshX, was shown to be regulated by BldD or activated by heat shock and osmotic stress, but are also autoregulated with different promoters [21,44,45]. Further studies showed σH^{SC} regulation by proteolysis and by a partner-switching mechanism of its anti-σ factor UshX and the anti-anti-σ factor BldG. Eight anti-σ factors were identified to be able to activate BldG and fifteen σ factors were identified to interact with BldG [46,47]. This intertwined large network of regulatory partners represents the complexity of σ factor regulation, especially of the stress-associated σB homologs, and reinforces the challenge of σ factor characterization and elucidation of its regulation. This becomes even more complex with *Actinoplanes* sp. SE50/110 carrying 78 σ factor genes (Genbank: NZ_LT827010.1).

Thereby, the adaption of molecular genetic tools like integrative vectors [48], CRISPR/Cas9 [49] and homologous recombination [50] for *Actinoplanes* sp. SE50/110 as well as advances in transcriptome analysis and gene annotation enables the genetic manipulation for gene characterization and investigation of regulatory networks in this organism. We here set out to manipulate sigma factor networks in this organism to increase acarbose yield and analyze the influence of *ACSP50_0507* on the *Actinoplanes* sp. SE50/110 phenotype. Therefore, we investigated the deletion and expression mutants of a systematically selected σ factor gene with respect to growth, cell morphology and acarbose production. We further analyzed its function using transcriptomic analysis for binding motif prediction and performed binding assays.

2. Materials and Methods

2.1. *Software and Databases Used for Bioinformatic Analysis*

Oligonucleotide design was performed with SnapGene 4.3 (GSL Biotech LLC, Chicago, IL, USA). EDGAR 3.0 [51], the Basic Local Alignment Search Tool [52–54] and CDD (Conserved Domain Database) [55–58] were used for comparative genome studies, homology prediction and functional domain search of proteins. Transcriptional analysis was performed with Readxplorer 2.2.3 and integrated DeSeq2 [59] and Improbizer [60] and WebLogo 3.7.10 [61] were used for binding motif prediction.

2.2. *Strains, Media and Supplements*

Cloning of expression and deletion plasmids was performed with *Escherichia coli* DH5αMCR, the donor strain *Escherichia coli* ET12567(pUZ8002) was used for conjugation into *Actinoplanes* sp. SE50/110, and plasmid replication for electro transformation into *Actinoplanes* sp. SE50/110 was performed using *Escherichia coli* ER2925 [62,63]. Protein expression was performed in *Escherichia coli* BL21(DE3) pLys.

E. coli was cultivated in liquid LB media (16 $g{\cdot}L^{-1}$ Luria/Miller broth (Carl Roth, GmbH & Co. KG, Karlsruhe, Germany) or grown on solid LB medium (supplemented with 16 $g{\cdot}L^{-1}$ agar–agar Kobel (Carl Roth, GmbH & Co. KG, Karlsruhe, Germany). The respective supplements were added to the media: apramycin (50 $\mu g{\cdot}mL^{-1}$), chloramphenicol (25 $\mu g{\cdot}mL^{-1}$), kanamycin (50 $\mu g{\cdot}mL^{-1}$), ampicillin (100 $\mu g{\cdot}mL^{-1}$).

For molecular genetic manipulation, *Actinoplanes* sp. SE50/110 was grown on solid soy flour medium (SFM: 20 $g{\cdot}L^{-1}$ soy flour, 20 $g{\cdot}L^{-1}$ mannitol, 20 $g{\cdot}L^{-1}$ agar, pH 8 adjusted with NaOH, tap water) or liquid NBS medium (11 $g{\cdot}L^{-1}$ glucose × $1H_2O$, 4 $g{\cdot}L^{-1}$ peptone, 4 $g{\cdot}L^{-1}$ yeast extract, 1 $g{\cdot}L^{-1}$ $MgSO_4{\cdot}7H_2O$, 2 $g{\cdot}L^{-1}$ KH_2PO_4, and 4 $g{\cdot}L^{-1}$ K_2HPO_4) with the respective supplements.

2.3. *Creation of Expression, Deletion and Complementation Strains*

2.3.1. Cloning of *ACSP50_0507* into the Integrative pSETT4tipA Expression Vector

The σ factor gene was cloned into the BbsI-linearized vector pSETT4*tipA* using Gibson Assembly [64] according to Schaffert et al. (2020) [48]. Oligonucleotides were ordered from metabion GmbH (Steinkirchen, Germany) and Sigma Aldrich (Merck KGaA, Darmstadt, Germany) (Table S1). The target genes were amplified by Polymerase Chain Reaction (PCR) using Phusion High-Fidelity PCR Master Mix with GC Buffer (Thermo Fisher Scientific, Waltham, MA, USA). The DNA constructs were transformed into *E. coli* DH5α according to Beyer et al. (2015) and selected on solid LB media, supplemented with 50 mg L^{-1} apramycin sulfate [65]. Plates were incubated overnight at 37 °C, and the obtained clones were screened using colony PCR and agarose gel electrophoresis. Positive clones were isolated using the GeneJET Plasmid-Miniprep-Kit (Thermo Fisher Scientific, Waltham, MA, USA) and verified using Sanger sequencing by our in-house sequencing facility.

2.3.2. Plasmid Transfer into *Actinoplanes* sp. SE50/110

Competent cells were obtained from 50 mL *Actinoplanes* sp. SE50/110 NBS culture, cultivated for 48 h at 140 rpm and 28 °C. After 15 min on ice, the cells were washed twice with TG buffer (10% glycerol (*v*/*v*), 1 mM Tris) and then washed in 10% (*v*/*v*) glycerin with centrifugation for 5 min at 5000× *g* and 4 °C in between. The cells were then resuspended in 3 mL of 10% (*v*/*v*) glycerin, and aliquots of 300 µL are prepared and subsequently frozen in liquid nitrogen and stored at −80 °C until further use.

Electro transformation of the pSETT4*tipA*-based expression plasmids into competent *Actinoplanes* sp. SE50/110 was performed with 100 ng expression plasmid, isolated from the methylase-deprived *E. coli* ER2925 strain, at 2500 kV, 25 µF and 200 Ω in a cuvette and incubated in 1 mL preheated CMR (10 g·L^{-1} glucose, 103 g·L^{-1} sucrose, 10.12 g·L^{-1} $MgCl_2 \times 6H_2O$, 15 g·L^{-1} TSB, 5 g·L^{-1} yeast extract) at 46 °C for 6 min before incubation at 28 °C and 1100 rpm overnight. The transformation broth was then grown on solid selective solid SFM agar, supplemented with 50 mg L^{-1} apramycin sulfate, and clones were screened by colony PCR and agarose gel electrophoresis. After gDNA isolation from NBS culture using the Quick-DNA Fungal/Bacterial Miniprep (Zymo Research Corporation, Irvine, CA, USA) kit, genomic integration was verified by Sanger Sequencing and Oxford Nanopore sequencing using the MinION (Oxford, UK), and minimap2, samtools and IGV were used for mapping, indexing and visualization (Figures S2 and S3) [55–57].

2.4. *Creation of the ACSP50_0507 Deletion Strain*

2.4.1. Construction of the Deletion Plasmid

The deletion strain was created using the CRISPR/Cas9-based vector pCRISPomyces-2 [66]. The spacer sequence was selected based on the CRISPy-web tool [67] and ordered as oligonucleotide, as well as its reversed complement, both including 4 bp overlap to the integration site of the vector (metabion GmbH, Steinkirchen, Germany and Sigma Aldrich—Merck KGaA, Darmstadt, Germany) (Table S1). Both oligonucleotides (5 µL, 10 µM) were annealed in 30 mM of HEPES at pH 7.8 under a temperature gradient of 98 °C to 4 °C with 1 °C min^{-1} and cloned into the BsaI (NEB, Ipswich, MA, USA) linearized vector by Golden Gate Assembly following the method outlined by Cobb et al. (2015) [66]. The DNA constructs were transformed into chemo-competent *E. coli* DH5α cells according to the method outlined by Beyer et al. and selected on solid LB media supplemented with 50 mg L^{-1} apramycin sulfate [65]. Plates were incubated overnight at 37 °C and blue-white screening was used to screen positive clones, which were then isolated using the GeneJET Plasmid-Miniprep-Kit (Thermo Fisher Scientific, Waltham, MA, USA). Homologous flanks of ~1.3 kb, operating as a template for the repair of the Cas9-induced double-strand break, were amplified by PCR using Phusion High-Fidelity PCR Master Mix with GC Buffer (Thermo Fisher Scientific, Waltham, MA, USA) and cloned in the XbaI linearized vector spacer construct by Gibson Assembly [64]. Transformation was performed as mentioned prior, and colony PCR and agarose gel electrophoresis were applied for screening. Positive clones were isolated using the GeneJET Plasmid-Miniprep-Kit (Thermo Fisher Scientific, Waltham, MA, USA) and verified by Sanger sequencing by our in-house sequencing core facility.

2.4.2. Conjugation

Conjugation was performed with competent *Actinoplanes* sp. SE50/110 cells and *E. coli* ET12567/pUZ8002 as a donor according to the protocol outlined by Schaffert et al. (2019) [68]. Plasmid curing of exconjugants was performed according to the method outlined by Wolf et al. (2016) [49], and the deletion was tested via PCR. Genome sequencing was performed via gDNA isolation of an NBS culture using Quick-DNA Fungal/Bacterial Miniprep (Zymo Research Corporation, Irvine, CA, USA) and ONT MinION of Oxford Nanopore (Oxford, UK) kits, and minimap2, samtools and IGV were used for mapping, indexing and visualization (Figure S1) [69–71].

2.5. Cultivation of Actinoplanes sp. SE50/110 Strains

For the pre-culture, 30 mL of NBS was inoculated from 200 µL of glycerol stock. After cultivation for 72 h at 28 °C and 140 rpm, 0.5 mL of culture was passaged to 30 mL of new media and cultivated for another 48 h. Shake flask cultivation was performed using 250 mL Erlenmeyer baffled cell culture flasks (Corning Inc., Corning, NY, USA) for 50 mL of culture at 28 °C and 140 rpm in maltose minimal medium ((1) 72.06 $g \cdot L^{-1}$ maltose$\cdot 1H_2O$, (2) 5 $g \cdot L^{-1}$ $(NH_4)_2SO_4$, 0.184 $g \cdot L^{-1}$ $FeCl_3 \cdot 4H_2O$, 5.7 $g \cdot L^{-1}$ $Na_3C_6H_5O_7 \cdot 2H_2O$, 1 $g \cdot L^{-1}$, 200 µL trace elements (stock concentration: 1 µM $CuCl_2$, 50 µM $ZnCl_2$, 7.5 µM $MnCl_2$), (3) $MgCl_2 \cdot 6H_2O$, 2 $g \cdot L^{-1}$ $CaCl_2 \cdot 2H_2O$, and (4) 5 $g \cdot L^{-1}$ each K_2HPO_4 and KH_2PO_4). Therefore, four medium components were separately prepared and autoclaved, whereas component (1) was sterile filtrated. Inoculation was performed using 30 mL of pre-culture, which was harvested by centrifugation at 5000 rpm for 5 min and washed twice with maltose minimal medium. The cell wet weight was measured, and the cells were diluted to 50 g mL^{-1}, of which 0.5 mL was used to inoculate 50 mL of maltose minimal medium. Cultivation was performed at 140 rpm and 28 °C. Growth was determined by a cell dry weight of 2 × 1 mL of culture sample in pre-weight 1.5 mL reaction tubes, after harvest by centrifugation at 14,000 rpm for 5 min and washing with H_2O (MilliQ) twice and a 48 h drying process at 60 °C.

2.6. Scanning Electron Microscopy (SEM)

Actinoplanes sp. SE50/110 strains were cultivated in liquid maltose minimal medium for 8 days. 1 mL cell culture sample was washed twice with 0.5 mL PBS with centrifugation at 1500 rpm for 1 min in between to remove remaining media. For purification, 0.5 mL cells solution was then transferred in a 5 mL glass vial and mixed with 5 mL of fixation reagent (2.5% (*v*/*v*) glutaraldehyde (Serva) in 100 mM phosphate buffer (pH 7.2) and incubated for 1 h on ice. Afterwards, the fixation solution was removed, and the samples were incubated in 5 mL of 100 mM phosphate buffer (pH 7.2) on ice for an additional 20 min. The solution was removed, and cells were then treated with 2-propanol (technical) in ascending concentrations, each for 20 min. The concentrations that were used were 30%, 40%, 50%, 60%, 70%, 80%, 90%, 98% of 2-propanol (technical) in 100 mM phosphate buffer (*v*/*v*). Lastly, the cells were incubated twice in 100% EtOH (HPLC grade) for 20 min. A measure of 100 µL of sample was then added to a 0.2 µm polycarbonate membrane filter (Poretics, OSMONICS, Livermore, CA, USA), dried and taped on a 12.7 mm × 3.1 mm carrier pin (MT7523, SMCAM, Science Services GmbH, Munich, Germany), before coating with gold for 150 s, at 30 mA (SCD 005 Sputter coater, BAL-TEC) under vacuum. A Hitachi S-450 Scanning Electron Microscope (High-Technologies Ltd., Tokyo, Japan) was used at 15 kV.

For the cell surface investigation, samples were alternative coated with a ~10 nm layer of ruthenium and imaged using the Dual-Beam FIB, FEI Helios Nanolab 600i (FEI Company, Hillsboro, OR, USA) at 5 kV and 0.17 nA.

2.7. Acarbose Quantification Using High-Performance Liquid Chromatography (HPLC)

For the quantification of acarbose, 2 × 1 mL of supernatant was collected and stored at −20 °C until methanol extraction and high-performance liquid chromatography (HPLC) was performed. Methanol extraction was performed after centrifugation of 1 mL supernatant for 2 min at 14,000 rpm to discard remaining solid particles. The supernatant was mixed at a ratio of 1:5 with methanol (100% v/v) and vortexed before centrifugation at 14,000× g for 2 min to remove the precipitate, and the methanol phase was transferred into HPLC vials. For the analysis, an Agilent 1100 HPLC system (G1312A Binary Pump, G1329A ALS autosampler, G1315A diode-array detector (DAD)) using a Hypersil APS-2 column, Thermo Fisher Scientific (Waltham, MA, USA) (125 × 4 mm, particle size: 3 μm) at 40 °C. As mobile phase containing 26% phosphate buffer (0.62 $g \cdot L^{-1}$ KH_2PO_4 and 0.38 $g \cdot L^{-1}$ $K_2HPO_4 \times H_2O$) (solvent A) and 74% acetonitrile (solvent B) was used with an isocratic flow rate of 1 mL min^{-1}. A measure of 40 μL of each sample was injected and separated over 15 min. Acarbose was detected at 210 nm using a DAD detector, and a standard curve allowed the acarbose quantification over the peak area.

2.8. RNA Isolation

During cultivation, samples of 2 × 1 mL were taken and subsequently frozen in liquid nitrogen and stored at −80 °C until further isolation using the Zymo, RNA Clean & Concentrator R1014 and Quick-DNA Fungal/Bacterial Miniprep Kit D6005 (Zymo Research Corporation, Irvine, CA, USA) according to the user protocol. Cell disruption was performed with 1 mL of cultivation sample at 2 × 2600 for 30 s with 5 min on ice in between. Disrupted samples were centrifuged for 4 min at 4000 rpm at 4 °C. DNaseI digestion was performed with 75 μL of digestion buffer and 5 μL of enzyme for 15 min at room temperature, and RNA was eluted with 87 μL RNase-free H_2O. Control PCR to exclude remaining DNA was performed with genomic context binding oligonucleotides and Phusion High-Fidelity PCR Master Mix with GC Buffer (Thermo Fisher Scientific, Waltham, MA, USA) at 62 °C for 30 s and following agarose gel electrophoresis. If required, an additional DNaseI treatment of 15 min at RT was performed with 85 μL samples, 10 μL Digestion Buffer and 5 μL DNaseI. RNA elution was performed using 30 μL RNase-free H_2O. RNA concentration and purity were analyzed using a NanoDrop 1000 spectrometer (Peqlab, Erlangen, Germany).

2.9. Reverse Transcription Quantitative PCR (RT-qPCR)

A Luna ® Universal Probe One-Step RT-qPCR Kit (New England BioLabs, Ipswich, MA, USA) was used for RT-qPCR with 100 ng RNA according to the manufacturer's protocol. Ninety-six-well LightCycler plates (Sarstedt, Nümbrecht, Germany) were used for the analysis with the LightCycler 96 System (Roche, Mannheim, Germany).

2.10. RNA-Sequencing and Data Processing

RNA quality was analyzed on an Agilent Bioanalyzer 2100 (Agilent Technologies, Santa Clara, CA, USA). For every sampling point, RNA of three biological replicates was isolated and analyzed. cDNA libraries were prepared according to [11] using the Illumina TruSeq stranded mRNA kit (Illumina, San Diego, CA, USA) and sequenced on Illumina NextSeq 500 in paired-end mode with a 2 × 74 nt read length. Reads were mapped on the reference genome (GenBank: LT827010.1). Differential gene expression analysis of triplicates including normalization was performed using Bioconductor package DESeq2 [72] included in the ReadXplorer v2.2 software [59]. The signal intensity value (*A*-value) was calculated by the log2 mean of normalized read counts and the signal intensity ratio (*M*-value) by the log2 fold change. The evaluation of the differential RNA-seq data was performed using an adjusted p-value (P_{adj}) cutoff of $P_{adj} \leq 0.05$ and a signal intensity ratio (*M*-value) cutoff of ≥ 1 or ≤ -1 if not mentioned otherwise. Genes with an *M*-value outside this range and $P_{adj} \leq 0.05$ were considered as differentially up- or downregulated.

Primary 5′-end-specific cDNA library was constructed and sequenced to include σ0507-specific TSSs. 5′-end cDNA library preparation was performed according to Busche et al. (2022) [73], purified and size-selected for fragments of approximately 100–1000 nt by gel electrophoresis, quantified and sequenced on a MiSeq system (Illumina, San Diego, CA, USA) in PE mode with 2 × 75 nt.

RNA-seq data were deposited in in the ArrayExpress database (www.ebi.ac.uk/arrayexpress, accessed on 15 February 2024) under accession number E-MTAB-13644.

2.11. Electrophoretic Mobility Shift Assays (EMSA)

The vector pJOE was used for protein expression with a C-terminal His_6 tag according to Nölting et al. (2023) [4]. Protein expression was verified via SDS PAGE and LC-MS/MS using a nano-LC Ultimate 3000 (Thermo Fisher Scientific, Waltham, MA, USA), coupled with ESI-Orbitrap MS/MS QExactive Plus (Thermo Fisher Scientific, Waltham, MA, USA), in negative ionization mode.

Complemental oligonucleotides of 61 bp were ordered (Sigma Aldrich—Merck KGaA, Darmstadt, Germany), with one oligonucleotide of each pair 5′-end biotin labeled. Annealing to a double-stranded DNA fragment was performed in 30 mM of HEPES at pH 7.8 under a temperature gradient from 98 °C to 4 °C with 1 °C min^{-1}. As a control, a 61 bp fragment of the luciferase coding sequence was used. Protein–DNA interaction was analyzed using a LightShift® Chemiluminescent EMSA Kit (Thermo Scientific, Waltham, MA, USA), following the manufacturer's protocol. For the reaction, 0.1 ug (3.40 μM) recombinant protein and 0.05 nM dsDNA were used. Blotting was performed on an Amersham Hybond™-N+ nylon transfer membrane (GE Healthcare, Chicago, IL, USA) with blot filter paper (Bio-Rad Laboratories, Inc., Hercules, CA, USA) and Trans-Blot® SD Semi-Dry (Bio-Rad Laboratories, Inc., Hercules, CA, USA).

3. Results

3.1. Sigma Factor Gene ACSP50_0507 Encodes a σH Homolog of Streptomyces coelicolor

Sigma factors influence a wide variety of cellular processes and form highly complex regulatory networks [74,75]. *Actinoplanes* sp. SE50/110 carries at least 78 sigma factor genes and 52 genes coding for anti-sigma factors and anti-sigma factor antagonists (GenBank: NZ_LT827010.1). Thereby, its sigma factor regulation seems to be of higher complexity than in *C. glutamicum* with 7 σ factors, or even *S. coelicolor*, with 65 sigma factor coding genes [38].

Actinoplanes sp. SE50/110 alternative sigma factor gene *ACSP50_0507* is strongly transcribed during the transition and early stationary growth phase, with a maximal transcript amount at the early stationary growth phase and therefore possibly influencing the cells during growth phase transition [12]. This σ B/F/G family protein (IPRO014322) of 270 amino acids features RNA-polymerase binding and −10 promoter recognition region 2 (IPRO07027), RNA polymerase and extended-10 binding region 3 (IPRO07624) and −35 promoter element recognition region 4 (cd06171). *ACSP50_0507* is a homolog of *A. missouriensis* (*AMIS_4630*) and *S. coelicolor* (*SCO5243*), with high protein sequence identity to *SCO5243* (56% identical amino acid residues) (Figure 1). *SCO5243* codes for σH^{Sc}, which is known to be involved in osmotic stress response, sporulation and morphological differentiation [21,44] and whose regulatory network was already elucidated [21,42,43,45–47,76]. Sequence alignment also shows an extended N-terminal σH^{Sc} without any domains. This protein region is not essential for its functionality since *sigH*^{Sc} is developmentally translated into three translational products, with the shortest one missing this extended region [77]. In *S. coelicolor*, σH^{Sc} is inactivated by anti-σ factor UshX, which itself can be inactivated by anti-anti-σ factor BldG [21]. Interestingly, homologs for both genes were identified in *Actinoplanes* sp. SE50/110, with ACSP50_7560 being homologous to UshX and ACSP50_0384 to BldG, respectively (Figure S7). Interestingly, ACSP50_0284 also shows a high similarity towards BldG but was shown in the phylogenetic analysis as more distantly related and is therefore a putative paralog.

Due to gene homology and further observations described below, ACSP50_0507 was designated σH^{As} and its gene as *sigH*^{As}. In *S. coelicolor*, σH^{Sc} is encoded in an operon with its anti-σ factor gene. However, in *Actinoplanes* sp. SE50/110, that is not the case. Its homolog *ACSP50_0507* is coded in an operon under the control of several promoters. Genes which are coded in the same operon code for a putative membrane protein (*ACSP50_0506*) carry a soluble N-ethylmaleimide-sensitive-factor attachment receptor (SNARE) domain and a putative secreted SDR family NAD(P)H-dependent oxidoreductase (*ACSP50_0505*).

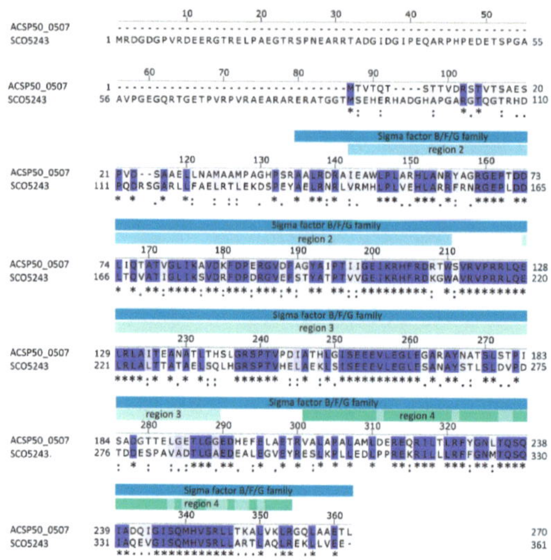

Figure 1. *Actinoplanes* sp. SE50/110 ACSP50_0507 protein sequence and domain annotation. Sequence alignment to σH (SCO5243) of *S. coelicolor* performed with Clustal Omega [78] and domains were predicted using InterPro [78,79]. ACSP50_0507 has a σ B/F/G family protein domain, containing three regions of different functions. Region 2 is an RNA-polymerase binding and −10 promoter recognition region, region 3 is an RNA-polymerase and extended −10-binding region and region 4 is a −35 promoter element-recognition region where DNA-binding residues are depicted in a lighter color. "*" indicates identical residues, ":" indicates conservation between groups of strongly similar properties and "." indicates conservation between groups of weakly similar properties.

3.2. The Cell Morphology Is Highly Affected by $sigH^{As}$ Expression

Actinoplanes sp. SE50/110 grows as dense mycelium, surrounded by an extended mesh of hyphae (Figure 2a). The long hyphae are occasionally branched and have a uniform shape (Figure 2b). Genomic insertion of empty vector control pSETT4*tipA* without gene of interest did not influence the strain's morphology (Figure 2c,d). Genomic insertion of an additional gene copy (pSETT4*tipA*-$sigH^{As}$) demonstrated a drastic influence on cell morphology, featuring a dense and irregularly formed mycelium, with no surrounding mesh of hyphae (Figure 2e). The expression strain has deformed hyphae with an irregular shape and exceeded branching (Figure 2f). Higher magnification shows that these hyphae are subdivided into parts of irregular sizes and shapes, with larger bulbous sections in between (Figure 2g). The cell surface seemingly shrank and is highly wrinkled (Figure 2h). While cell morphology of the expression strain was drastically influenced, no changes could be observed with the $sigH^{As}$ deletion (Δ$sigH^{As}$) and gene complementation (CΔ$sigH^{As}$) strains (Figure S5), making the expression strain especially interesting for further characterization.

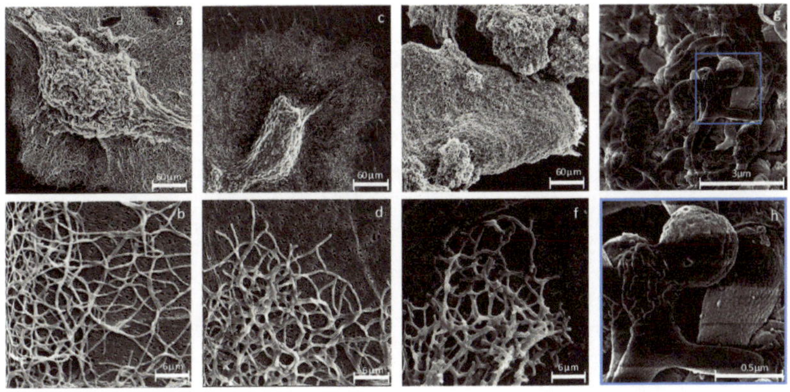

Figure 2. Scanning electron microscopy (SEM) of *Actinoplanes* sp. SE50/110. Strains were cultivated in maltose minimal medium for 5 days before sample preparation for microscopic imaging. The wild-type (**a**) and empty vector control strain pSETT4*tipA* (**c**) grow as large, dense mycelium, surrounded by a large web of hyphae. Hyphae of the wild-type (**b**) and pSETT4*tipA* (**d**) have a uniform shape and are occasionally branched. The morphology of the *sigH*^{As} expression strain is drastically influenced to a dense mycelium without a hyphal mesh surrounding it (**e**). Hyphae are deformed, with an irregular shape (**f**,**g**), and closer magnification of the blue marked area shows a highly wrinkled cell surface (**h**).

3.3. The *sigH*^{As} Expression Strain Achieves a Two-Fold Acarbose Yield Due to Prolonged Acarbose Production during Stationary Growth

Cultivation of *Actinoplanes* sp. SE50/110 wild type, empty vector control strain pSETT4*tipA* and expression strain pSETT*tipA-sigH*^{As} was performed to analyze growth and acarbose production. Growth was determined by cell dry weight determination. The expression strain has similar growth towards the end of cultivation than wild-type and empty vector control strain (Figure 3a). The expression strain achieved an increasing acarbose concentration beyond stationary growth at 118 h of cultivation, whereas the wild-type (WT) and empty vector control strain pSETT*tipA* showed a stagnating acarbose concentration from this period of growth. A maximal acarbose concentration of 0.853 $g \cdot L^{-1}$ at 166 h of cultivation by the *sigH*^{As} expression strain was observed, whereas WT and empty vector control had ~0.471 $g \cdot L^{-1}$. At this time, the expression strain achieved its optimal biomass-related acarbose yield of 63.95 $mg_{acab} \cdot g_{cdw}{}^{-1}$, a doubling of yield as compared to 31.75 $mg_{acab} \cdot g_{cdw}{}^{-1}$ achieved by the WT (Figure 3b).

The *sigH*^{As} deletion strain did show a slight growth advantage, and a similar acarbose end-concentration was achieved (Figure S4a). When *sigH*^{As} was reintroduced by genomic integration using pSETT4*tipA*, the complementation strain CΔ*sigH*^{As} showed similar growth behavior and acarbose end-concentration as the reference strain (Figure S4b).

Gene transcription analysis using RT-qPCR revealed a 2.5-fold higher transcript level of *sigH*^{As} in the expression strain (Figure 4). This unexpectedly low differential gene transcription was further examined by a gene copy-specific amplification, using 5′UTR-specific oligonucleotides. It revealed downregulation of the chromosomal *sigH*^{As} gene copy in the expression strain, thereby indicating a much higher transcript level of the gene copy on the integration vector. It further indicates negative regulation, when σH^{As} levels are increased due to the expression of a second gene copy. The relative transcript level of the deletion strain is at the lower limit of detection, and the transcriptional levels of *sigH*^{As} were apparently restored to the wild-type level by gene complementation.

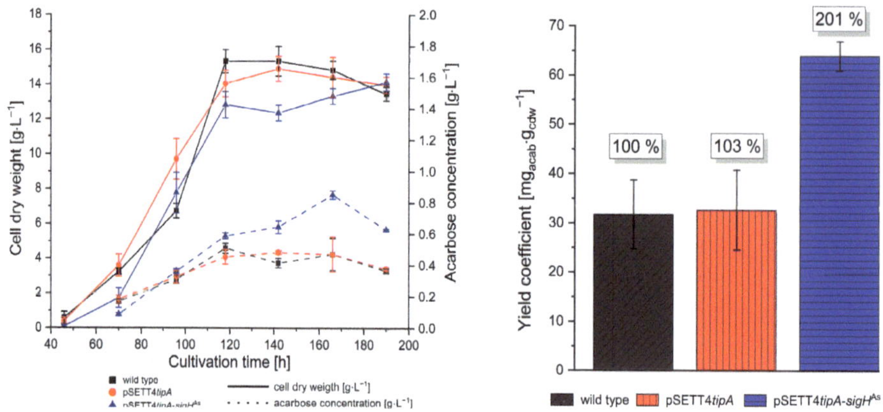

(**a**) (**b**)

Figure 3. Characterization of growth and acarbose production of *Actinoplanes* sp. SE50/110 $sigH^{As}$ expression and control strains. (**a**) Cell dry weight (dense lines) and acarbose concentration (bracket lines) from supernatant during cultivation in maltose minimal medium are shown for *Actinoplanes* sp. SE50/110 wild type, empty vector strain pSETT4*tipA* and σH^{As} expression strain pSETT4*tipA*-$sigH^{As}$ (n = 3). (**b**) Optimal biomass-related acarbose yield of *Actinoplanes* sp. SE50/110 wild type, empty vector strain pSETT4*tipA* and expression strain pSETT4*tipA*-$sigH^{As}$ at 166 h of cultivation (n = 3).

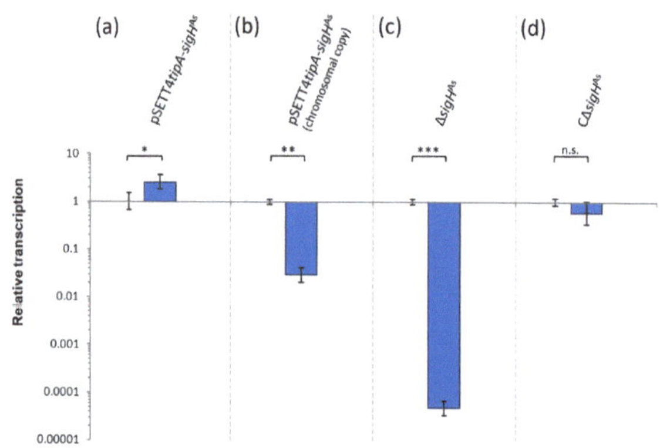

Figure 4. Relative $sigH^{As}$ transcript amounts of $sigH^{As}$ expression, deletion and complementation strains. RNA was isolated from cell pellet of 1 mL of cultivation sample at 96 h of cultivation in maltose minimal medium. RT-qPCR was performed for three biological replicates per condition with two technical replicates each. (**a**) $sigH^{As}$ overexpression detected for the chromosomal and ectopic $sigH^{As}$ gene copy of the expression strain (pSETT4*tipA*-$sigH^{As}$) compared to the empty vector control strain. The gene expression strain (pSETT4*tipA*-$sigH^{As}$) has a 2.5-fold increased transcript level. (**b**) Repression of transcription of the chromosomal gene copy. The expression strain (pSETT4*tipA*-$sigH^{As}$) was compared to the empty vector control strain with 5′UTR specific oligonucleotides to differentiate between chromosomal and ectopic gene copy. The genomic integrated additional gene copy must be highly transcribed to achieve 2.5-fold upregulation detected for chromosomal and ectopic gene copy, despite the downregulation of the chromosomal gene copy. (**c**) Transcription of the deletion strain $\Delta sigH^{As}$ was compared to *Actinoplanes* sp. SE50/110 wild type and is at the lower limit of detection. (**d**) Gene transcription of the complementation strain ($C\Delta sigH^{As}$) was compared to the empty vector control strain. Asterisks indicate *p*-value of a two-sided *t*-test with n.s. $p \geq 0.05$, * $p < 0.05$; ** $p < 0.01$ and *** $p < 0.001$.

3.4. Increased Transcription of Acarbose Biosynthesis Genes during Growth and Late Stationary Growth Phase

Comparative whole transcriptome analysis of empty vector control strain and *sigH*^{As} expression strain revealed a high influence on gene transcription in general, with varying differential gene transcription during cultivation. Interestingly, besides 739 significantly upregulated genes (M-value ≥ 1; $P_{adj} < 0.01$), we also observed 657 downregulated genes (M-value ≤ -1; $P_{adj} < 0.01$) during the growth phase (96 h). With 2975 genes, differential transcription is maximal during late stationary growth, where 1707 genes are upregulated (M-value ≥ 1; $P_{adj} < 0.01$) and 1268 genes are downregulated (M-value ≤ -1; $P_{adj} < 0.01$) (Figure S6). During late stationary growth, the highest number of regulatory genes, totaling 247, are differential transcribed.

The genes of the acarbose gene cluster were found to be upregulated during growth and late stationary growth phases at 96 h and 166 h of cultivation (Figure 5a), respectively, with a maximal average upregulation with an M-value of ~1.96 during the late stationary phase (166 h) and maximal upregulation of the acarbose export coding *acbWXY* operon with M-values up to 4.44 (*acbY*) (Figure 5b). During the growth phase (96 h), *acbABD* are the highest upregulated genes of the gene cluster, with M-values of 3.35 and 2.50 for the biosynthesis genes *acbA* (dTDP-glucose synthase) and *acbB* (dTDP-4-keto-6-deoxyglucose dehydratase) and 4.10 for the extracellular acarbose metabolism gene *acbD* (acarviose transferase). Interestingly, during the transition phase and early stationary growth, the *acb* genes are downregulated to an average M-value of −0.78. Thereby, *acb* gene transcription is influenced in similar levels, except the monocistronically transcribed *acbZ* (extracellular α-amylase) and *acbE* (α-amylase) genes. *acbZ* transcription is downregulated during all sampling points during stationary growth (118 h–166 h), whereas transcription of *acbE* remains upregulated even when *acb* genes are downregulated in general.

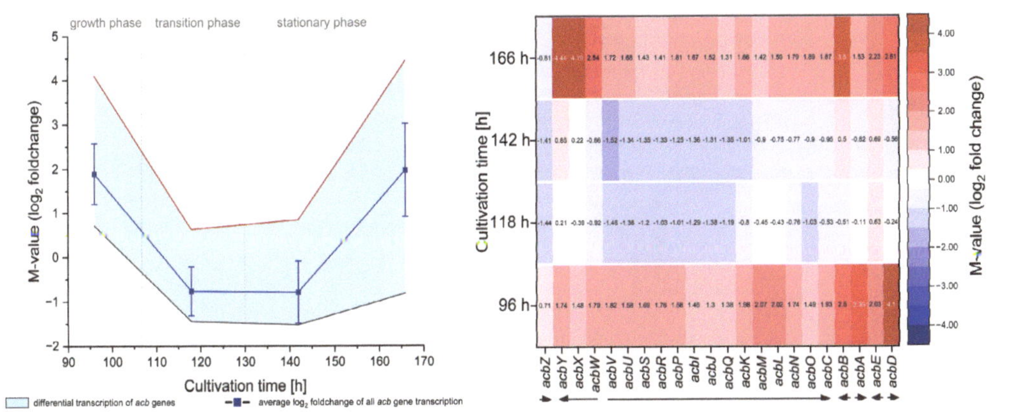

(a) (b)

Figure 5. Differential transcription of *acb* genes. Transcriptome sequencing data of expression strain pSETT4*tipA*-*sigH*^{As} compared to the empty vector control strain pSETT4*tipA* during growth. (**a**) Average differential *acb* gene transcription and standard deviation of *sigH*^{As} expression strain is shown in dark blue. Minimal and maximal differential gene transcription is shown within the colored area, defined by red (maximal) or black (minimal) lines. (**b**) Differential gene transcription of single *acb* genes is shown as a heatmap with the respective M-values within the color-marked areas. Arrows indicate operon structure and gene orientation.

3.5. Altogether, 185 Genes Are Permanently Upregulated in the *sigH*^{As} Expression Mutant

Since gene overexpression resulted in a 2.5-fold increased transcription and sigma factors act as transcriptional activators, the focus was set on genes with increased transcription. Data filtering for genes with an increased transcription (M-value ≥1; P_{adj} < 0.05) at all analyzed sampling times left 185 genes.

Of these 185 genes, 20 genes belong to a putative NRPS biosynthesis gene cluster [11]. This cluster of 38 genes (*ACSP50_3026-3063*) is upregulated in the expression strain. Six genes of this putative biosynthesis cluster are among the 25 highest permanently upregulated genes (Table 1). It is suspected that biosynthesis cluster-specific regulators strengthened the cluster upregulation. The cluster contains an AraC family transcriptional regulator gene (*ACSP50_3061*) and an Lrp/AsnC family transcriptional regulator gene (*ACSP50_3031*). Both regulatory genes are affected within the expression strain. While the putative AraC family regulator gene is both up- and downregulated during cultivation, the Lrp/AsnC family regulator gene is only upregulated during the end of stationary growth. Since the transcription of both regulators is not continuously up- or downregulated, they are probably not involved in the permanent upregulation of the putative biosynthesis gene cluster alone.

Table 1. Genes of *sigH^As^* expression mutant showing the strongest average upregulation at all sampling times (96 h, 118 h, 142 h and 166 h) with M-value ≥ 1 and P_{adj} < 0.05. Genes are sorted by locus tag. The gene annotation is based on GenBank file NZ_LT827010.1 and further bioinformatic analysis.

Locus Tag	Function Annotation (NZ_LT827010.1)	Highest M-Value	P_{adj}	Cultivation Time [h]
ACSP50_0208	Putative secreted protein	8.93	1.09×10^{-14}	166
ACSP50_0424	PadR family transcriptional regulator	4.48	9.69×10^{-15}	166
ACSP50_2587	HAD (haloacid dehalogenase) family hydrolase	5.03	2.184×10^{-50}	142
ACSP50_2935	Protein of unknown function	4.79	3.81×10^{-19}	142
ACSP50_3039 [2]	oligosaccharide flippase, MATE like superfamily	4.78	2.63×10^{-44}	142
ACSP50_3041 [1,2]	Putative secreted protein	4.19	9.58×10^{-23}	166
ACSP50_3042 [1,2]	protein of unknown function (DUF 4082)	3.89	1.67×10^{-23}	142
ACSP50_3043 [1,2]	Glycosyltransferase family 2 protein	5.52	9.44×10^{-21}	166
ACSP50_3047 [2]	MbtH-like protein	5.26	6.89×10^{-34}	166
ACSP50_3058 [2]	Gfo/Idh/MocA family oxidoreductase	4.63	1.66×10^{-40}	96
ACSP50_3379	Putative secreted protein	4.29	3.26×10^{-13}	142
ACSP50_3380	Membrane protein of unknown function (DUF4235)	4.8	1.40×10^{-12}	142
ACSP50_3381	Putative transmembrane protein (DUF 3618)	5.24	3.93×10^{-28}	118
ACSP50_3382	Putative Actinobacterial Holin-X, holin superfamily III	5.33	1.93×10^{-37}	118
ACSP50_3383	Uncharacterized BrkB/YihY/UPF0761 family membrane protein	4.95	7.99×10^{-32}	118
ACSP50_3888	Putative secreted protein	8.35	1.54×10^{-48}	118
ACSP50_4135	Putative secreted protein	9.53	1.63×10^{-54}	118
ACSP50_4921	Putative membrane protein (DUF3618)	4.6	9.46×10^{-16}	142
ACSP50_4922 [1]	Putative Actinobacterial Holin-X, holin superfamily III	4.24	1.24×10^{-12}	142
ACSP50_4923 [1]	Protein of unknown function	4.44	1.69×10^{-11}	118
ACSP50_5880	Putative secreted protein	10.17	3.46×10^{-20}	118
ACSP50_6038	MFS transporter DHA2 family	6.54	2.72×10^{-44}	166
ACSP50_6083	Putative transmembrane protein	6.98	6.50×10^{-72}	118
ACSP50_6593	Putative transmembrane protein	4.91	1.81×10^{-35}	142
ACSP50_7588	Protein of unknown function	5.36	4.39×10^{-86}	96

[1] Coded in the same operon; [2] members of an uncharacterized putative NRPS biosynthesis gene cluster.

Aligning with the highly affected cell morphology, the majority of maximal upregulated genes code for membrane-associated proteins or carry a signal peptide and are therefore putative secreted proteins. *ACSP50_3382* and *ACSP50_4922* code for putative actinobacterial holin proteins. Furthermore, the PadR transcriptional regulator coding gene *ACSP50_0424* and a haloacid dehalogenase family hydrolase coding gene *ACSP50_2587* are strongly upregulated genes. Among the 185 permanent upregulated genes are several genes related to transport, amino acid supply, regulation and trehalose biosynthesis (Table S3). Interestingly, none of the five homologs of σH^{Sc} dependent genes *SCO3557, ssgB, dpsA, gltB* and *gltD* are found among the 185 permanent upregulated genes.

3.6. *σH^{As} Binds to a Determined Promoter Region, Upstream of Genes Permanently Upregulated in the Expression Strain*

sigH^{As} expression resulted in the permanent upregulation of 185 genes, belonging to 178 transcription units, which were used for binding motif prediction. TSSs were assigned according to a native 5′-end transcript dataset, obtained for the *sigH*^{As} expression and deletion strains. Upstream sequences of these genes, comprising −60 to +10 nucleotides (nt) relative to the transcription start site (TSS, +1), were used for the analysis using Improbizer and visualization using WebLogo 3.7.10 [60,61]. The emerging promoter motif contains a −35 region with 5′-G/CGTTTA/C-3′ and a −10 region with 5′-GGGTAN$_3$T-3′. The −35 region and −10 region have an average distance of 14.6 ± 1.9 nt (Figure 6b). Interestingly, binding motif prediction with four known binding sites of the σH^{As} homolog σH^{Sc} [43,46] identified a highly similar binding motif of a −35 region with 5′-GTTTC/GC/G-3′ with an average distance of 14.2 nt to a −10 region of 5′-GGGT/CA-3′ (Figure 6c).

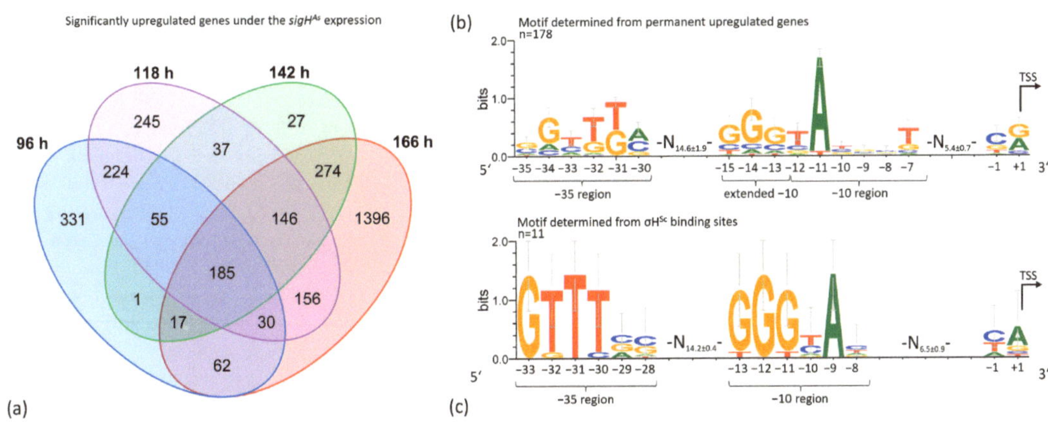

Figure 6. σH^{As} binding motif prediction. (**a**) Venn diagram of significantly upregulated genes (M-value ≥1; P_{adj} < 0.05) of the *sigH*^{As} expression mutant from growth to late stationary phase (96 h, 142 h, 118 h, 166 h). (**b**) Predicted binding motif of −35 and −10 region and transcription start site (TSS) for σH^{As}. The motif was predicted based on upstream sequences (−60 to +10 nt) according to the TSS, determined by 5′-end specific transcript sequencing. The analysis was performed with Improbizer [61] and visualized with WebLogo 3.7.10 [60] for 178 upstream sequences of permanently upregulated genes. (**c**) Determined binding motif of the homologous σH^{Sc} of *S. coelicolor*, based on 11 known binding sites [43,46].

To demonstrate in vitro binding of σH^{As} to a promoter region of strong differentially transcribed *ACSP50_4135* (M-value 9.5 ± 1.64 × 10^{-54}), C-terminal hexa-histidine tagged σH^{As} was recombinantly expressed in *E. coli* BL21(DE3) pLysS, verified by SDS-PAGE (Figure S8) and analysis of tryptic peptide digestion by Orbi-trap-MS/MS. Electrophoretic mobility shift assay (EMSA) was performed with biotinylated double-stranded DNA of the respective promoter region of −60 to +1 bp, which were determined based on 5′-end specific transcriptome data. σH^{As} binds to the predicted promoter region 5′-TGATTA-3′

(−35 region) and 5′-GGGGATTAG-3′ ((extended) −10 region) of *ACSP50_4135*, proven by the shift in labeled DNA and protein complex (Figure 7a). The addition of unlabeled specific dsDNA in excess proved the binding specificity of σH^{As} to the DNA sequence.

Since RT-qPCR indicated an autoregulatory effect, the promoter region of *sigH*^{As} was also examined for the predicted binding motif. A putative −35 region with 5′-CTTTCG-3′ and −10 region with 5′-GGCTACGGT-3′ was determined according to the TSS and tested for σH^{As} binding using EMSA (Figure 7b). Binding of σH^{As} to the biotinylated −60 to +1 sequence was confirmed and was reversed by the addition of unlabeled specific dsDNA in excess (Figure 7b). No band shift and, therefore, no unspecific binding of σH^{As} to an intergenic dsDNA sequence, lacking the presence of a promoter, were detected (Figure 7c).

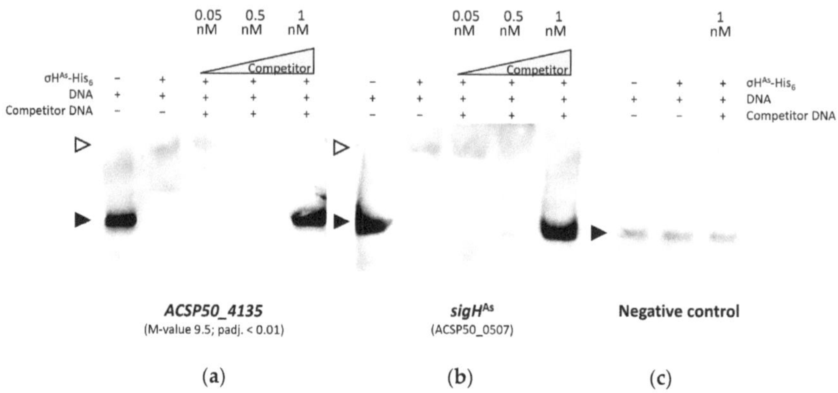

Figure 7. Electrophoretic mobility shift assay (EMSA) with recombinantly expressed C-terminal His$_6$ tagged σH^{As}. Binding was tested for upstream region of ACSP50_4135 and its own promoter. All binding reactions contained Poly (dI-dC) as a non-specific competitor and 0.05 nM of the respective biotinylated dsDNA fragment of 61 base pairs. To the reaction, 3.40 uM of protein and 0.05–1 nM of specific unlabeled competitor DNA were either added (+) or not (−). Filled arrows indicate biotinylated DNA and open arrows indicate the shifted protein–DNA complex. (**a**) EMSA with the promoter region (−60 to +1) of ACSP50_4135 confirms specific binding by σH^{As}-His$_6$. (**b**) Binding to the promoter region (−60 to +1) of *sigH*^{As} was also confirmed. (**c**) σH^{As}-His6 does not show unspecific binding to DNA, which was confirmed by a lack of binding to the intragenic region of luciferase DNA (GenBank: EU239244.1).

4. Discussion

The investigation of the influence of sigma factor on cellular processes and acarbose production in particular is of interest since several studies already highlight the influence of sigma factor on growth and metabolite production in other bacteria [16,17,26]. This becomes even more promising since acarbose production in *Actinoplanes* sp. SE50/110 is coupled to growth [14]. In this work, we characterized the influence and regulatory mechanism of alternative sigma factor σH^{As} (ACSP50_0507).

Protein sequence alignment revealed high similarity to σH^{Sc} (SCO5243) of *S. coelicolor*, where its complex regulatory network is comprehensively characterized. It is a member of the alternative σHBF family and has a dual role in the transcriptional regulation of *S. coelicolor*, controlling osmotic stress response and septation of aerial hyphae into spores [21,45]. *SigH*^{Sc} transcription is activated upon heat shock and salt stress. σH^{Sc} is further developmentally regulated in the course of differentiation by BldD, alternative translation, and proteolysis, and transcription is autoregulated under osmotic stress through one of four known upstream promoters [21,44,77]. Among other sigma factors, σH^{Sc} is part of a complex regulatory network of σB^{Sc}, the master regulator for osmotic and oxidative stress response in *S. coelicolor* and further σB homologues [22,80,81]. Thereby, the regulation of σB homologues is closely interconnected by promoter cross-recognition with both their networks and their own genes [43,46]. On the protein level, σH^{Sc} is negatively regulated by the anti-σ factor UshX, which itself can be inactivated by anti-anti σ factor BldG [45,82]. Interestingly, homologs of the direct interaction partners are present in *Actinoplanes* sp. SE50/110, with ACSP50_7560 homologous to *S. coelicolor* anti-σH^{Sc} UshX and ACSP50_0384 homologous to anti-anti-σH^{Sc} BldG. It is therefore highly likely that a similar regulatory cascade exists for σH^{As}.

σB homologs are stress-associated alternative σ factors with diverse roles in morphological differentiation. Strikingly, the σH^{As} expression strain features a drastically deformed cell morphology. The hyphae show severe branching and have an irregular shape with visible compartmentalization. The cell surface is severely wrinkled and appears dehydrated. This is also reflected at the transcriptional level, where the majority of permanently highly upregulated genes are coding for membrane-associated or secreted proteins. Interestingly, *ACSP50_3382* and *ACSP50_4922* encode putative actinobacterial holin superfamily II proteins. Holin proteins promote cell lysis through their accumulation, but their involvement in differentiation, biofilm formation, stress response, non-lytic transport of toxins, gene transfer agents and proteins has also been shown [83–86]. Since cell morphology is highly affected and the cells appear dehydrated, the upregulation of two holin coding genes and further membrane-associated proteins aligns with the highly affected cell morphology.

σH^{As} expression leads to the global and permanent upregulation of a putative NRPS biosynthesis gene cluster, comprising 38 genes (*ACSP50_3026-3063*), of which 6 genes are among the 25 highest permanently upregulated genes. Two transcriptional regulator coding genes (*ACSP50_3061* and *ACSP50_3031*) are located within the putative biosynthesis gene cluster. Since both are not permanently up- or downregulated in the expression strain, it is not likely that any of them cause permanent cluster upregulation alone; therefore, transcriptional activation by σH^{As} is suspected. However, upstream sequences of transcriptional units within this gene cluster do not show similarity to the predicted binding motif. Thus, indirect cluster regulation by σH^{As} through the activation of further putative regulatory elements can be assumed. Within the *sigH^As^* expression strain, a maximal number of genes is transcriptionally upregulated at late stationary growth, with increasing upregulation throughout the stationary phase. During late stationary growth, the influence on regulatory genes is highest with 247 differentially transcribed regulatory genes, among them, 12 encoding σ factors, 3 anti-σ factors and 5 anti-anti-σ factors. If these factors are not themselves regulated post-transcriptionally, they will cause a broad range of secondary regulatory effects at the late stationary growth phase but probably also at the other sampled cultivation time points, where regulatory genes are also found to be up- or downregulated (Table S3).

Since the *S. coelicolor* homolog σH^{Sc} is transcriptionally activated under osmotic stress and involved in osmotic stress response, it prompted an investigation into transcriptional indications of functional similarity of σH^{As} in response to osmotic stress [87–89]. One important mechanism is to prevent cell dehydration by increasing the intracellular osmotic potential. This can be facilitated by the production or import of ions, peptides, free amino acids or other osmolytes [90–92]. Within the $sigH^{As}$ expression mutant, a high number of transport-related genes are permanently significantly upregulated. Among them are peptidases, amino acid importer coding genes (*ACSP50_3876* and *ACSP50_7166*) and the operon *ACSP50_0791-0795*, coding for a putative peptide transport system similar to the known DppABCDF or OppABCDEF systems [93–96]. Such upregulation could increase the amino acid availability, which is used to decrease osmotic pressure [97–99]. Besides the upregulation of acarbose biosynthesis genes and an unknown NRPS biosynthesis gene cluster, the trehalose biosynthesis genes *ACSP50_5263*, coding for trehalose synthase/amylase TreS, *ACSP50_6610*, coding for maltooligosyl trehalose synthase, and the trehalose-6-phosphate synthase coding *ACSP50_3343*, are permanently upregulated. Trehalose is an organic solute that is produced or imported under osmotic stress [100–102]. Since no trehalose importer is currently known in *Actinoplanes* sp. SE50/110, trehalose biosynthesis is required. It was further shown that an osmolality upshift results in increased branching of hyphae in *S.*

venezuelae and *S. coelicolor* [87,88], similar to the observed hyperbranching of the *sigH*^{As} expression mutant.

Acarbose production in different *Actinoplanes* species is increased under a certain osmolality, which is assumed to result from increased membrane permeability and higher maltose transport [103,104]. It was further shown that by-product formation is also reduced under specific osmolality conditions [104–106]. Altogether, the upregulation of transport, trehalose biosynthesis and membrane-associated genes within the *sigH*^{As} expression strain and its hyperbranching and dehydrated cell shape points to its putative involvement in osmotic stress response, similar to its *S. coelicolor* homolog.

A σH^{As} binding motif was predicted and has high similarity with the known binding sites of the *S. coelicolor* homolog. In this work, binding of recombinantly expressed σH^{As}-His$_6$ to the upstream sequence of the highest average upregulated gene *ACSP50_4135* was experimentally confirmed. Despite ectopic overexpression under the medium–strong *tipA* promoter, *sigH*^{As} gene transcription is only increased 2.5-fold. Gene copy-specific RT-qPCR determined the downregulation of the chromosomal gene copy, which can be compensated by ectopic overexpression and therefore results in low overall gene upregulation. Transcriptional inactivation of the chromosomal gene copy is most likely caused by an autoregulatory effect, since σH^{As} binding to its own upstream sequence was confirmed in vitro. Interestingly, the homologous σH^{Sc} also binds within its own promoter region [20,21,45]. Despite the binding motif similarity and sequence similarity of both σ factor proteins, none of the five homologs of σH^{Sc}-dependent genes are found to be permanently upregulated. Even considering their phenotypic and functional similarity, the regulatory networks of σH^{Sc} and σH^{As} seem different. It can be speculated that the expansion and functional differentiation of the σ factor family occurred during evolution after separation of the genera *Streptomyces* and *Actinoplanes*.

The expression strain achieved a doubled biomass-related acarbose yield at 166 h of cultivation. Strikingly, an increase in acarbose concentration continued during stationary growth. This is extraordinary for *Actinoplanes* sp. SE50/110, where acarbose production ends with stationary growth in the wildtype strain. σH^{As} gene deletion did not affect the acarbose yield; thus, it is not essential for acarbose production. *Acb* gene transcription is increased during the growth and late stationary phase in the *sigH*^{As} expression mutant, with the exceptions of *acbZ* and *acbE*. Both gene products are not involved in the acarbose synthesis pathway itself but code for α-amylases involved in extracellular acarbose metabolism and are transcribed monocistronically [12]. Transcription of *acbZ* remains downregulated during stationary growth and *acbE* transcription remains upregulated during the start of stationary growth, when other acarbose biosynthesis genes are downregulated within the strain. In previous studies, it was shown that the upregulation of *acbE* or *acbD* did not achieve an increased acarbose concentration [13]; therefore, it is not suspected that *acbE* upregulation alone is the reason for and increased acarbose production. Since no σH^{As} binding motif was found within the *acb* gene cluster, no direct transcriptional regulation by σH^{As} is assumed. However, general *acb* gene upregulation during later cultivation times indicates at least one additional transcriptional regulatory level, highlighting the challenges to elucidate the regulation of acarbose biosynthesis in *Actinoplanes* sp. SE50/110. Acarbose production is closely connected to the cell membrane, with Acb proteins and acarbose biosynthesis located at the inner cell membrane [107]. According to the carbophore model, acarbose could function as a shuttle for the importation of malto-oligosaccharides [3,108]. Therefore, acarbose is exported and reimported as higher acarbose variants, which enables the uptake, intracellular release and metabolism of additional glucose [109,110]. A significant morphological change is likely to influence not only cell membrane-associated acarbose production but also its use as a putative carbophore.

5. Conclusions

We show that sigma factor engineering is a useful tool to achieve a coordinated phenotype change. Ectopic expression of the alternative sigma factor gene *ACSP50_0507* (coding for σH[As]), a homolog of σH[Sc] of *Streptomyces coelicolor*, achieved a two-fold increased acarbose yield in *Actinoplanes* sp. SE50/110. The genetic network, regulated by σH[As], comprises membrane-associated functions, and its cellular morphology was deformed accordingly. A binding motif for the σH[As] promoter was determined und experimentally validated for highly upregulated genes and the σH[As] promoter region, showing an autoregulatory function.

Supplementary Materials: The following supporting information can be downloaded at: https://www.mdpi.com/article/10.3390/microorganisms12061241/s1, Table S1: Oligonucleotides used in this study; Table S2: Venn diagram dataset; Table S3: Permanently upregulated genes; Figure S1: *sigH[As]* deletion (*ΔsigH[As]*) is verified by nanopore whole-genome sequencing; Figure S2: *sigH[As]* complementation of *ΔsigH[As]*, based on the integrative pSET4*tipA* vector, is verified by nanopore whole-genome sequencing; Figure S3: *sigH[As]* expression, based on the integrative pSET4*tipA* vector, is verified by nanopore whole-genome sequencing; Figure S4: Characterization of growth and acarbose production of *Actinoplanes* sp. SE50/110 sigma factor deletion and complementation mutants; Figure S5: Scanning electron microscopy (SEM) of *Actinoplanes* sp. SE50/110 *sigH[As]* deletion and complementation strain; Figure S6: Transcriptomic analysis of *Actinoplanes* sp. SE50/110 *sigH[As]* expression strain whole transcriptome; Figure S7: Phylogenetic tree of *Streptomyces coelicolor* σH[Sc] regulatory proteins and their *Actinoplanes* sp. SE50/110 and *Actinoplanes missouriensis* homologs; Figure S8: Purity of recombinant expressed σH[As]-His_6 analyzed using SDS-PAGE.

Author Contributions: Conceptualization, L.S.; methodology, L.S.; validation, L.S.; formal analysis L.S.; investigation, L.S.; resources, A.H., K.N. and J.K.; data curation, L.S. and T.B.; writing—original draft preparation, L.S.; writing—review and editing, S.S.-B. and J.K.; visualization, L.S., L.B. and K.N.; supervision, J.K. and A.P.; project administration, J.K. and A.P.; funding acquisition, J.K. and A.P. All authors have read and agreed to the published version of the manuscript.

Funding: This research was funded by Bayer AG (Leverkusen, Germany).

Data Availability Statement: Data are contained within the article and Supplementary Materials.

Acknowledgments: We thank IIT Biotech GmbH for support in RNA sequencing, Levin Joe Klages (CeBiTec, Bielefeld) for support in nanopore sequencing, and Svenja Vinke (Harvard Medical School, Boston) for proofreading of the manuscript. We further thank Olga Blifernez-Klassen and Birgit Gerlinde Katharina Mittmann (CeBiTec, Bielefeld) for the introduction to EMSA and Carola Eck and Thomas Patschkowski for the MS protein analysis. We acknowledge support for the publication costs by the Open Access Publication Fund of Bielefeld University and the Deutsche Forschungsgemeinschaft (DFG).

Conflicts of Interest: The authors declare no conflicts of interest. The funders had no role in the design of the study; in the collection, analyses, or interpretation of data; in the writing of the manuscript; or in the decision to publish the results.

References

1. Parenti, F.; Coronelli, C. Members of the genus Actinoplanes and their antibiotics. *Annu. Rev. Microbiol.* **1979**, *33*, 389–411. [CrossRef] [PubMed]
2. Truscheit, E.; Frommer, W.; Junge, B.; Müller, L.; Schmidt, D.D.; Wingender, W. Chemistry and Biochemistry of Microbialα-Glucosidase Inhibitors. *Angew. Chem. Int. Ed. Engl.* **1981**, *20*, 744–761. [CrossRef]
3. Wehmeier, U.F.; Piepersberg, W. Biotechnology and molecular biology of the?-glucosidase inhibitor acarbose. *Appl. Microbiol. Biotechnol.* **2004**, *63*, 613–625. [CrossRef] [PubMed]
4. Nölting, S.; März, C.; Jacob, L.; Persicke, M.; Schneiker-Bekel, S.; Kalinowski, J. The 4-α-Glucanotransferase AcbQ Is Involved in Acarbose Modification in *Actinoplanes* sp. SE50/110. *Microorganisms* **2023**, *11*, 848. [CrossRef]
5. Tsunoda, T.; Samadi, A.; Burade, S.; Mahmud, T. Complete biosynthetic pathway to the antidiabetic drug acarbose. *Nat Commun* **2022**, *13*, 3455. [CrossRef] [PubMed]
6. Zhang, C.-S.; Podeschwa, M.; Altenbach, H.-J.; Piepersberg, W.; Wehmeier, U.F. The acarbose-biosynthetic enzyme AcbO from *Actinoplanes* sp. SE 50/110 is a 2-epi-5-epi-valiolone-7-phosphate 2-epimerase. *FEBS Lett.* **2003**, *540*, 47–52. [CrossRef] [PubMed]
7. Zhang, C.-S.; Podeschwa, M.; Block, O.; Altenbach, H.-J.; Piepersberg, W.; Wehmeier, U.F. Identification of a 1-epi-valienol 7-kinase activity in the producer of acarbose, *Actinoplanes* sp. SE50/110. *FEBS Lett.* **2003**, *540*, 53–57. [CrossRef] [PubMed]

8. Zhang, C.-S.; Stratmann, A.; Block, O.; Brückner, R.; Podeschwa, M.; Altenbach, H.-J.; Wehmeier, U.F.; Piepersberg, W. Biosynthesis of the C(7)-cyclitol moiety of acarbose in Actinoplanes species SE50/110. 7-O-phosphorylation of the initial cyclitol precursor leads to proposal of a new biosynthetic pathway. *J. Biol. Chem.* **2002**, *277*, 22853–22862. [CrossRef] [PubMed]
9. Stratmann, A.; Mahmud, T.; Lee, S.; Distler, J.; Floss, H.G.; Piepersberg, W. The AcbC protein from Actinoplanes species is a C7-cyclitol synthase related to 3-dehydroquinate synthases and is involved in the biosynthesis of the alpha-glucosidase inhibitor acarbose. *J. Biol. Chem.* **1999**, *274*, 10889–10896. [CrossRef]
10. Drepper, A.; Pape, H. Acarbose 7-phosphotransferase from *Actinoplanes* sp.: Purification, properties, and possible physiological function. *J. Antibiot.* **1996**, *49*, 664–668. [CrossRef]
11. Wolf, T.; Schneiker-Bekel, S.; Neshat, A.; Ortseifen, V.; Wibberg, D.; Zemke, T.; Pühler, A.; Kalinowski, J. Genome improvement of the acarbose producer *Actinoplanes* sp. SE50/110 and annotation refinement based on RNA-seq analysis. *J. Biotechnol.* **2017**, *251*, 112–123. [CrossRef] [PubMed]
12. Droste, J.; Ortseifen, V.; Schaffert, L.; Persicke, M.; Schneiker-Bekel, S.; Pühler, A.; Kalinowski, J. The expression of the acarbose biosynthesis gene cluster in *Actinoplanes* sp. SE50/110 is dependent on the growth phase. *BMC Genom.* **2020**, *21*, 818. [CrossRef] [PubMed]
13. Wolf, T.; Droste, J.; Gren, T.; Ortseifen, V.; Schneiker-Bekel, S.; Zemke, T.; Pühler, A.; Kalinowski, J. The MalR type regulator AcrC is a transcriptional repressor of acarbose biosynthetic genes in *Actinoplanes* sp. SE50/110. *BMC Genom.* **2017**, *18*, 3. [CrossRef] [PubMed]
14. Wendler, S.; Ortseifen, V.; Persicke, M.; Klein, A.; Neshat, A.; Niehaus, K.; Schneiker-Bekel, S.; Walter, F.; Wehmeier, U.F.; Kalinowski; et al. Carbon source dependent biosynthesis of acarviose metabolites in *Actinoplanes* sp. SE50/110. *J. Biotechnol.* **2014**, *191*, 113–120. [CrossRef]
15. Yuan, B.; Keller, N.P.; Oakley, B.R.; Stajich, J.E.; Wang, C.C.C. Manipulation of the Global Regulator mcrA Upregulates Secondary Metabolite Production in *Aspergillus wentii* Using CRISPR-Cas9 with In Vitro Assembled Ribonucleoproteins. *ACS Chem. Biol.* **2022**, *17*, 2828–2835. [CrossRef]
16. Zhuo, Y.; Zhang, W.; Chen, D.; Gao, H.; Tao, J.; Liu, M.; Gou, Z.; Zhou, X.; Ye, B.-C.; Zhang, Q.; et al. Reverse biological engineering of hrdB to enhance the production of avermectins in an industrial strain of Streptomyces avermitilis. *Proc. Natl. Acad. Sci. USA* **2010**, *107*, 11250–11254. [CrossRef] [PubMed]
17. Taniguchi, H.; Henke, N.A.; Heider, S.A.E.; Wendisch, V.F. Overexpression of the primary sigma factor gene sigA improved carotenoid production by *Corynebacterium glutamicum*: Application to production of β-carotene and the non-native linear C50 carotenoid bisanhydrobacterioruberin. *Metab. Eng. Commun.* **2017**, *4*, 1–11. [CrossRef] [PubMed]
18. Viollier, P.H.; Kelemen, G.H.; Dale, G.E.; Nguyen, K.T.; Buttner, M.J.; Thompson, C.J. Specialized osmotic stress response systems involve multiple SigB-like sigma factors in *Streptomyces coelicolor*. *Mol. Microbiol.* **2003**, *47*, 699–714. [CrossRef] [PubMed]
19. Paget, M.S.; Kang, J.G.; Roe, J.H.; Buttner, M.J. sigmaR, an RNA polymerase sigma factor that modulates expression of the thioredoxin system in response to oxidative stress in *Streptomyces coelicolor* A3(2). *EMBO J.* **1998**, *17*, 5776–5782. [CrossRef]
20. Kormanec, J.; Sevcíková, B.; Halgasová, N.; Knirschová, R.; Rezuchová, B. Identification and transcriptional characterization of the gene encoding the stress-response sigma factor sigma(H) in *Streptomyces coelicolor* A3(2). *FEMS Microbiol. Lett.* **2000**, *189*, 31–38.
21. Sevciková, B.; Benada, O.; Kofronova, O.; Kormanec, J. Stress-response sigma factor sigma(H) is essential for morphological differentiation of *Streptomyces coelicolor* A3(2). *Arch. Microbiol.* **2001**, *177*, 98–106. [CrossRef] [PubMed]
22. Lee, E.-J.; Karoonuthaisiri, N.; Kim, H.-S.; Park, J.-H.; Cha, C.-J.; Kao, C.M.; Roe, J.-H. A master regulator sigmaB governs osmotic and oxidative response as well as differentiation via a network of sigma factors in *Streptomyces coelicolor*. *Mol. Microbiol.* **2005**, *57*, 1252–1264. [CrossRef] [PubMed]
23. Yamazaki, H.; Ohnishi, Y.; Horinouchi, S. An A-factor-dependent extracytoplasmic function sigma factor (sigma(AdsA)) that is essential for morphological development in Streptomyces griseus. *J. Bacteriol.* **2000**, *182*, 4596–4605. [CrossRef] [PubMed]
24. Gallagher, K.A.; Schumacher, M.A.; Bush, M.J.; Bibb, M.J.; Chandra, G.; Holmes, N.A.; Zeng, W.; Henderson, M.; Zhang, H.; Findlay, K.C.; et al. c-di-GMP Arms an Anti-σ to Control Progression of Multi-cellular Differentiation in Streptomyces. *Mol. Cell* **2020**, *77*, 586–599.e6. [CrossRef]
25. Chater, K.F.; Bruton, C.J.; Plaskitt, K.A.; Buttner, M.J.; Méndez, C.; Helmann, J.D. The developmental fate of *S. coelicolor* hyphae depends upon a gene product homologous with the motility sigma factor of *B. subtilis*. *Cell* **1989**, *59*, 133–143. [PubMed]
26. Taniguchi, H.; Busche, T.; Patschkowski, T.; Niehaus, K.; Pátek, M.; Kalinowski, J.; Wendisch, V.F. Physiological roles of sigma factor SigD in *Corynebacterium glutamicum*. *BMC Microbiol.* **2017**, *17*, 158. [CrossRef] [PubMed]
27. Wang, H.; Yang, L.; Wu, K.; Li, G. Rational selection and engineering of exogenous principal sigma factor (σ(HrdB)) to increase teicoplanin production in an industrial strain of *Actinoplanes teichomyceticus*. *Microb. Cell Factories* **2014**, *13*, 10. [CrossRef]
28. Kang, J.G.; Paget, M.S.; Seok, Y.J.; Hahn, M.Y.; Bae, J.B.; Hahn, J.S.; Kleanthous, C.; Buttner, M.J.; Roe, J.H. RsrA, an anti-sigma factor regulated by redox change. *EMBO J.* **1999**, *18*, 4292–4298. [CrossRef] [PubMed]
29. Li, W.; Bottrill, A.R.; Bibb, M.J.; Buttner, M.J.; Paget, M.S.B.; Kleanthous, C. The Role of zinc in the disulphide stress regulated anti-sigma factor RsrA from *Streptomyces coelicolor*. *J. Mol. Biol.* **2003**, *333*, 461–472. [CrossRef]
30. Walsh, N.P.; Alba, B.M.; Bose, B.; Gross, C.A.; Sauer, R.T. OMP peptide signals initiate the envelope stress response by activating DegS protease via relief of inhibition mediated by its PDZ domain. *Cell* **2003**, *113*, 61–71. [CrossRef]

31. Schöbel, S.; Zellmeier, S.; Schumann, W.; Wiegert, T. The Bacillus subtilis sigmaW antisigma factor RsiW is degraded by intramembrane proteolysis through YluC. *Mol. Microbiol.* **2004**, *52*, 1091–1105. [CrossRef] [PubMed]
32. Homerova, D.; Sevcikova, B.; Rezuchova, B.; Kormanec, J. Regulation of an alternative sigma factor σI by a part-ner switching mechanism with an anti-sigma factor PrsI and an anti-anti-sigma factor ArsI in Streptomyces coe-licolor A3(2). *Gene* **2012**, *492*, 71–80. [CrossRef] [PubMed]
33. Grossman, A.D.; Straus, D.B.; Walter, W.A.; Gross, C.A. Sigma 32 synthesis can regulate the synthesis of heat shock proteins in Escherichia coli. *Genes Dev.* **1987**, *1*, 179–184. [CrossRef]
34. Missiakas, D.; Mayer, M.P.; Lemaire, M.; Georgopoulos, C.; Raina, S. Modulation of the Escherichia coli sigmaE (RpoE) heat-shock transcription-factor activity by the RseA, RseB and RseC proteins. *Mol. Microbiol.* **1997**, *24*, 355–371. [CrossRef] [PubMed]
35. Wiegeshoff, F.; Beckering, C.L.; Debarbouille, M.; Marahiel, M.A. Sigma L is important for cold shock adaptation of Bacillus subtilis. *J. Bacteriol.* **2006**, *188*, 3130–3133. [CrossRef] [PubMed]
36. Hecker, M.; Pané-Farré, J.; Völker, U. SigB-dependent general stress response in Bacillus subtilis and related gram-positive bacteria. *Annu. Rev. Microbiol.* **2007**, *61*, 215–236. [CrossRef] [PubMed]
37. Gruber, T.M.; Gross, C.A. Multiple sigma subunits and the partitioning of bacterial transcription space. *Annu. Rev. Microbiol.* **2003**, *57*, 441–466. [CrossRef] [PubMed]
38. Bentley, S.D.; Chater, K.F.; Cerdeño-Tárraga, A.-M.; Challis, G.L.; Thomson, N.R.; James, K.D.; Harris, D.E.; Quail, M.A.; Kieser, H.; Harper, D.; et al. Complete genome sequence of the model actinomycete *Streptomyces coelicolor* A3(2). *Nature* **2002**, *417*, 141–147. [CrossRef] [PubMed]
39. Helmann, J.D.; Moran, C.P. RNA Polymerase and Sigma Factors. In *Bacillus Subtilis and Its Closest Relatives*; Sonenshein, A.L., Hoch, J.A., Losick, R., Eds.; ASM Press: Washington, DC, USA, 2001; pp. 287–312.
40. Kunst, F.; Ogasawara, N.; Moszer, I.; Albertini, A.M.; Alloni, G.; Azevedo, V.; Bertero, M.G.; Bessières, P.; Bolotin, A.; Borchert, S.; et al. The complete genome sequence of the gram-positive bacterium Bacillus subtilis. *Nature* **1997**, *390*, 249–256. [CrossRef]
41. Bentley, S.D.; Brown, S.; Murphy, L.D.; Harris, D.E.; Quail, M.A.; Parkhill, J.; Barrell, B.G.; McCormick, J.R.; San-tamaria, R.I.; Losick, R.; et al. SCP1, a 356,023 bp linear plasmid adapted to the ecology and developmental biology of its host, *Streptomyces coelicolor* A3(2). *Mol. Microbiol.* **2004**, *51*, 1615–1628. [CrossRef]
42. Kormanec, J.; Sevcikova, B.; Novakova, R.; Homerova, D.; Rezuchova, B.; Mingyar, E. The Complex Roles and Regulation of Stress Response σ Factors in *Streptomyces coelicolor*. In *Stress and Environmental Regulation of Gene Expression and Adaptation in Bacteria*; De Bruijn, F.J., Ed.; Wiley: Hoboken, NJ, USA, 2016; pp. 328–343.
43. Sevcikova, B.; Rezuchova, B.; Mazurakova, V.; Homerova, D.; Novakova, R.; Feckova, L.; Kormanec, J. Cross-Recognition of Promoters by the Nine SigB Homologues Present in *Streptomyces coelicolor* A3(2). *Inter-Natl. J. Mol. Sci.* **2021**, *22*, 7849. [CrossRef] [PubMed]
44. Kelemen, G.H.; Viollier, P.H.; Tenor, J.; Marri, L.; Buttner, M.J.; Thompson, C.J. A connection between stress and development in the multicellular prokaryote *Streptomyces coelicolor* A3(2). *Mol. Microbiol.* **2001**, *40*, 804–814. [CrossRef] [PubMed]
45. Sevcikova, B.; Kormanec, J. Activity of the *Streptomyces coelicolor* stress-response sigma factor sigmaH is regulated by an anti-sigma factor. *FEMS Microbiol. Lett.* **2002**, *209*, 229–235. [PubMed]
46. Mazurakova, V.; Sevcikova, B.; Rezuchova, B.; Kormanec, J. Cascade of sigma factors in streptomycetes: Identi-fication of a new extracytoplasmic function sigma factor sigmaJ that is under the control of the stress-response sigma factor sigmaH in *Streptomyces coelicolor* A3(2). *Arch. Microbiol.* **2006**, *186*, 435–446. [CrossRef] [PubMed]
47. Sevcikova, B.; Rezuchova, B.; Mingyar, E.; Homerova, D.; Novakova, R.; Feckova, L.; Kormanec, J. Pleiotropic an-ti-anti-sigma factor BldG is phosphorylated by several anti-sigma factor kinases in the process of activating multiple sigma factors in *Streptomyces coelicolor* A3(2). *Gene* **2020**, *755*, 144883. [CrossRef] [PubMed]
48. Schaffert, L.; Jacob, L.; Schneiker-Bekel, S.; Persicke, M.; März, C.; Rückert, C.; Pühler, A.; Kalinowski, J. pSETT4, an Improved φC31-Based Integrative Vector System for *Actinoplanes* sp. SE50/110. *Microbiol. Resour. Announc.* **2020**, *9*, 10–1128. [CrossRef] [PubMed]
49. Wolf, T.; Gren, T.; Thieme, E.; Wibberg, D.; Zemke, T.; Pühler, A.; Kalinowski, J. Targeted genome editing in the rare actinomycete *Actinoplanes* sp. SE50/110 by using the CRISPR/Cas9 System. *J. Biotechnol.* **2016**, *231*, 122–128. [CrossRef] [PubMed]
50. Zhao, Q.; Xie, H.; Peng, Y.; Wang, X.; Bai, L. Improving acarbose production and eliminating the by-product component C with an efficient genetic manipulation system of *Actinoplanes* sp. SE50/110. *Synth. Syst. Biotechnol.* **2017**, *2*, 302–309. [PubMed]
51. Dieckmann, M.A.; Beyvers, S.; Nkouamedjo-Fankep, R.C.; Hanel, P.H.G.; Jelonek, L.; Blom, J.; Goesmann, A. EDGAR3.0: Comparative genomics and phylogenomics on a scalable infrastructure. *Nucleic Acids Res.* **2021**, *49*, W185–W192. [CrossRef]
52. Altschul, S.F.; Gish, W.; Miller, W.; Myers, E.W.; Lipman, D.J. Basic local alignment search tool. *J. Mol. Biol.* **1990**, *215*, 403–410. [CrossRef]
53. Altschul, S.F.; Madden, T.L.; Schäffer, A.A.; Zhang, J.; Zhang, Z.; Miller, W.; Lipman, D.J. Gapped BLAST and PSI-BLAST: A new generation of protein database search programs. *Nucleic Acids Res.* **1997**, *25*, 3389–3402. [CrossRef] [PubMed]
54. Altschul, S.F.; Wootton, J.C.; Gertz, E.M.; Agarwala, R.; Morgulis, A.; Schäffer, A.A.; Yu, Y.-K. Protein database searches using compositionally adjusted substitution matrices. *FEBS J.* **2005**, *272*, 5101–5109. [CrossRef] [PubMed]
55. Marchler-Bauer, A.; Bryant, S.H. CD-Search: Protein domain annotations on the fly. *Nucleic Acids Res.* **2004**, *32*, W327–W331. [CrossRef] [PubMed]

56. Marchler-Bauer, A.; Lu, S.; Anderson, J.B.; Chitsaz, F.; Derbyshire, M.K.; DeWeese-Scott, C.; Fong, J.H.; Geer, L.Y.; Geer, R.C.; Gonzales, N.R.; et al. CDD: A Conserved Domain Database for the functional annotation of proteins. *Nucleic Acids Res.* **2011**, *39*, D225–D229. [CrossRef]
57. Marchler-Bauer, A.; Derbyshire, M.K.; Gonzales, N.R.; Lu, S.; Chitsaz, F.; Geer, L.Y.; Geer, R.C.; He, J.; Gwadz, M.; Hurwitz, D.I.; et al. CDD: NCBI's conserved domain database. *Nucleic Acids Res.* **2015**, *43*, D222–D226. [CrossRef] [PubMed]
58. Marchler-Bauer, A.; Bo, Y.; Han, L.; He, J.; Lanczycki, C.J.; Lu, S.; Chitsaz, F.; Derbyshire, M.K.; Geer, R.C.; Gonza-les, N.R.; et al. CDD/SPARCLE: Functional classification of proteins via subfamily domain architectures. *Nucleic Acids Res.* **2017**, *45*, D200–D203. [CrossRef] [PubMed]
59. Hilker, R.; Stadermann, K.B.; Schwengers, O.; Anisiforov, E.; Jaenicke, S.; Weisshaar, B.; Zimmermann, T.; Goesmann, A. ReadXplorer 2-detailed read mapping analysis and visualization from one single source. *Bioinformatics* **2016**, *32*, 3702–3708. [CrossRef] [PubMed]
60. Ao, W.; Gaudet, J.; Kent, W.J.; Muttumu, S.; Mango, S.E. Environmentally induced foregut remodeling by PHA-4/FoxA and DAF-12/NHR. *Science* **2004**, *305*, 1743–1746. [CrossRef] [PubMed]
61. Crooks, G.E.; Hon, G.; Chandonia, J.-M.; Brenner, S.E. WebLogo: A sequence logo generator. *Genome Res.* **2004**, *14*, 1188–1190. [CrossRef]
62. Militello, K.T.; Simon, R.D.; Qureshi, M.; Maines, R.; VanHorne, M.L.; Hennick, S.M.; Jayakar, S.K.; Pounder, S. Conservation of Dcm-mediated cytosine DNA methylation in Escherichia coli. *FEMS Microbiol. Lett.* **2012**, *328*, 78–85. [CrossRef]
63. Kieser, T.; Bibb, M.; Buttner, M.; Chater, K. *Practical Streptomyces Genetics*; Innes: Norwich, UK, 2000.
64. Gibson, D.G.; Young, L.; Chuang, R.-Y.; Venter, J.C.; Hutchison, C.A.; Smith, H.O. Enzymatic assembly of DNA molecules up to several hundred kilobases. *Nat. Methods* **2009**, *6*, 343–345. [CrossRef] [PubMed]
65. Beyer, H.M.; Gonschorek, P.; Samodelov, S.L.; Meier, M.; Weber, W.; Zurbriggen, M.D. AQUA Cloning: A Versatile and Simple Enzyme-Free Cloning Approach. *PLoS ONE* **2015**, *10*, e0137652. [CrossRef]
66. Cobb, R.E.; Wang, Y.; Zhao, H. High-efficiency multiplex genome editing of Streptomyces species using an engi-neered CRISPR/Cas system. *ACS Synth. Biol.* **2015**, *4*, 723–728. [CrossRef] [PubMed]
67. Blin, K.; Pedersen, L.E.; Weber, T.; Lee, S.Y. CRISPy-web: An online resource to design sgRNAs for CRISPR appli-cations. *Synth. Syst. Biotechnol.* **2016**, *1*, 118–121. [CrossRef]
68. Schaffert, L.; März, C.; Burkhardt, L.; Droste, J.; Brandt, D.; Busche, T.; Rosen, W.; Schneiker-Bekel, S.; Persicke, M.; Pühler, A.; et al. Evaluation of vector systems and promoters for overexpression of the acarbose biosynthesis gene acbC in *Actinoplanes* sp. SE50/110. *Microb. Cell Fact* **2019**, *18*, 195. [CrossRef] [PubMed]
69. Li, H. New strategies to improve minimap2 alignment accuracy. *Bioinformatics* **2021**, *37*, 4572–4574. [CrossRef]
70. Danecek, P.; Bonfield, J.K.; Liddle, J.; Marshall, J.; Ohan, V.; Pollard, M.O.; Whitwham, A.; Keane, T.; McCarthy, S.A.; Davies, R.M.; et al. Twelve years of SAMtools and BCFtools. *GigaScience* **2021**, *10*, giab008. [CrossRef] [PubMed]
71. Robinson, J.T.; Thorvaldsdóttir, H.; Winckler, W.; Guttman, M.; Lander, E.S.; Getz, G.; Mesirov, J.P. Integrative genomics viewer. *Nat. Biotechnol.* **2011**, *29*, 24–26. [CrossRef]
72. Love, M.I.; Huber, W.; Anders, S. Moderated estimation of fold change and dispersion for RNA-seq data with DESeq2. *Genome Biol.* **2014**, *15*, 550. [CrossRef]
73. Busche, T.; Dostálová, H.; Rucká, L.; Holátko, J.; Barvík, I.; Štěpánek, V.; Pátek, M.; Kalinowski, J. Overlapping SigH and SigE sigma factor regulons in Corynebacterium glutamicum. *Front. Microbiol.* **2022**, *13*, 1059649. [CrossRef]
74. Paget, M.S. Bacterial Sigma Factors and Anti-Sigma Factors: Structure, Function and Distribution. *Biomolecules* **2015**, *5*, 1245–1265. [CrossRef]
75. Campbell, E.A.; Westblade, L.F.; Darst, S.A. Regulation of bacterial RNA polymerase sigma factor activity: A structural perspective. *Curr. Opin. Microbiol.* **2008**, *11*, 121–127. [CrossRef]
76. Sevcikova, B.; Kormanec, J. The ssgB gene, encoding a member of the regulon of stress-response sigma factor sigmaH, is essential for aerial mycelium septation in *Streptomyces coelicolor* A3(2). *Arch. Microbiol.* **2003**, *180*, 380–384. [CrossRef] [PubMed]
77. Viollier, P.H.; Weihofen, A.; Folcher, M.; Thompson, C.J. Post-transcriptional regulation of the Streptomyces coe-licolor stress responsive sigma factor, SigH, involves translational control, proteolytic processing, and an an-ti-sigma factor homolog. *J. Mol. Biol.* **2003**, *325*, 637–649. [CrossRef] [PubMed]
78. Sievers, F.; Higgins, D.G. Clustal Omega for making accurate alignments of many protein sequences. *Protein Sci. A Publ. Protein Soc.* **2018**, *27*, 135–145. [CrossRef] [PubMed]
79. Paysan-Lafosse, T.; Blum, M.; Chuguransky, S.; Grego, T.; Pinto, B.L.; Salazar, G.A.; Bileschi, M.L.; Bork, P.; Bridge, A.; Colwell, L.; et al. InterPro in 2022. *Nucleic Acids Res.* **2023**, *51*, D418–D427. [CrossRef] [PubMed]
80. Lee, E.-J.; Cho, Y.-H.; Kim, H.-S.; Ahn, B.-E.; Roe, J.-H. Regulation of sigmaB by an anti- and an anti-anti-sigma factor in *Streptomyces coelicolor* in response to osmotic stress. *J. Bacteriol.* **2004**, *186*, 8490–8498. [CrossRef] [PubMed]
81. Lee, E.-J.; Cho, Y.-H.; Kim, H.-S.; Roe, J.-H. Identification of sigmaB-dependent promoters using consen-sus-directed search of *Streptomyces coelicolor* genome. *J. Microbiol.* **2004**, *42*, 147–151. [PubMed]
82. Sevcikova, B.; Rezuchova, B.; Homerova, D.; Kormanec, J. The anti-anti-sigma factor BldG is involved in activa-tion of the stress response sigma factor σ(H) in *Streptomyces coelicolor* A3(2). *J. Bacteriol.* **2010**, *192*, 5674–5681. [CrossRef]

83. Palmer, T.; Finney, A.J.; Saha, C.K.; Atkinson, G.C.; Sargent, F. A holin/peptidoglycan hydrolase-dependent pro-tein secretion system. *Mol. Microbiol.* **2021**, *115*, 345–355. [CrossRef]
84. Brüser, T.; Mehner-Breitfeld, D. Occurrence and potential mechanism of holin-mediated non-lytic protein translocation in bacteria. *Microb. Cell* **2022**, *9*, 159–173. [CrossRef]
85. Govind, R.; Dupuy, B. Secretion of Clostridium difficile toxins A and B requires the holin-like protein TcdE. *PLoS Pathog.* **2012**, *8*, e1002727. [CrossRef]
86. Real, G.; Pinto, S.M.; Schyns, G.; Costa, T.; Henriques, A.O.; Moran, C.P. A gene encoding a holin-like protein in-volved in spore morphogenesis and spore germination in Bacillus subtilis. *J. Bacteriol.* **2005**, *187*, 6443–6453. [CrossRef] [PubMed]
87. Bhowmick, S.; Shenouda, M.L.; Tschowri, N. Osmotic stress responses and the biology of the second messenger c-di-AMP in Streptomyces. *microLife* **2023**, *4*, uqad020. [CrossRef]
88. Fuchino, K.; Flärdh, K.; Dyson, P.; Ausmees, N. Cell-Biological Studies of Osmotic Shock Response in *Streptomyces* spp. *J. Bacteriol.* **2017**, *199*, 10–1128. [CrossRef]
89. de Jong, W.; Vijgenboom, E.; Dijkhuizen, L.; Wösten, H.A.B.; Claessen, D. SapB and the rodlins are required for development of *Streptomyces coelicolor* in high osmolarity media. *FEMS Microbiol. Lett.* **2012**, *329*, 154–159. [CrossRef]
90. Wood, J.M. Bacterial osmoregulation: A paradigm for the study of cellular homeostasis. *Annu. Rev. Microbiol.* **2011**, *65*, 215–238. [CrossRef] [PubMed]
91. Kol, S.; Merlo, M.E.; Scheltema, R.A.; de Vries, M.; Vonk, R.J.; Kikkert, N.A.; Dijkhuizen, L.; Breitling, R.; Takano, E. Metabolomic characterization of the salt stress response in *Streptomyces coelicolor*. *Appl. Environ. Microbiol.* **2010**, *76*, 2574–2581. [PubMed]
92. Ignatova, Z.; Gierasch, L.M. Inhibition of protein aggregation in vitro and in vivo by a natural osmoprotectant. *Proc. Natl. Acad. Sci. USA* **2006**, *103*, 13357–13361. [CrossRef]
93. Masulis, I.S.; Sukharycheva, N.A.; Kiselev, S.S.; Andreeva, Z.S.; Ozoline, O.N. Between computational predic-tions and high-throughput transcriptional profiling: In depth expression analysis of the OppB trans-membrane subunit of Escherichia coli OppABCDF oligopeptide transporter. *Res. Microbiol.* **2020**, *171*, 55–63. [CrossRef]
94. Perego, M.; Higgins, C.F.; Pearce, S.R.; Gallagher, M.P.; Hoch, J.A. The oligopeptide transport system of Bacillus subtilis plays a role in the initiation of sporulation. *Mol. Microbiol.* **1991**, *5*, 173–185. [CrossRef] [PubMed]
95. Payne, J.W.; Grail, B.M.; Marshall, N.J. Molecular recognition templates of peptides: Driving force for molecular evolution of peptide transporters. *Biochem. Biophys. Res. Commun.* **2000**, *267*, 283–289.
96. Pletzer, D.; Braun, Y.; Dubiley, S.; Lafon, C.; Köhler, T.; Page, M.G.P.; Mourez, M.; Severinov, K.; Weingart, H. The Pseudomonas aeruginosa PA14 ABC Transporter NppA1A2BCD Is Required for Uptake of Peptidyl Nucleoside Antibiotics. *J. Bacteriol.* **2015**, *197*, 2217–2228. [CrossRef] [PubMed]
97. Zaprasis, A.; Brill, J.; Thüring, M.; Wünsche, G.; Heun, M.; Barzantny, H.; Hoffmann, T.; Bremer, E. Osmoprotec-tion of Bacillus subtilis through import and proteolysis of proline-containing peptides. *Appl. Environ. Microbiol.* **2013**, *79*, 576–587. [CrossRef]
98. Detmers, F.J.; Lanfermeijer, F.C.; Poolman, B. Peptides and ATP binding cassette peptide transporters. *Res. Microbiol.* **2001**, *152*, 245–258. [CrossRef]
99. Slamti, L.; Lereclus, D. The oligopeptide ABC-importers are essential communication channels in Gram-positive bacteria. *Res. Microbiol.* **2019**, *170*, 338–344.
100. Rimmele, M.; Boos, W. Trehalose-6-phosphate hydrolase of Escherichia coli. *J. Bacteriol.* **1994**, *176*, 5654–5664. [CrossRef]
101. Strøm, A.R.; Kaasen, I. Trehalose metabolism in Escherichia coli: Stress protection and stress regulation of gene expression. *Mol. Microbiol.* **1993**, *8*, 205–210. [CrossRef] [PubMed]
102. Pospísl, S.; Halada, P.; Petrícek, M.; Sedmera, P. Glucosylglycerate is an osmotic solute and an extracellular metabolite produced by Streptomyces caelestis. *Folia Microbiol.* **2007**, *52*, 451–456. [CrossRef]
103. Jürgen, B.; Michael, S.; Ulrich, S. Osmotically Controlled Fermentation Process for the Preparation of Acarbose. Patent US6130072A, 10 October 2000.
104. Wang, Y.-J.; Liu, L.-L.; Wang, Y.-S.; Xue, Y.-P.; Zheng, Y.-G.; Shen, Y.-C. Actinoplanes utahensis ZJB-08196 fed-batch fermentation at elevated osmolality for enhancing acarbose production. *Bioresour. Technol.* **2012**, *103*, 337–342. [CrossRef]
105. Choi, B.T.; Shin, C.S. Reduced formation of byproduct component C in acarbose fermentation by *Actinoplanes* sp. CKD485-16. *Biotechnol. Prog.* **2003**, *19*, 1677–1682. [CrossRef]
106. Cheng, X.; Peng, W.-F.; Huang, L.; Zhang, B.; Li, K. A novel osmolality-shift fermentation strategy for im-proving acarbose production and concurrently reducing byproduct component C formation by *Actinoplanes* sp. A56. *J. Ind. Microbiol. Biotechnol.* **2014**, *41*, 1817–1821. [CrossRef]
107. Wendler, S.; Otto, A.; Ortseifen, V.; Bonn, F.; Neshat, A.; Schneiker-Bekel, S.; Walter, F.; Wolf, T.; Zemke, T.; Wehmeier, U.F.; et al. Comprehensive proteome analysis of *Actinoplanes* sp. SE50/110 highlighting the location of proteins encoded by the acarbose and the pyochelin biosyn-thesis gene cluster. *J. Proteom.* **2015**, *125*, 1–16. [CrossRef]
108. Wehmeier, U.F.; Piepersberg, W. Chapter 19 Enzymology of Aminoglycoside Biosynthesis—Deduction from Gene Clusters. In *Complex Enzymes in Microbial Natural Product Biosynthesis*; Hopwood, D., Ed.; Academic Press: London, UK, 2009; Volume 459, pp. 459–491.

109. Wendler, S.; Hürtgen, D.; Kalinowski, J.; Klein, A.; Niehaus, K.; Schulte, F.; Schwientek, P.; Wehlmann, H.; Wehmeier, U.F.; Pühler, A. The cytosolic and extracellular proteomes of *Actinoplanes* sp. SE50/110 led to the identification of gene products involved in acarbose metabolism. *J. Biotechnol.* **2013**, *167*, 178–189. [CrossRef]
110. Licht, A.; Bulut, H.; Scheffel, F.; Daumke, O.; Wehmeier, U.F.; Saenger, W.; Schneider, E.; Vahedi-Faridi, A. Crystal structures of the bacterial solute receptor AcbH displaying an exclusive substrate preference for β-D-galactopyranose. *J. Mol. Biol.* **2011**, *406*, 92–105. [CrossRef]

MDPI AG
Grosspeteranlage 5
4052 Basel
Switzerland
Tel.: +41 61 683 77 34

Microorganisms Editorial Office
E-mail: microorganisms@mdpi.com
www.mdpi.com/journal/microorganisms

www.ingramcontent.com/pod-product-compliance
Lightning Source LLC
Chambersburg PA
CBHW041346210926
51850CB00013B/3009
9783725829163